LabVIEW® For Everyone

Graphical Programming Made Even Easier

Lisa K. Wells
Jeffrey Travis

Prentice Hall PTR, Upper Saddle River, NJ 07458
http://www.prenhall.com

D1533787

Library of Congress Cataloging-in-Publication Data

Wells, Lisa K.
    LabVIEW for everyone : graphical programming made even easier /
Lisa Wells and Jeffrey Travis.
       p.   cm.
    Includes index.
    ISBN 0-13-268194-3 (paper)
    1. Laboratories--Computer programs. 2. LabVIEW. I. Travis,
Jeffrey. II. Title.
Q183.A1W45 1997
006--dc20
                           96-19491
                               CIP

© 1997 by Prentice Hall PTR
Prentice-Hall, Inc.
A Division of Simon and Schuster
Upper Saddle River, NJ 07458

The publisher offers discounts on this book when ordered
in bulk quantities. For more information, contact:

Corporate Sales Department
Prentice Hall PTR
One Lake Street
Upper Saddle River, NJ 07458
Phone: 800-382-3419
Fax: 201-236-7141;
e-mail: corpsales@prenhall.com

Printed in the United States of America

10 9 8 7 6 5 4

ISBN: 0-13-268194-3

Prentice-Hall International (UK) Limited, London
Prentice-Hall of Australia Pty. Limited, Sydney
Prentice-Hall of Canada Inc., Toronto
Prentice-Hall Hispanoamericana, S.A., Mexico
Prentice-Hall of India Pte. Ltd., New Delhi
Prentice-Hall of Japan, Inc., Tokyo
Simon & Schuster Asia Pte. Ltd., Singapore
Editora Prentice-Hall do Brasil, Ltda., Rio de Janeiro

To our families—
for all of their support and understanding

Contents

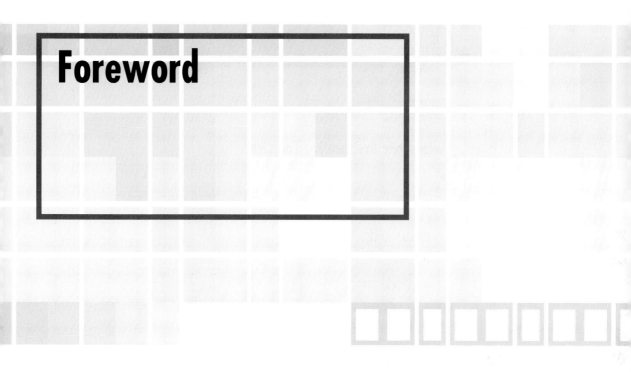

Foreword

LabVIEW celebrates its tenth birthday this year, and although the development team is larger than ever and I am still actively designing and coding the next version along with them, the effort is just as exciting, challenging, and rewarding as always. The excitement comes from our continuously expanding base of enthusiastic customers and the unique systems they build using LabVIEW. The challenge is still how to deploy our limited resources to provide the most valuable enhancements as quickly as possible. And the reward is hearing a customer say that he not only got his job done faster, cheaper and better using LabVIEW, but he had more fun in the process.

With more than two man-centuries of development behind us, it is easy to have confidence in the future of LabVIEW, but that was not the case when it was first introduced—there was the much more significant risk in convincing people that programming graphically was viable.

We had combined two concepts, dataflow and structured programming, and presented the result as a tool for constructing "virtual instruments"—software modules that acquire, process, and display measurements much like conventional bench-top instruments. Dataflow is an elegant and rigorous programming model closely related to functional programming, yet easy to understand and ideally suited to graphical representation. Structured programming is a well-proven technique for constructing maintainable and understandable programs, and it is supported by all the modern textual programming languages. The breakthrough we made was to combine these two methodologies and invent the first practical graphical language, G, which is the heart of LabVIEW. Because G is based on sound computer science principles, it is well suited to a broad range of programming tasks: from low-level computational programming to high-level user-interface programming, from real-time embedded systems to widely distributed networked systems. Ten years after its introduction, there is still no other graphical language that comes close to G, and it's unlikely there ever will be.

But back in 1986 it wasn't clear at all that LabVIEW would be taken seriously. LabVIEW 1 was a huge application, possibly the largest personal computer application of its time, yet it delivered the barest minimum functionality—but it implied so much more. And the innovation inspired our first customers to attempt—and largely succeed at—applications that to us seemed impossible. We shuddered at the thought of megabytes of virtual instrument diagrams being interpreted, but these early applications proved that structured dataflow was a robust model and that graphical programming was a viable technology.

Our first customers gave us the motivation to redouble our efforts and embark on the second and subsequent versions. Once it had been demonstrated that you could construct software by wiring icons together, it was a simple matter to imagine improvements of all kinds: make it possible to move icons that have been wired, make the software execute as fast as if it were a compiled C program, make it easier to edit and debug diagrams, run on all the popular computers, provide more data types, and display styles and graphs, and on and on.

LabVIEW has grown and matured during the past decade, but the powerful core concepts and principles have remained unchanged. I believe this is a key ingredient in the success of

LabVIEW. This book will be yet another key ingredient in the future success of LabVIEW, detailing the benefits of the G graphical programming language for countless applications. Lisa and Jeff have done a great job making LabVIEW easy for beginners, teaching experts helpful tidbits, and describing the latest version of LabVIEW. It is a pleasant journey through the core concepts of G, as well as many interesting details of the entire LabVIEW environment. Enjoy!

JEFF KODOSKY
Inventor of LabVIEW
Vice President of R&D, National Instruments Corporation

Preface: Good Stuff to Know Before You Get Started

Welcome to LabVIEW

LabVIEW®, or Laboratory Virtual Instrument Engineering Workbench, is a graphical programming language that has been widely adopted throughout industry, academia, and research labs as the standard for data acquisition and instrument control software. Currently running on Macintoshes (and Mac clones), Sun SPARCstations, HP 9000/700 Series workstations, and personal computers (PCs) running Windows 3.1, Windows NT, or Windows 95, LabVIEW is a powerful and flexible instrumentation and analysis software system. Computers are much more flexible than standard instruments, and creating your own LabVIEW program, or *virtual instrument* (VI), is simple. LabVIEW's intuitive user interface makes writing and using programs exciting and fun!

Dr. Viola Vogel investigates the structure and function of membrane mimetic interfaces using LabVIEW software on a Macintosh, an epifluorescent microscope, a Langmuir trough, and a high-power laser system. (Photo courtesy of University of Washington.)

LabVIEW departs from the sequential nature of traditional programming languages and features an easy-to-use graphical programming environment, including all of the tools necessary for data acquisition (DAQ), data analysis, and presentation of results. With its graphical programming language, called "G," you program using a graphical block diagram that compiles into machine code. Ideal for a countless number of science and engineering applications, LabVIEW helps you solve many types of problems in only a fraction of the time and hassle it would take to write "conventional" code.

■ Beyond the Lab

LabVIEW has found its way into such a broad spectrum of virtual instrumentation applications that it is hard to know where to begin. As its name implies, it began in the laboratory and still remains very popular in many kinds of laboratories—from major research and development laboratories around the world (such as Lawrence Livermore, Argonne, Batelle, Sandia, Jet Propulsion Laboratory, White Sands, and Oak Ridge in the United States, and CERN in Europe), to R&D laboratories in many industries, and to teaching laboratories in universities all over the world, especially in the disciplines of electrical and mechanical engineering and physics.

The spread of LabVIEW beyond the laboratory has gone in many directions—up (aboard the space shuttle), down (aboard U.S. Navy submarines), and around the world (from oil wells in the North Sea to factories in New Zealand). It is still moving in all directions, headed for new horizons! Virtual instrumentation systems are known for their low cost, both in hardware and development time, and their great flexibility. Is it any wonder that they are so popular?

■ The Expanding Universe of Virtual Instrumentation

Perhaps the best way to describe the expansion (or perhaps explosion) of LabVIEW applications is to generalize it. There are niches in many industries where measurements of some kind are required—most often of temperature, whether it be in an oven, a refrigerator, a greenhouse, a clean room, or a vat of soup. Beyond temperature, users measure pressure, force, displacement, strain, pH, and so on, ad infinitum. Personal computers are used virtually everywhere. LabVIEW is the catalyst that links the PC with measuring things, not only because it makes it easy, but also because it brings along the ability to analyze what you have measured and display it and communicate it halfway around the world if you so choose.

After measuring and analyzing something, the next logical step often is to change (control) something based upon the results. For example, measure temperature and then turn on either a furnace or a chiller. Again, LabVIEW makes this easy to do; monitoring and control have become LabVIEW strengths. Sometimes it is direct monitoring and control, or it may be through communicating with a programmable logic controller (PLC) in what is commonly called supervisory control and data acquisition (SCADA).

■ The Results

You will find descriptions of LabVIEW applications inter-spersed throughout this book. They are written by users in a very brief style to give you the essence of the application and chosen from industry segments in which LabVIEW tends to be very popular—automated electronics testing, semiconductor manufacturing, medical instrumentation, automotive testing, and industrial automation (two applications featured). Of the multitude of successful LabVIEW applications, these are particularly interesting current examples of virtual instrumentation at its finest!

A few of LabVIEW's many uses include:

◆ *Simulating heart activity*

◆ *Controlling an ice cream-making process*

◆ *Detecting hydrogen gas leaks on the space shuttle*

◆ *Monitoring feeding patterns of baby ostriches*

◆ *Modeling power systems to analyze power quality*

◆ *Measuring physical effects of exercise in lab rats*

◆ *Controlling motion of servo and stepper motors*

◆ *Testing circuit boards in computers and other electronic devices*

◆ *Simulating motion in a virtual reality system*

◆ *Supervisor control of the heating ventilation and air conditioning (HVAC) system in National Instruments' manufacturing facility*

Objectives of this Book

LabVIEW for Everyone will help you get LabVIEW up and running quickly and easily, and will start you down the road to becoming an expert G programmer. The book offers additional examples and activities to demonstrate techniques, identifies other sources of information about LabVIEW, and features descriptions of cool LabVIEW applications. You are invited to open, inspect, use, and modify any of the programs on the accompanying CD-ROM (although in many cases you'll need the full version of LabVIEW, not the included sample software, to be able to do much with them).

This book expects you to have basic knowledge of your computer's operating system. If you don't have much computer experience, you may want to spend a little time with your operating system manual and familiarize yourself with your computer. For example, you should know how to access menus, open and save files, make backup disks, and use a mouse.

After reading this book and working through the exercises, you should be able to do the following, and much more, with the greatest of ease:

◆ *Write LabVIEW programs, called virtual instruments, or VIs.*

◆ *Employ various debugging techniques.*

◆ *Manipulate both built-in LabVIEW functions and library VIs.*

◆ *Create and save your own VIs so that you can use them as subVIs, or subroutines.*

◆ *Design custom graphical user interfaces (GUIs).*

◆ *Save your data in a file and display it on a graph or chart.*

◆ *Build applications that use General Purpose Interface Bus (GPIB) or serial instruments.*

◆ *Create applications that use plug-in DAQ boards.*

◆ *Use built-in analysis functions to process your data.*

◆ *Optimize the speed and performance of your LabVIEW programs.*

◆ *Employ advanced techniques such as globals, locals, and attribute nodes.*

◆ *Use LabVIEW to create your instrumentation applications.*

LabVIEW from National Instruments has extensive graphics and analysis capabilities that are used to control and monitor the operation of process control applications, such as this wind tunnel experiment.

Organization

LabVIEW for Everyone helps you get started quickly with LabVIEW to develop your instrumentation and analysis applications. The book is divided into two main sections: Fundamentals and Advanced Topics. The Fundamentals section contains nine chapters and teaches you the fundamentals of G programming in LabVIEW. The Advanced Topics section contains six chapters that further develop your skills and introduce helpful techniques and optimizing strategies. We suggest that you work through the beginning section to master the basics; then, if you're short on time, skip around to what you really want to learn in the advanced section.

In both sections, chapters have a special structure to facilitate learning:

♦ *Overview*, *goals*, and *key terms* describe the main ideas covered in that chapter.

♦ Discussion of the featured topics.

♦ *Activities* reinforce the information presented in the discussion.

♦ *Wrap It Up!* summarizes important concepts and skills taught in the chapter.

♦ Additional activities in many chapters give you more practice with the new material.

▓ Fundamentals

Chapter 1 describes LabVIEW and introduces you to some of LabVIEW's features and uses.

In Chapter 2, you will get an overview of virtual instrumentation: how data acquisition, GPIB, serial port communication, and data analysis are performed with LabVIEW. You will also learn about LabVIEW's history.

In Chapter 3, you will get acquainted with the LabVIEW environment, including the essential parts of a virtual instrument (or VI), the Help window, menus, tools, palettes, and subVIs.

In Chapters 4 and 5, you will become familiar with the basics of G programming in LabVIEW—using controls and indicators (such as numerics, Booleans, and strings); wiring, creating, editing, debugging, and saving VIs; creating subVIs; and documenting your work. You will also begin to understand why G is considered a dataflow programming language.

Chapter 6 describes the basic G programming structures in LabVIEW: while loops, for loops, shift registers, case structures, sequence structures, and formula nodes. It also teaches you how to introduce timing into your programs.

In Chapter 7, you will learn how to use two important data structures—arrays and clusters—in your programs. You will also explore LabVIEW's built-in functions for manipulating arrays and clusters.

Chapter 8 details the variety of charts and graphs available in LabVIEW and teaches you how to use them for animated and informative data presentation.

Chapter 9 discusses string data types, string functions, and tables. It also talks a little about how to save data in and read data from a file, using LabVIEW's easy File I/O VIs.

▤ Advanced Topics

Chapter 10 teaches you more about data acquisition, GPIB, and serial communication. You will learn a bit of theory and some hardware considerations, and you will find a valuable guide to many common acronyms used in instrumentation. Chapter 10 also discusses software setup for data acquisition.

Chapter 11 discusses a few basics on how to use LabVIEW to acquire data using plug-in DAQ boards, as well as a brief overview on communicating with other instruments using GPIB and serial protocols.

Chapter 12 covers some invaluable features like local and global variables, attribute nodes, occurrences, data type conversions, and much more.

Chapter 13 shows you how to configure VI behavior and appearance using VI Setup options and how to access front panel controls using the keyboard. It also introduces LabVIEW's very useful Find function and Profile window.

Chapter 14 covers input/output (I/O) considerations such as printing, advanced file I/O, and networking with other computers.

In Chapter 15, you will learn how to create your own online help files and how to add a customized look to your applications by importing pictures and using the Control Editor. Chapter 15 also describes some good programming techniques that you can use to make your programs run faster, use less memory, port more easily to other platforms, and behave more efficiently overall.

You will find a glossary, index, and appendices at the end of the book.

Appendix A tells you how to contact National Instruments, the maker of LabVIEW, and points you toward other resources that can help you with LabVIEW. It also describes add-on toolkits available to enhance LabVIEW's functionality.

Appendix B talks a little bit about troubleshooting your VIs and answers some common questions.

Pyramid Ranch Research uses LabVIEW to integrate and control its PC-based system that measures, analyzes, and records the dynamic lift or vertical forces generated by the feet during a golf swing.

Conventions Used in this Book

The following table describes the conventions used in this book:

Convention	**Definition**
bold	Bold text denotes VI names, function names, menus, and menu items. In addition, bold text denotes VI input and output parameters. For example, "Choose **New** from the **File** menu to create a new document."
italic	Italic text denotes emphasis, a cross reference, or an introduction to a key term or concept. For example, VIs have three main parts: the *front panel*, the *block diagram*, and the *icon/connector*.
Courier	Courier type denotes text or characters that you enter using the keyboard, and folder or directory names. Sections of code, programming examples, syntax examples, and messages and responses that the computer automatically prints to the screen also appear in this font. For example, "Type Digital Indicator inside the bordered box."
< >	Angle brackets enclose names of keys on the keyboard. For example, <shift>.
Underline	Underlining denotes the name of a control or indicator in your LabVIEW program.
	Note. This icon marks information to which you should pay special attention.
	Watch out! This icon flags a common pitfall or special information that you should be aware of in order to keep out of trouble.
	Tips and Hints. This icon calls your attention to useful tips and hints on how to do something efficiently.

Convention	Definition
	This information applies only to the Macintosh platform.
	This information applies only to UNIX platforms.
	This information applies only to Windows platforms.

A Note About Paths

Different platforms have different conventions for specifying path names. For example, Windows paths take the form X:\LABVIEW\MINE.LLB\BINGO.VI. *The same path on a Macintosh would be denoted* Hard Drive Name:LabVIEW:Mine.llb:Bingo.vi. *On UNIX machines, it would be* /USR/LABVIEW/MINE.LLB/BINGO.VI. *Rather than specifying a full path, this book will list the default path from the LabVIEW directory or folder when telling you where to find an example VI. To simplify notation, we will use the Windows standard to describe paths; if you use Mac OS or UNIX machines, please substitute colons or forward slashes where necessary.*

LabVIEW Installation Instructions (Full Version)

If you have the full version of LabVIEW and need instructions on how to install it, please see the release notes that came with your software.

In addition, you will need to install the Everyone directory or folder from the CD-ROM in the back of this book. It contains the activities in this book and their solutions. To do so, open the appropriate directory for your platform and copy the Everyone directory or folder into your LabVIEW directory. Or, on Windows machines, you can run the setup program on the CD, tell it you have LabVIEW, and tell it to install the Everyone directory in your LabVIEW directory for you.

The EXAMPLES *directory or folder installed by the full version of LabVIEW contains great example programs that you can use as is or modify to suit your needs.*

Your Very Own Demo Disk

The CD included with this book contains a demonstration version of LabVIEW as well as solutions to the activities in the book. It also contains a few example programs for your enjoyment. You can use this sample software to work through almost all of the activities in the Fundamentals section (Chapters 1–9); exceptions are clearly marked. *Most of the activities in the Advanced Topics section of the book are beyond the scope of the sample software and may not load into it.* The sample software does not have the ability to do data acquisition or instrument control using GPIB or the serial port. In addition, most of the analysis functions are not available.

To Accompany
LabVIEW®
FOR EVERYONE

Lisa K. Wells
Jeffrey Travis

ISBN 0-13-268194-3

Software compiled by Crystal Drumheller, Jeffrey Travis, and Lisa Wells

NATIONAL INSTRUMENTS®
The Software is the Instrument®

PTR
PH

LabVIEW © 1996 National Instruments Corp. All rights reserved.
Published by Prentice Hall PTR, Upper Saddle River, NJ 07458

Please note that you can only save VIs temporarily using the sample software. Once you quit the LabVIEW demonstration version, the VIs saved during that session are lost.

To access the part of the sample software that lets you build VIs or investigate existing examples, start the application and select the Explore LabVIEW for Your Own Applications button in the lower right corner of the LabVIEW Demo main menu. Click OK on the first dialog box that pops up. Under Windows and UNIX, select New VI in the next dialog box to start building a fresh VI or choose Open VI to open an existing one. On Macintosh, choose **New** from the **File** menu to open a new VI or **Open** to bring up one that's already built.

The installation process places a directory or folder called vi.lib *in the* LABVIEW *directory or folder.* vi.lib *contains support files for LabVIEW; you should not change the contents of this folder in any way or you might cause yourself much suffering.*

■ Minimum Specifications for LabVIEW and the Sample Software

The basic engine of the sample software is the same as the full version of LabVIEW. The system requirements are also about the same, with the exception of hard disk space. You need about 50 MB of disk storage space to install the entire LabVIEW package (custom installations can be smaller). You can run the sample software directly from the CD-ROM if disk space is tight.

■ Windows

The Windows 3.1 version of LabVIEW 4.0 runs under Windows 3.1 and Windows for Workgroups 3.1 in 386 enhanced mode. You should have a minimum of 8 MB of RAM. LabVIEW can run on an 80386-based PC, but we strongly recommend a computer with an 80486 CPU. In addition, LabVIEW for Windows requires a math coprocessor.

The Windows 95 version of LabVIEW 4.0 runs under any system that supports Windows 95. You should have a minimum of 8 to 12 MB of RAM for this version to run smoothly.

The Windows NT version of LabVIEW 4.0 runs only under Windows NT version 3.5.1 or greater. You should have a minimum of 12 to 16 MB of RAM for this version to run smoothly. LabVIEW for Windows NT only runs on Windows NT 80x86 computers.

The Windows 95 and NT sample versions include online help and require about 13 MB of hard disk space; the Windows 3.1 version does not have online help and needs only about 6.5 MB.

■ MacOS

LabVIEW for the 680x0-based Macintosh requires a math coprocessor. LabVIEW requires a minimum of 5 MB of available RAM, in addition to the RAM requirements for your system software and any other applications that you want to run simultaneously. We recommend that you have at least 8 MB of RAM. You will need about 13 MB of hard disk space.

LabVIEW requires System 7 and will not run under earlier versions of the Macintosh operating system.

■ Sun and HP

LabVIEW for Sun runs on SPARCstations under Solaris 1.1 and 1.1.1 (SunOS 4.1.3) and Solaris 2.3 or later. LabVIEW for HP-UX runs on Hewlett-Packard Model 9000 Series 700 computers under HP-UX 9.0.3 or later.

The workstation should have 32 MB of RAM, with 32 MB or more of swap space storage. LabVIEW can run on less than 24 MB of RAM, but performance will suffer. You will need 8–10 MB of hard disk space.

■ All

LabVIEW uses a directory to store temporary files. Some of the temporary files are large, so we recommend that you have several megabytes of disk space available for this temporary directory. The default for the temporary directory is /tmp on Windows and UNIX, and inside the trash can on MacOS.

■ How to Install the Sample Software

■ Windows

Run setup.exe on the top level of the CD-ROM. It will ask you if you already have a copy of LabVIEW 4.0 or greater. If you do, you can install only the Everyone directory, or you can install the LabVIEW demo software, including Everyone (which takes a lot more space, but has some neat features).

If you do not already have LabVIEW (or if you choose to install the demo software), the setup program will install the correct version of the sample software (Windows 3.1 or Windows 95-NT) in a directory of your choice.

The setup program also gives you the choice of running the demo software from the CD. If you choose this option, setup creates a shortcut to the CD.

■ MacOS

Copy the appropriate folder (either 68KMac or PowerMac) onto your hard drive. Rename the folder LabVIEW if you would like to. If you want to access the LabVIEW demo guide tutorial, install the Adobe Acrobat viewer as well.

■ Sun

SunOS 4.x:

Use the command, `mount -rt hsfs /dev/sr0/cdrom`, then run `/cdrom/lvdemo` to run from the CD-ROM.

If you want to install the software on your hard disk, copy the appropriate directory onto your disk, then launch the file `labview`.

Solaris 2:

The CD mounts automatically as `/cdrom/cdrom0`. Run `/cdrom/cdrom0/lvdemo`.

If you want to install the software on your hard disk, copy the appropriate directory onto your disk, then launch the file `labview`.

■ HP

Use `sam` to mount the CD-ROM. Run `/cdrom/LVDEMO`.

If you want to install the software on your hard disk, copy the appropriate directory onto your disk, then launch the file `LABVIEW`.

Purchasing LabVIEW

LabVIEW runs on the following platforms: Macintosh, Power Macintosh, PCs running Windows 3.1, Windows NT, or Windows 95, Sun SPARCstations, and HP 9000/700-series workstations running HP-UX. If you would like information on how to purchase LabVIEW, contact National Instruments.

National Instruments
6504 Bridge Point Parkway
Austin, Texas 78730

Telephone: (512) 794-0100
Fax: (512) 794-8411
E-mail: `info@natinst.com`
Worldwide Web: `http://www.natinst.com`

Or contact the local National Instruments branch office for your country. See *Appendix A* for branch contact information.

About the Authors

Lisa Wells has extensive experience both using and teaching LabVIEW. During the two years she taught LabVIEW classes, Lisa became attuned to the unique learning needs of the LabVIEW beginner. Lisa first learned LabVIEW as a student and wrote her undergraduate thesis on using LabVIEW for monitoring electrical power quality. She joined the National Instruments (Austin, TX) team in August 1992, after earning a B.S. in electrical engineering and a B.A. in liberal arts from the University of Texas at Austin. Her first 18 months at the company were focused on technical support for LabVIEW and teaching LabVIEW courses. In early 1994, Lisa diverted her attentions to developing and promoting LabVIEW and related products for use by students and instructors in the classroom and laboratory. Lisa is also the author of the *LabVIEW Student Edition User's Guide* and the *LabVIEW Instructor's Guide*. When she's not playing with LabVIEW, Lisa enjoys traveling, snow skiing, water sports, and camping.

Jeffrey Travis is an electrical engineer who first learned to use LabVIEW while obtaining his Master's degree at the University of Texas. He currently works as a systems integration consultant at VI Technology, Inc. in Austin, Texas. He specializes in LabVIEW-based applications for automated testing and signal processing. Jeffrey has had a chance to manage LabVIEW projects for applications ranging from virtual reality environments to testing circuit boards at nuclear power plants. His professional interests include instrumentation system design, biomedical engineering, signal processing, circuit design, and LabVIEW programming. Since there's more to life than LabVIEW, Jeffrey enjoys preparing gourmet dinners for his wife, playing classical guitar, skiing, traveling, and reading literature classics. Jeffrey can best be reached via e-mail at *travis@vi-tech.com*.

Acknowledgments

Although only Jeffrey and Lisa's names appear on this book, many people's contributions were indispensable in the creation and production of this work. Without their support, *LabVIEW for Everyone* would still be just an idea.

Gary Johnson, Sandy Bartnett, Monnie Anderson, Joe Savage, and Andres Thorarinsson deserve much appreciation for their invaluable comments and suggestions on how to improve our manuscript. A big thank-you also goes to Greg McKaskle, Trevor Petruk, Gregg Fowler, and the rest of the LabVIEW development team for their never-ending commitment to LabVIEW and support for this book.

Much appreciation is due all the folks at National Instruments and Prentice Hall who handled the sales, marketing, publications, and various other aspects of *LabVIEW for Everyone.*. Dudley Baker, Kathy Brown, Lara Farwell, Leesa Levesque, Tamra Pringle, and Jeff Kodosky deserve special recognition.

We owe our undying gratitude to Crystal Drumheller for her help writing and compiling the software for the activities in this book.

Lisa sends a special thanks to Dan Phillips, the other half of the National Instruments in Education team, for his patience when this book took priority over everything else .

Jeffrey would like to thank Alex Backus, for supporting him in this endeavor and allowing him to take those "book days" off work to write. A special thanks as well to those engineers at VI Technology, Ashutosh Bhalerao, Mehar Gangishetti, Niranjan Ravulapalli, JP Lugo, and Jorge Rodríguez, whose encouragement and LabVIEW ideas were well received. Least but not last, Nestlé, Jeffrey's cat, is grudgingly acknowledged for walking over the keyboard while he was trying to write the book at his home computer.

Last but by no means least, Jeffrey's most humble gratitude goes to his wife, Stephanie, for being there and putting up with the late nights, a stressed husband and everything else that went into writing this book

Finally, we would like to thank Bernard Goodwin, Diane Spina, and Joe Czerwinski from Prentice Hall for their hard work, flexibility, and devotion to producing a timely and successful publication.

LabVIEW® For Everyone

Graphical Programming Made Even Easier

OVERVIEW

Welcome to the world of LabVIEW! This chapter gives you a basic explanation of LabVIEW, its capabilities, and how it can make your life easier.

YOUR GOALS

- Develop an idea of what LabVIEW really is
- Learn what a graphical language and dataflow programming mean
- Peruse the introductory examples
- Get a feel for the LabVIEW environment

KEY TERMS

- LabVIEW
- G
- virtual instrument (VI)
- dataflow
- graphical language
- front panel
- block diagram
- icon
- connector
- Toolbar
- hierarchy

What in the World Is LabVIEW?

1

What Exactly is LabVIEW, and What Can it Do for Me?

You'd probably like to know what exactly LabVIEW is before you go much further. What can you do with it and what can it do for you? LabVIEW®, short for **Lab**oratory **V**irtual **I**nstrument **E**ngineering **W**orkbench, is a programming environment in which you create programs with graphics; in this regard it differs from traditional programming languages like C, Pascal, or BASIC, in which you program with text. However, LabVIEW is much more than a language. It is a program development and execution system designed for people, like scientists and engineers, who need to program as part of their jobs. LabVIEW works on PCs running Microsoft Windows, Mac OS, Sun SPARCstations, and HP 9000/700 series workstations running HP-UX.

Using the very powerful G programming language, LabVIEW can increase your productivity by orders of magnitude. Programs that take weeks or months to write using conventional programming languages can be completed in hours using LabVIEW, because it is specifically designed to take measurements, analyze data, and present results to the user. And because LabVIEW has such a versatile graphical user interface and is so easy to program with, it is also ideal for simulations, presentation of ideas, general programming, or even teaching basic programming concepts.

LabVIEW offers more flexibility than standard laboratory instruments because it is software-based. You, not the instrument manufacturer, define instrument functionality. Your computer, plug-in hardware, and LabVIEW comprise a completely configurable *virtual instrument* to accomplish your tasks. Using LabVIEW, you can create exactly the type of virtual instrument you need, when you need it, at a fraction of the cost of traditional instruments. When your needs change, you can modify your virtual instrument in moments.

The Space Industries Sheet Float Zone Furnace is used for high-temperature superconductor materials processing research in a microgravity environment aboard the NASA KC-135 parabolic aircraft. LabVIEW controls the industrialized Macintosh-based system.

LabVIEW tries to make your life as hassle-free as possible. It has extensive libraries of functions and subroutines to help you with most programming tasks, without the fuss of pointers, memory allocation, and other arcane programming problems found in conventional programming languages. LabVIEW also contains application-specific libraries of code for data acquisition

(DAQ), General Purpose Interface Bus (GPIB) and serial instrument control, data analysis, data presentation, and data storage. The Analysis Library contains a multitude of useful functions, including signal generation, signal processing, filters, windows, statistics, regression, linear algebra, and array arithmetic.

Because of LabVIEW's graphical nature, it is inherently a data presentation package. Output appears in any form you desire. Charts, graphs, and user-defined graphics comprise just a fraction of available output options. This book will show you how to present data in all of these forms.

LabVIEW's programs are portable across platforms, so you can write a program on a Macintosh and then load and run it on a Windows machine without changing a thing in most applications. You will find LabVIEW applications improving operations in any number of industries, from every kind of engineering and process control to biology, farming, psychology, chemistry, physics, teaching, and many others.

■ Dataflow and the Graphical Language 'G'

The LabVIEW program development environment is different from commercial C or BASIC development systems in one important respect. While other programming systems use text-based languages to create lines of code, LabVIEW uses a *graphical* programming language called *G* to create programs in a pictorial form called a block diagram, eliminating a lot of the syntactical details. With this method, you can concentrate on the flow of data within your application; the simpler syntax doesn't obscure what the program is doing. The following figure shows a simple LabVIEW user interface and the code behind it.

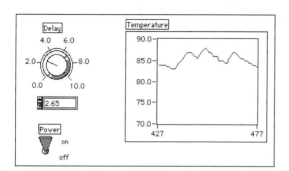
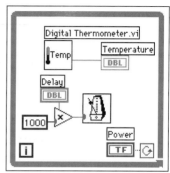

User Interface   Graphical Code

LabVIEW uses terminology, icons, and ideas familiar to scientists and engineers. It relies on graphical symbols rather than textual language to describe programming actions. The principle of *dataflow*, in which functions execute only after receiving the necessary data, governs execution in a straightforward manner. You can learn LabVIEW even if you have little or no programming experience, but you will find knowledge of programming fundamentals very helpful.

■ How Does LabVIEW Work?

LabVIEW programs are called *virtual instruments* (*VIs*) because their appearance and operation imitate actual instruments. However, behind the scenes they are analogous to main programs, functions, and subroutines from popular programming languages like C or BASIC. Hereafter, we will refer to a LabVIEW program as a 'VI' (pronounced "vee eye", NOT the Roman numeral six as we've heard some people say).

A VI has three main parts:

• The *front panel* is the interactive user interface of a VI, so named because it simulates the front panel of a physical instrument. The front panel can contain knobs, push buttons, graphs, and many other controls (which are user inputs) and indicators (which are program outputs). Input data using a mouse and keyboard, and then view the results produced by your program on the screen.

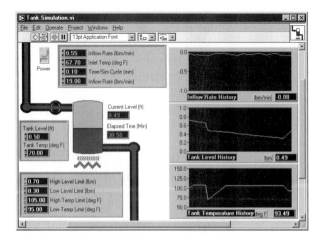

Front Panel

• The *block diagram* is the VI's source code, constructed in Lab-VIEW's graphical programming language, G. *The block diagram is the actual executable program.* The components of a block diagram are lower-level VIs, built-in functions, constants, and program execution control structures. You draw wires to connect the appropriate objects together to indicate the flow of data between them. Front panel objects have corresponding terminals on the block diagram so data can pass from the user to the program and back to the user.

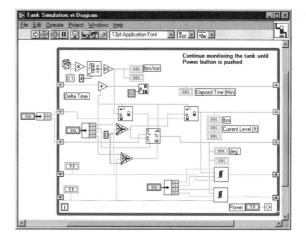

Block Diagram

• In order to use a VI as a subroutine in the block diagram of another VI, it must have an *icon* and a *connector.* A VI that is used within another VI is called a *subVI* and is analogous to a subroutine. The icon is a VI's pictorial representation and is used as an object in the block diagram of another VI. A VI's connector is the mechanism used to wire data into the VI from other block diagrams when the VI is used as a subVI. Much like parameters of a subroutine, the connector defines the inputs and outputs of the VI.

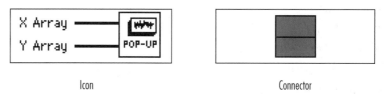

Icon   Connector

Virtual instruments are *hierarchical* and *modular*. You can use them as top-level programs or subprograms. With this architecture, LabVIEW promotes the concept of *modular programming*. First, you divide an application into a series of simple subtasks.

Next, build a VI to accomplish each subtask and then combine those VIs on a top-level block diagram to complete the larger task.

Modular programming is a plus because you can execute each subVI by itself, which facilitates debugging. Furthermore, many low-level subVIs often perform tasks common to several applications and can be used independently by each individual application.

Just so you can keep things straight, we've listed a few common LabVIEW terms with their conventional programming equivalents.

LabVIEW Terms and Their Conventional Equivalents

LabVIEW	Conventional Language
VI	program
function	function
subVI	subroutine, subprogram
front panel	user interface
block diagram	program code
G	C, Pascal, BASIC, etc.

Demonstration Examples

Okay, enough reading for now. To get an idea of how LabVIEW works, you can open and run a few existing LabVIEW programs.

If you are using the full version of LabVIEW, just launch it. Make sure you have already installed the Everyone directory as described in the Preface; it contains the activities for this book. After launching LabVIEW, a new, untitled window will appear. To open an example, select **Open** from the **File** menu and choose the one you want.

If you are using the LabVIEW sample software, the Everyone directory will be installed when you install the sample version. To access the part of the evaluation software that lets you build VIs or investigate existing examples, start the application and select the Explore LabVIEW for Your Own Applications button in the lower right corner of the demo's main menu. Click OK on the first dialog box that pops up. If you are using Windows or UNIX, select New VI on the next dialog box to start building a fresh VI or Open VI to access an existing one. On Macintosh machines, select **New** from the **File** menu to create a new VI or **Open** to bring up one that's already built.

Throughout this book, use the left mouse button (if you have more than one) unless we specifically tell you to use the right one. On Macintosh computers, <command>-click when right-mouse functionality is necessary. In most LabVIEW situations, the <control> key on PCs will correspond to <command> on Macs, <meta> on Suns, and <alt> on HP machines.

Activity 1-1
Temperature System Demo

Open and run the VI called **Temperature System Demo.vi** by following these steps:

1. Launch LabVIEW if you haven't already. If you are using the sample software, make sure you are in the practice area.

2. Select **Open** from the **File** menu, or click the Open VI button if you have a dialog box that says "There are no VIs open."

3. Next, open the Everyone directory or folder by double-clicking on it. Then select CH1.LLB. Finally, open **Temperature System Demo.vi**. (If you have the full version of LabVIEW, you can also find this example under examples/apps/tempsys.llb). After a few moments, the Temperature System Demo front panel window appears, as shown in the next illustration. The front panel contains numeric controls, Boolean switches, slide controls, knob controls, charts, graphs, and a thermometer indicator.

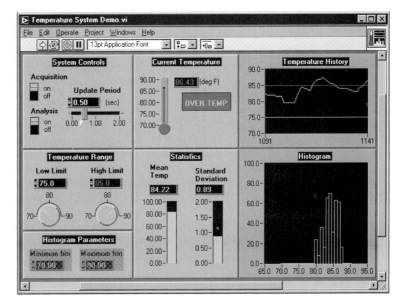

Run Button

Run Button (active)

Abort Button

4. Run the VI by clicking on the Run button. The button changes appearance to indicate that the VI is running. The *Toolbar*, which is the row of icons on the top bar of the screen, also changes, since editing functionality won't be necessary while the VI is running.

Notice also that the Abort button becomes active in the Toolbar. You can press it to abort program execution.

Temperature System Demo.vi simulates a temperature monitoring application. The VI makes temperature measurements and displays them in the thermometer indicator and on the chart. Although the readings are simulated in this example, you can easily modify the program to measure real values. The Update Period slide controls how fast the VI acquires the new temperature readings. LabVIEW also plots high and low temperature limits on the chart; you can change these limits using the Temperature Range knobs. If the current temperature reading is out of the set range, LEDs light up next to the thermometer.

This VI continues to run until you click the Acquisition switch to *off*. You can also turn the data analysis on and off. The Statistics section shows you a running calculation of the mean and standard deviation, and the Histogram plots the frequency with which each temperature value occurs.

■ Tweaking Values

Operating Tool

Enter Button

5. Use the cursor, which takes on the personality of the Operating tool while the VI is running, to change the values of the high and low limits. Highlight the old high or low value, either by clicking twice on the value you want to change, or by clicking and dragging across the value with the Operating tool. Then type in the new value and click on the enter button, located next to the run button on the Toolbar.

6. Change the Update Period slide control by placing the Operating tool on the slider, then clicking and dragging it to a new location.

You can also operate slide controls using the Operating tool by clicking on a point on the slide to snap the slider to that location, by clicking on a scroll button to move the slider slowly toward the arrow, or by clicking in the slide's digital display and entering a number.

Even though the display changes, LabVIEW does not accept the new values in digital displays until you press the enter button, or click the mouse in an open area of the window.

7. Try adjusting the other controls in a similar manner.

8. Stop the VI by clicking on the <u>Acquisition</u> switch.

■ Examine the Block Diagram

The block diagram shown in the following illustration represents a complete LabVIEW application. You don't need to understand all of these block diagram elements right now—we'll deal with them later. Just get a feel for the nature of a block diagram. If you already do understand this diagram, you'll probably fly through the first part of this book!

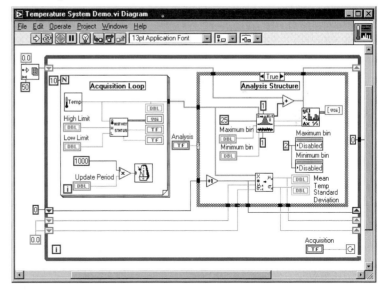

9. Open the block diagram of **Temperature System Demo.vi** by choosing **Show Diagram** from the **Windows** menu.

10. Examine the different objects in the diagram window. Don't panic at the detail shown here! These structures are explained step by step later in this book.

11. Open the Help window by choosing **Show Help** from the **Help** menu. Position the cursor over different objects in the block diagram and watch the Help window change to show descrip-

tions of the objects. If the object is a function or subVI, the Help window will describe the inputs and outputs as well.

Hierarchy

LabVIEW's power lies in the hierarchical nature of its VIs. After you create a VI, you can use it as a subVI in the block diagram of a higher-level VI, and you can have as many layers of hierarchy as you need. To demonstrate this versatile ability, look at a subVI of **Temperature System Demo.vi**.

12. Open the **Temperature Status** subVI by double-clicking on its icon.

The front panel shown in the following illustration springs to life.

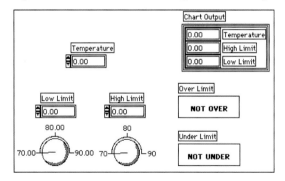

Icon and Connector

The icon and connector provide the graphical representation and parameter definitions needed if you want to use a VI as a subroutine or function in the block diagrams of other VIs. They reside in the upper right corner of the VI's front panel window. The icon graphically represents the VI in the block diagram of other VIs, while the connector terminals are where you must wire the inputs and outputs. These terminals are analogous to parameters of a subroutine or function. You need one terminal for each front panel control and indicator through which you want to pass data to the VI. The icon sits on top of the connector pattern until you choose to view the connector.

Temperature ──────── Chart Output
High Limit ──── Over Limit
Low Limit ──── Under Limit

Temperature Status icon   connector

By using subVIs, you can make your block diagrams modular and more manageable. This modularity makes VIs easy to maintain, understand, and debug. In addition, you can often create one sub-VI to accomplish a function required by many different VIs.

Now run the top level VI with both its window and the **Temperature Status** subVI window visible. Notice how the subVI values change as the main program calls it over and over.

13. Select **Close** from the **File** menu of the **Temperature Status** subVI. Do not save any changes.

14. Select **Close** from the **File** menu of **Temperature System Demo.vi**, and do not save any changes.

*Selecting **Close** from the **File** menu of a VI diagram closes the block diagram window only. Selecting **Close** on a front panel window closes both the panel and the diagram.*

Activity 1-2
Frequency Response Example

This example measures the frequency response of an unknown "black box." A function generator supplies a sinusoidal input to the black box (hint: it contains a bandpass filter, which lets only certain signal components through it). A digital multimeter measures the output voltage of the black box. While this VI uses subVIs to simulate a function generator and a digital multimeter, real instruments could easily be hooked up to a real black box to provide real-world data. You would then use sub-VIs to control data acquisition, GPIB transfers, or serial port communication to bring in or send out real data instead of simulating it.

You will open, run, and observe the VI in this activity.

1. Select **Open** from the **File** menu to open the VI or click the Open VI button if you have the "There are no VIs open" dialog box.

2. Select the EVERYONE directory, then CH1.LLB. Finally, double-click on **Frequency Response.vi**. (If you have the full version of

LabVIEW, you can also find this example under
examples/apps/freqresp.llb). The front panel shown in
the next illustration should appear.

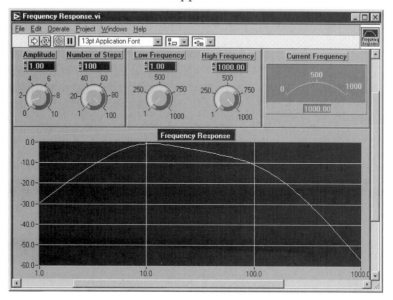

Run Button

3. Run the VI by clicking on the run button. You can specify the
amplitude of the input sine wave and the number of steps the VI
uses to find the frequency response by changing the Amplitude
control and the Number of Steps control, then run the VI again.
You can also specify the frequency sweep by inputting the upper
and lower limits with the Low Frequency and High Frequency
knobs. Play with these controls and observe the effect they have
on the output of the "black box."

4. Open and examine the block diagram by choosing **Show Diagram** from the **Windows** menu.

5. Close the VI by selecting **Close** from the **File** menu. These exercises should give you a basic feel for LabVIEW's 'G' programming environment. With the G language, you'll find writing powerful applications (and debugging them) to be a snap! Read on to learn how!

Wrap It Up!

LabVIEW is a powerful and flexible instrumentation and
analysis software system. It uses the graphical programming
language, G, to create programs called *virtual instruments*, or

VIs. The user interacts with the program through the *front panel.* Each front panel has an accompanying *block diagram,* which is the VI's source code. LabVIEW has many built-in functions to facilitate the programming process; components are wired together to show the flow of data within the block diagram. Stay tuned—the next chapters will teach you how to effectively use LabVIEW's many features.

You will find the solutions to every activity in the upcoming chapters in the EVERYONE *directory on the CD that accompanies this book. We'll trust you not to cheat!*

Additional Activities

■ Activity 1-3 **Neat Demos**

Even if you have the full version of LabVIEW, install the demo version that comes with this book (installation instructions can be found in the Preface). Run the demo, but do not shell out to the development environment like you did in the previous activities. Instead, explore the demo and see the different types of things LabVIEW can do. Throughout this book you will find that you can learn a lot just by observing existing programs.

After you've looked at what the different buttons can show you, press the Explore LabVIEW for Your Own Applications button to get to the practice area. Open and play with the VIs found in Demos.11b. Make sure to **Show Diagram** from the **Windows** menu so you can get a feel for how these neat programs work.

■ Activity 1-4 **Examples**

In this activity, you will look at some example programs that ship with LabVIEW. *You will need the full version of LabVIEW to do this activity.*

1. Open the **readme.vi** located in LabVIEW's examples directory.

Run Button

2. Run the VI by clicking on its Run button. The **readme.vi** allows you to browse through directories, libraries (.llb), and VIs to display their description.

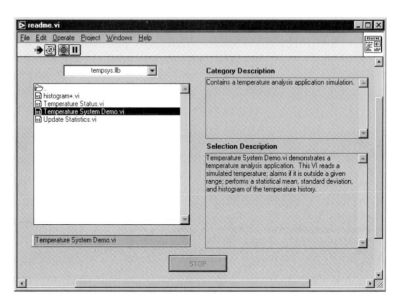

If you highlight a directory or VI name, this VI will display information about the directory's contents or the VI's functions. You can double-click to open directories and access the directories or VIs inside. Press the <u>Stop</u> button when you've finished exploring.

3. Choose **Show Diagram** from the **Windows** menu to see what the program looks like.

4. Now look through and run other VIs in the `examples` directory to get an idea of the LabVIEW environment and what you can do with it. You might use **readme.vi** to choose which ones interest you. Although all of the examples are extremely educational, you should investigate a few particularly interesting directories: `apps`, `appbuild.llb`, `analysis`, and `daq` . Feel free to browse through any VIs that strike your fancy; you can learn a lot just by watching how they work. Also feel free to modify and use these examples for your own applications.

5. When you're done, select **Close** from the **File** menu to close each VI. Do not save any changes you may have made.

OVERVIEW

Virtual instrumentation is the foundation for the modern laboratory. A virtual instrument consists of a computer, software, and a plug-in board simulating the function of traditional hardware instrumentation; it's also what we call a LabVIEW program. Because their functionality is software-defined by the user, virtual instruments are extremely flexible, powerful, and cost-effective. This chapter explains how to communicate with the outside world (e.g., take measurements, 'talk' to an instrument, send data to another computer) using LabVIEW. We're only giving you a very brief overview here; you can learn more about acquiring data, controlling instruments, and networking your computer with LabVIEW in the second half of this book. You'll also learn a little about how LabVIEW has changed over the years.

YOUR GOALS

- Understand the nature of data acquisition and GPIB
- Be able to describe the components of a typical DAQ or GPIB system
- Learn about your computer's built-in serial port
- Appreciate the usefulness of analysis functions
- Learn a little about VXI
- See how LabVIEW can exchange data with other computers and applications
- Know about some of the toolkits that enhance LabVIEW's capabilities

KEY TERMS

- data acquisition (DAQ)
- General Purpose Interface Bus (GPIB)
- Institute of Electrical and Electronic Engineers (IEEE) 488 standard
- serial port

- VXI
- networking
- dynamic link library (DLL)
- code interface node (CIN)
- object linking and embedding (OLE)
- toolkit

Virtual Instrumentation: Hooking Your Computer Up to the Real World

2

The Evolution of LabVIEW

In 1983, National Instruments began to search for a way to minimize the time needed to program instrumentation systems. Through this effort, the LabVIEW virtual instrument concept evolved—intuitive front panel user interfaces combined with an innovative block diagram programming methodology to produce an efficient, software-based graphical instrumentation system.

LabVIEW version 1 was released in 1986 on the Macintosh only. Although the Mac was not widely used for measurement and instrumentation applications, its graphical nature best accommodated the LabVIEW technology until the more common operating systems could support it.

By 1990, National Instruments had completely rewritten LabVIEW, combining new software technology with years of customer feedback. More importantly, LabVIEW 2 featured a compiler that made execution speeds of VIs comparable with

programs created in the C programming language. The United States Patent Office issued several patents recognizing the innovative LabVIEW technology.

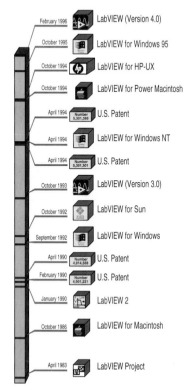

February 1996	LabVIEW (Version 4.0)
October 1995	LabVIEW for Windows 95
October 1994	LabVIEW for HP-UX
October 1994	LabVIEW for Power Macintosh
April 1994	U.S. Patent Number 5,301,336
April 1994	LabVIEW for Windows NT
April 1994	U.S. Patent Number 5,301,301
October 1993	LabVIEW (Version 3.0)
October 1992	LabVIEW for Sun
September 1992	LabVIEW for Windows
April 1990	U.S. Patent Number 4,914,568
February 1990	U.S. Patent Number 4,901,221
January 1990	LabVIEW 2
October 1986	LabVIEW for Macintosh
April 1983	LabVIEW Project

As new graphical operating systems like the MacOS appeared, National Instruments ported the now mature LabVIEW technology to the other platforms: PCs and workstations. In 1992, they introduced LabVIEW for Windows and LabVIEW for Sun based on the new portable architecture.

LabVIEW 3 arrived in 1993 for Macintosh, Windows, and Sun operating systems. LabVIEW 3 programs written on one platform can run on another. This multiplatform compatibility gives users the opportunity to choose the development platform while ensuring that they can run their VIs on other platforms.

In 1994, the list of LabVIEW-supported platforms grew to include Windows NT, Power Macintoshes, and HP workstations. 1995 brought about an adaptation to Windows 95.

LabVIEW 4, the most recent release as of 1996, features a more customizable development environment so users can create their own workspace to match their industry, experience level, and development habits. In addition, LabVIEW 4 adds high-powered editing and debugging tools for advanced instrumentation systems, as well as OLE-based connectivity and distributed execution tools.

Although LabVIEW is a very powerful simulation tool, it is most often used to gather data from an external source, and it contains many VIs built especially for this purpose. For example, LabVIEW can command plug-in *data acquisition*, or *DAQ*, boards to acquire or generate analog and digital signals. You might use DAQ boards and LabVIEW to monitor a temperature, send signals to an external system, or determine the frequency of an unknown signal. LabVIEW also facilitates data transfer over the *General Purpose Interface Bus (GPIB)* and through your computer's built-in serial port. GPIB is frequently used to communicate with oscilloscopes, scanners, and multimeters, and to drive instruments from remote locations. LabVIEW software can also control sophisticated VXI hardware instrumentation systems. Once you have acquired or received your data, you can use LabVIEW's many analysis VIs to process and manipulate it.

Often you will find it useful to share data with other applications or computers in addition to an instrument. LabVIEW has built-in functions that simplify this process, supporting several networking protocols, external calls to existing code or dynamic link libraries (DLLs), and object linking and embedding (OLE). We'll spend the rest of this chapter talking about some of the tasks LabVIEW was designed to accomplish.

What Is Data Acquisition?

Data acquisition, or *DAQ*, is simply the process of measuring a real-world signal, such as a voltage, and bringing that information into the computer for processing, analysis, storage, or other data manipulation. The following illustration shows the components of a DAQ system. Physical phenomena represent the real-world signals you are trying to measure, such as speed, temperature, humidity, pressure, flow, pH, start-stop, radioactivity, light intensity, and so on. Transducers, a fancy word for 'sensors,' evaluate the physical phenomena and produce electrical signals proportionately. For example, thermocouples, a type of transducer, convert temperature into a voltage that an A/D (analog to digital) converter can measure. Other examples of transducers include strain gauges, flowmeters, and pressure transducers, which measure force, rate of flow, and pressure, respectively. In each case, the electrical signal produced by the transducer is directly related to the phenomenon it monitors.

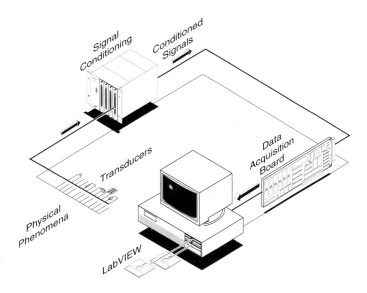

DAQ System

LabVIEW can command DAQ boards to read analog input signals (A/D conversion), generate analog output signals (D/A conversion), read and write digital signals, and manipulate the on-board counters for frequency measurement, pulse generation, etc., to interface with the transducers. In the case of analog input, the voltage data from the sensor goes into the plug-in DAQ board in the computer, which sends the data into computer memory for storage, processing, or other manipulation.

Signal conditioning modules "condition" the electrical signals generated by transducers so that they are in a form that the DAQ board can accept. For example, you would want to isolate a high-voltage input such as lightning, lest you fry both your board and your computer—a costly mistake! Signal conditioning modules can apply to many different types of conditioning: amplification, linearization, filtering, isolation, and so on. Not all applications will require signal conditioning, but many do, and you should pay attention to your specifications to avoid a potential disaster. In addition, information loss can be even worse than equipment loss! Noise, nonlinearity, overload, aliasing, etc. can hopelessly corrupt your data and LabVIEW will not save you. Signal conditioning is often *not* optional—it's best to check before you start!

The AT-MIO-16E-2, AT-MIO-16XE-50, AT-MIO-16E-10, AT-MIO-16DE-10, and AT-MIO-64E-3, from National Instruments, are completely software-configurable, software-calibrated plug-in DAQ boards for PC AT and EISA computers.

To acquire data in your lab using the virtual instrumentation approach, you will need a DAQ board, a computer configured with LabVIEW and DAQ driver software, and some method of connecting your transducer signal to the board, such as a connector block, breadboard, cable, or wire. You may also need signal conditioning equipment, depending on the specifications of your application.

For example, if you wanted to measure a temperature, you would need to wire the temperature sensor to an analog input channel on the DAQ board in your computer (often via signal conditioning equipment, depending on the sensor). Then use LabVIEW's DAQ VIs to read the channel on the board, display the temperature on the screen, record it in a data file, and analyze it any way you need to.

The built-in LabVIEW data acquisition VIs only work with National Instruments DAQ boards. If you are using a board from another vendor, you will have to get a driver from them (if they have one), or you will have to write your own driver code and call it from LabVIEW using code interface nodes or dynamic link libraries—not a trivial task.

What Is GPIB?

Hewlett Packard developed the *General Purpose Interface Bus*, or *GPIB*, in the late 1960s to facilitate communication between computers and instruments. A bus is simply the means by which computers and instruments transfer data, and GPIB provided a much-needed specification and protocol to govern this communication. The Institute of Electrical and Electronic Engineers (IEEE) standardized GPIB in 1975, and it became known

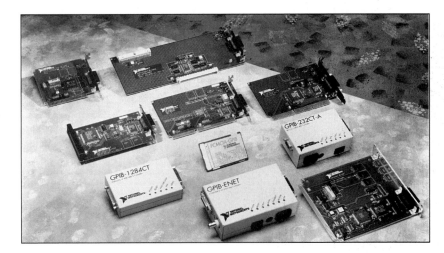

as the *IEEE 488* standard. GPIB's original purpose was to provide computer control of test and measurement instruments. However, its use has expanded beyond these applications into other areas, such as computer-to-computer communication and control of multimeters, scanners, and oscilloscopes.

GPIB is a digital, 24-conductor parallel bus. It consists of eight data lines, five bus management lines (ATN, EOI, IFC, REN, and SRQ), three handshake lines, and eight ground lines. GPIB uses an eight-bit parallel, byte-serial, asynchronous data transfer scheme. In other words, whole bytes are sequentially moved across the bus at a speed determined by the slowest participant in the transfer. Because GPIB sends data in bytes (one byte = eight bits), the messages transferred are frequently encoded as ASCII character strings. Your computer can only perform GPIB communication if it has a GPIB board (or external GPIB box) and the proper drivers installed.

You can have many instruments and computers connected to the same GPIB bus. Every device, including the computer interface board, must have a unique GPIB address between 0 and 30, so that the data source and destinations can be specified by this number. Address 0 is normally assigned to the GPIB interface board. Instruments connected to the bus can use addresses 1 through 30. The GPIB has one Controller, usually your computer, that controls the bus management functions. To transfer instrument commands and data on the bus, the Controller addresses one Talker and one or more Listeners. The data strings are then sent across the bus from the Talker to the Listener(s). The LabVIEW GPIB VIs automatically handle the addressing and most other bus management functions, saving you the hassle of low-level programming. The following illustration shows a typical GPIB system.

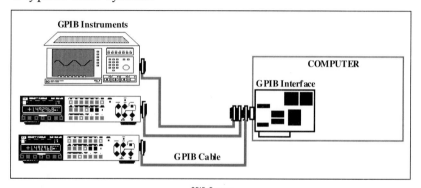

GPIB System

Although using GPIB is one way to bring data into a computer, it is fundamentally different from performing data acquisition, even though both use boards that plug into the computer. Using a special protocol, GPIB talks to another computer or instrument to bring in data acquired by that device, while data acquisition involves connecting a signal directly up to a DAQ board in the computer.

To use GPIB as part of your virtual instrumentation system, you need a GPIB board or external box, a GPIB cable, LabVIEW and a computer, and an IEEE 488-compatible instrument with which to communicate (or another computer containing a GPIB board). You also need to install the GPIB driver software on your computer as well, according to the directions that accompany LabVIEW or the board.

LabVIEW's GPIB VIs communicate with National Instruments GPIB boards, but not those from other manufacturers. If you have another vendor's board, you can either get driver software from them (if it's available) or write your own driver code and integrate it into LabVIEW. As with DAQ drivers, it's not an easy thing to do!

We'll talk more about DAQ and GPIB in Chapters 10 and 11.

Communica- tion Using the Serial Port

Serial communication is another popular means of transmitting data between a computer and another computer, or a peripheral device such as a programmable instrument. It uses the built-in *serial port* (either RS-232 or RS-422 standard) in your computer. Serial communication uses a transmitter to send data one bit at a time over a single communication line to a receiver. You can use this method when data transfer rates are low, or when you must transfer data over long distances. It is slower and less reliable than the GPIB, but you do not need a board in your computer to do it and your instrument does not need to conform to the IEEE 488 standard. The following illustration shows a typical serial communication system.

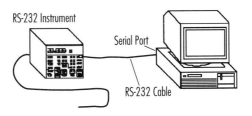

RS-232 Instrument

Serial Port

RS-232 Cable

Serial Communication System

Serial communication is handy because most computers have one or two serial ports built in—you can send and receive data without buying any special hardware. Many GPIB instruments also have built-in serial ports. However, unlike GPIB, a serial port can communicate with only one device, which can be limiting for some applications. Serial port communication is also painstakingly slow and has no built-in error checking capabilities. However, serial communication has its uses (it's certainly economical!), and the LabVIEW Serial library contains ready-to-use functions for serial port operations. If you have a cable and a device to 'talk' to, you are all set to try out serial communication!

Real-World Applications: Why We Analyze

Once you get data into your computer, you may want to process your data somehow. Modern, high-speed floating-point digital signal processors have become increasingly important to real-time and analysis systems. A few of the many possible analysis applications for LabVIEW include biomedical data processing, speech synthesis and recognition, and digital audio and image processing.

The importance of integrating analysis libraries into laboratory stations is obvious: The raw data collected from your DAQ board or GPIB instrument does not always immediately convey useful information. Often you must transform the signal, remove noise perturbations, correct for data corrupted by faulty equipment, or compensate for environmental effects such as temperature and humidity. This picture shows data that epitomizes the need for the analysis functions.

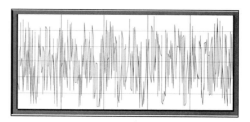

By analyzing and processing the digital data, you can extract the useful information from the noise and present it in a form more comprehensible than the raw data. The processed data looks more like this:

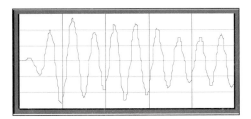

The LabVIEW block diagram programming method and the extensive set of LabVIEW analysis VIs simplify the development of analysis applications. The following sample block diagram illustrates the LabVIEW programming concept.

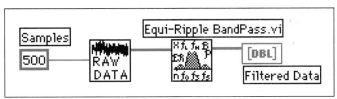

Because the LabVIEW analysis functions give you popular data analysis techniques in discrete VIs, you can wire them together, as shown in the previous picture, to analyze data. Instead of worrying about implementation details for analysis routines as you do in most programming languages, you can concentrate on solving your data analysis problems. LabVIEW's analysis VIs are powerful enough for experts to build sophisticated analysis applications using digital signal processing (DSP), digital filters, statistics, or numerical analysis. At the same time, they are simple enough for novices to perform sophisticated calculations.

The LabVIEW analysis VIs efficiently process blocks of information represented in digital form. They cover the following major processing areas:

◆ *Pattern generation*   ◆ *Curve fitting*

◆ *Digital signal processing*   ◆ *Linear algebra*

◆ *Digital filtering*   ◆ *Numerical analysis*

◆ *Smoothing windows*   ◆ *Measurement-based analysis*

◆ *Statistical analysis*

A Little Bit about VXI

The *VXI*bus, an acronym for VMEbus eXtensions for Instrumentation, is an instrumentation standard for instrument-on-a-card systems. First introduced in 1987 and based on the VMEbus (IEEE 1014) standard, the VXIbus is an exciting and fast-growing platform for instrumentation systems. VXI consists of a mainframe chassis with slots holding modular instruments on plug-in boards. A variety of instrument and mainframe sizes is available from numerous vendors, and you can also use VME modules in VXI systems. VXI has a wide array of uses in traditional test and measurement and ATE (automated test equipment) applications. VXI's popularity is also growing as a platform for data acquisition and analysis in research and industrial control applications. Many users are migrating to VXI by integrating it into existing systems along with traditional GPIB instruments and/or plug-in DAQ boards.

An engineer uses LabVIEW and VXI to test frequency response.

VXI combines some of the best technology from GPIB instruments, plug-in DAQ boards, and modern computers. Like GPIB, VXI offers a wealth of powerful instruments from leading vendors. Like plug-in DAQ boards, VXI offers modularity, flexibility, and much higher performance. Because VXI combines a sophisticated instrument environment with a modern computer backplane, VXI instruments have the ability to communicate at very high speeds using the best techniques of both GPIB instruments and plug-in DAQ boards.

VXI*plug&play* is a name used in conjunction with VXI products that have additional standardized features beyond the scope of the baseline specifications. VXI*plug&play*-compatible instruments include standardized software, which provides soft front panels, instrument drivers, and installation routines to take full advantage of instrument capabilities and make your programming task as easy as possible. LabVIEW software for VXI is fully compatible with VXI*plug&play* specifications.

Connectivity

In some applications, you will want to share data with other programs or computers. LabVIEW has built-in functions that simplify this process. These VIs facilitate communication over a network, call *dynamic link libraries (DLLs)* or external code, and support *object linking and embedding (OLE)*. Using the add-on SQL Toolkit—DatabaseVIEW, LabVIEW can also communicate with most SQL (structured query language) databases.

■ Networking

For our purposes, *networking* refers to communication between multiple processes that can, optionally, run on separate computers. This communication usually occurs over a hardware network such as ethernet or LocalTalk. One main use for networking in software applications is to allow one or more applications to use the services of another application.

For communication between processes to work, the processes must use a common communications language, referred to as a protocol. Several protocols have emerged as accepted standards, and they are not generally compatible with one another.

◆ *TCP—available on all computers*

◆ *UDP—available on all computers*

◆ *DDE—available on the PC, for communication between Windows applications*

◆ *AppleEvents—available on the Macintosh, for sending messages between Mac applications*

◆ *PPC—available on the Macintosh, for sending and receiving data between Mac applications*

This book will talk more about networking in Chapter 14, *Communications and Advanced File I/O.*

■ DLLs and CINs

For increased flexibility, LabVIEW can call external text-based code routines or *dynamic link libraries (DLLs)* and integrate these routines into its own program execution. In case you were wondering, a dynamic link library is a library of shared functions that an application can link to at runtime, instead of at compile time. LabVIEW uses a special block diagram structure called a *code interface node (CIN)* to link conventional, text-based code to a VI. LabVIEW calls the executable code when the node executes, passing input data from the block diagram to the executable code, and returning data from the executable code to the block diagram. Similarly, you can use the Call Library Function to call a DLL if you are running under Windows.

Most applications never require the use of a CIN or DLL. Although the LabVIEW compiler can usually generate code that is fast enough for most tasks, CINs and DLLs are useful for some tasks that are time-critical, require a great deal of data manipulation, or that you've already written specific code for. They are also useful for tasks that you can't perform directly from the diagram, such as calling system routines for which LabVIEW functions do not exist.

■ OLE

Object linking and embedding (OLE) is a Microsoft tool for embedding objects from one application into another application. For example, a spreadsheet might be included on a word processing document. When the text document is loaded, the current values that are found in the spreadsheet are automatically included in the document. LabVIEW includes OLE automation client functions, which allow it to control OLE automation servers such as ACCESS, Excel, and Word. This functionality is only available on Windows machines.

LabVIEW Add-on Toolkits

You can use the following special add-on *toolkits* with Lab-VIEW to increase your flexibility and capabilities. For more information about them, see Appendix A or contact National Instruments.

- ◆ *Application Builder*
- ◆ *Test Executive Toolkit*
- ◆ *JTFA Toolkit*
- ◆ *Digital Filter Design Toolkit*
- ◆ *Third Octave Analyzer Toolkit*
- ◆ *Signal Processing Suite*
- ◆ *Picture Control Toolkit*
- ◆ *SPC Toolkit*
- ◆ *PID Control Toolkit*
- ◆ *SQL Toolkit—DatabaseVIEW*

Many third-party developers, often National Instruments Alliance Members, make other add-ons to LabVIEW that do all sorts of things. If you have a specific task and you want to know if someone's already done it, we suggest posting to the `info-labview` user forum (see Appendix A for details).

Wrap It Up!

LabVIEW's built-in functions facilitate hardware communication with external devices so you don't have to write involved programs. LabVIEW virtual instruments can work with several types of hardware to gather or exchange data: plug-in DAQ boards, GPIB boards, your computer's built-in serial port, or VXI hardware. You can use National Instruments DAQ boards, managed by LabVIEW, to read and generate analog input, analog output, and digital signals and also to perform counter/timer operations. LabVIEW can also control communication over the GPIB bus (assuming you have a GPIB board) or command a VXI instrumentation system. If you don't have any special hardware, LabVIEW can communicate with other devices through your computer's serial port.

LabVIEW analysis VIs make it easy for you to process and manipulate data once you have brought it into your computer. Rather than working through tricky algorithms by hand or trying to write your own low-level code, you can simply access the built-in LabVIEW functions that suit your needs.

You can use LabVIEW's built-in connectivity functions to communicate with other computers linked by a network or with other applications on the same computer. LabVIEW includes VIs to support several networking protocols, OLE compatibility, and to call DLLs and external text-based code.

If you want to expand LabVIEW's considerable functionality, you can buy add-on toolkits to accomplish specific tasks. Toolkits are available to build LabVIEW applications, design digital filters, automate test execution, analyze certain signals, create custom pictures, perform statistical process and PID control, and communicate with an SQL database.

The next chapters will teach you the fundamentals of LabVIEW programming, so get ready to write some code!

Radio-Linked Environmental Monitoring And Display System

By Andres Tho-
rarinsson,
Chief Engineer,
Vista Engineers

The Challenge

Developing an auto-
mated environmental
monitoring system
with a graphical user
interface in a reason-
able time.

The Solution

Using LabVIEW to de-
velop a supervisory
control and data ac-
quisition (SCADA) sys-
tem for monitoring
programmable logic
controllers (PLCs) and
data loggers.

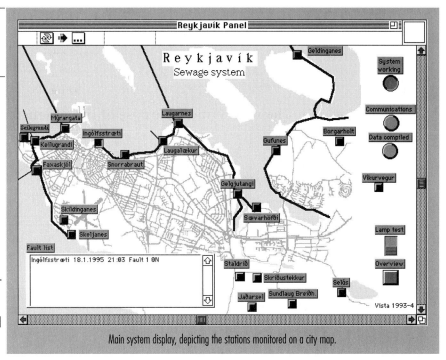

Main system display, depicting the stations monitored on a city map.

Introduction

The City of Reykjavik, Iceland, has built a LabVIEW-based automated system for monitoring its sewer system. The primary job of the system is to monitor several pumping stations as well as stations that measure flow rates and rainfall. The sites were previously monitored during weekly visits. Currently, we have interconnected five pumping stations, five measuring stations, and four monitoring stations.

The Design Phase

At the start of the project, two key questions emerged:

1. Will communication occur via wire or radio?

After investigation, we found that radio communications would have several benefits, such as lower operating costs, easy installation without wiring, and easy expandability.

2. Which software should we select for the monitoring stations?

After contacting some of the larger manufacturers of software for acquisition and control, we felt that only LabVIEW had the capabilities required—especially its flexibility in writing drivers for communications, string handling, and downloading to hard disk—as well as its easy user interface and its system integration capability. Although LabVIEW offers much more than the engineers asked for, we chose it for developing the monitoring system.

The most critical purpose of the system is to monitor the alarms in the *pumping stations*, although we also monitor operating conditions. For reliability, we chose Telemecanique TSX-17 PLCs for data collection and analysis. We created a standard information mailbox, suitable for a maximum of eight pumps. The mailbox has numerous data fields, such as the total running hours of each pump, number of starts, energy usage (kWh) of each set of pumps, level in sump, flow, temperature, 16 alarm states, and 16 on/off states. When transmitted, the mailbox contains the name of the station with a timestamp.

For the *measuring stations*, we chose Kantronics KTU data loggers, which are built to interface to radio modems. Because we needed to measure and transmit only analog and digital signals at these stations, without overhead items such as units and time stamps, we chose not to use the standard mailbox of the data logger. To minimize air time, we wrote drivers to communicate with each measuring station—each with parameters specially adjusted for that station.

The Software

The LabVIEW program in the monitoring station has a great deal of flexibility. From the city map, the operator selects any station for monitoring. By selecting one side of the mark on the screen, the newest data is displayed. By selecting the other side of the mark, a trend display for the last 24 hours is shown . For the biggest pump stations, the trend has up to 12 lines, divided between four trend displays. An experienced operator can see at once if the pump station is working well.

An alarm situation at one of the pump stations is indicated by an audio alarm and a text screen, along with an acknowledge button that tells which pump station is giving the alarm and which type of alarm it is. The last 20 alarms are retained on the city map.

It is very easy to add a new pump station to the system. To set up communications, the engineer types the call sign of the new station in the Callsign VI and makes a new set of instructions for the driver. Then, the pump station is automatically included in the monitoring station's call list, and information will be collected, analyzed, and saved to disk. Buttons are added to the city map; the display page for that station is copied from another pump station, then changed accordingly.

The VI for communication to a pump station receives the call sign and type of station, and then makes the call using Mac<->XX vi. The received string is manipulated and then stored in (1) an Excel file, (2) as a Global Variable, and (3) as a log file for trending. The Excel file is kept automatically in a folder named by the pumping station—the file itself has the name of the station and the current date. The monitor station contacts each pumping station and measurement station every 15 minutes, as directed by the clock function.

The LabVIEW program, because PLCs are used for data collection and calculations, is a supervisory control and data acqui-

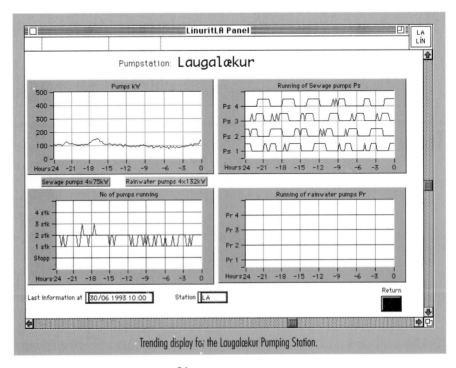

Trending display for the Laugalækur Pumping Station.

sition (SCADA) application written on the Macintosh. The program is divided in two main parts. Part 1, the Communications Engine, contacts the stations, sends commands, receives data, manipulates it, and saves to disk and to data files. Part 2 controls the city map with all its buttons, trend charts, and so on.

A very important part of the system is its fault handling. A 16-bit fault word is included in the mailbox from each pump station. This 16-bit word is extracted and checked each time data is received. If a fault is found, the operator is notified with a text screen and an audio alarm.

Summary

Over a period of one year, a three-man team worked part time to design and install this radio monitoring system. The system has operated for about two years—we have already upgraded it to include weather stations for monitoring road conditions. The biggest improvement we made was implementing one main radio station for all communications to pumping stations and measurement stations and distributing the data on a LAN with a server. For this purpose, we designed a different front end, so the operator can browse through data from whatever source or date desired. Another improvement was upgrading the original program written with LabVIEW 2 to LabVIEW 3. It was later ported to Windows on the PC with little difficulty.

Using LabVIEW, which has proved very stable, we created a professional solution in a very reasonable time frame. The system is readily expandable with an easy user interface.

OVERVIEW

In this chapter, you will investigate the LabVIEW environment and learn how its three parts—the front panel, block diagram, and icon/connector—work together. When all three main components are properly developed, you have a VI that can stand alone or be used as a subVI in another program. You will also learn about the LabVIEW environment: pull-down and pop-up menus, floating palettes and subpalettes, the Toolbar, and how to get help. To finish up, we will discuss the power of subVIs and why you should use them.

GOALS

- Understand and practice using the front panel, block diagram, and icon/connector
- Learn the difference between controls and indicators
- Be able to recognize the difference between the block diagram terminals of controls and indicators
- Understand the principle of dataflow programming
- Become familiar with LabVIEW menus, both pop-up and pull-down
- Learn about the capabilities and uses of the Toolbar, **Tools** palette, **Controls** palette, **Functions** palette, and subpalettes
- Learn why the Help window can be your most valuable ally
- Understand what a subVI is and why it's useful
- Work through the activities to get a feel for how LabVIEW works

KEY TERMS

- control
- indicator
- wire
- subVI
- terminal

- node
- dataflow
- pop-up menus
- Toolbar
- thumbtack

- palette
- subpalette
- Help window

The LabVIEW Environment: Building Your Own Workbench

3

Front Panels

Simply put, the *front panel* is the window through which the user interacts with the program. When you run a VI, you must have the front panel open so that you can input data to the executing program. You will also find the front panel indispensable because that's where you see your program's output. The following illustration shows an example of a LabVIEW front panel.

■ Controls and Indicators

The front panel is primarily a combination of *controls* and *indicators*. **Controls** simulate typical input objects you might find on a conventional instrument, such as knobs and switches. Controls allows the user to input values; they supply data to the block diagram of the VI. **Indicators** show output values produced by the

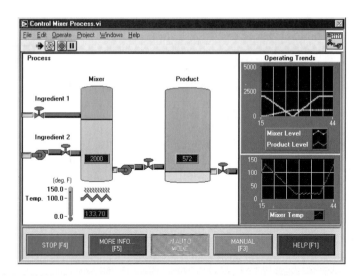

LabVIEW Front Panel

program. Consider this simple way to think about controls and indicators:

Controls = Inputs from the User = Source Terminals

Indicators = Outputs to the User = Destinations or "Sinks"

They are generally not interchangeable, so make sure you understand the difference.

You "drop" controls and indicators onto the front panel by selecting them from a *subpalette* of the floating **Controls** palette window and placing them in a desired spot. Once an object is on the front panel, you can easily adjust its size, shape, position, color, and other attributes.

Block Diagrams

The *block diagram* window holds the graphical source code of a LabVIEW VI, written in the graphical language G. LabVIEW's block diagram corresponds to the lines of text found in a more conventional language like C or BASIC—it is the actual executable code. You construct the block diagram by wiring together objects that perform specific functions. In this section, we will discuss the various components of a block diagram: *terminals*, *nodes*, and *wires*.

The following simple VI computes the sum of two numbers. Its diagram shows examples of terminals, nodes, and wires.

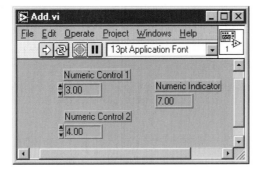

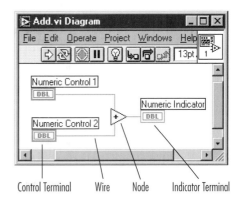

Control Terminal   Wire   Node   Indicator Terminal

■ Terminals

When you place a control or indicator on the front panel, Lab-VIEW automatically creates a corresponding *terminal* on the block diagram. You cannot delete a block diagram terminal that belongs to a control or indicator, although you may try to your heart's content. The terminal disappears only when you delete its corresponding control or indicator on the front panel.

Control terminals have THICK borders, while indicator terminal borders are THIN. It is VERY important to distinguish between the two since they are NOT functionally equivalent (Control=Input, Indicator=Output, so they are NOT interchangeable).

You can think of terminals as entry and exit ports in the block diagram, or as sources and destinations. Data that you enter into <u>Numeric Control 1</u> (shown in the previous illustration) exits the front panel and enters the block diagram through the <u>Numeric Control 1</u> terminal on the diagram. The data from <u>Numeric Control 1</u> follows the wire and enters the **Add** function input terminal. When the **Add** function completes its internal calculations, it produces new data values at its exit terminal. The data flows to the <u>Numeric Indicator</u> terminal and reenters the front panel, where it is displayed for the user.

■ Nodes

A *node* is just a fancy word for a program execution element. Nodes are analogous to statements, operators, functions, and subroutines in standard programming languages. The **Add** and **Subtract** functions represent one type of node. A structure is another type of node. Structures can execute code repeatedly or conditionally, similar to loops and Case statements in traditional programming languages. LabVIEW also has special nodes, called Formula Nodes, which are useful for evaluating mathematical formulas or expressions.

■ Wires

A LabVIEW VI is held together by *wires* connecting nodes and terminals. Wires are the data paths between source and destination terminals; they deliver data from one source terminal to one or more destination terminals. If you connect more than one source or no source at all to a wire, LabVIEW disagrees with what you're doing, and the wire will appear broken.

> *This principle of wires connecting source and destination terminals explains why controls and indicators are not interchangeable. Controls are source terminals, while indicators are destinations, or "sinks."*

Each wire has a different style or color, depending on the data type that flows through the wire. The block diagram shown previously depicts the wire style for a numeric scalar value—a thin, solid line. The following chart shows a few wires and corresponding types.

	Scalar	1D Array	2D Array	Color
Floating-Point Number	————	————	————	Orange
Interger Number	————	————	————	BLue
Boolean	··············	wwwwwwww	wwwwwwww	Green
String	∿∿∿∿∿∿	∘∘∘∘∘∘∘∘	ℛℛℛℛℛℛ	Purple

Basic wires styles used in block diagrams.

To avoid confusing your data types, simply match up the colors and styles!

■ Dataflow Programming—Going with the Flow

Since LabVIEW is not a text-based language, its code cannot execute "line by line." The principle that governs G program execution is called *dataflow*. Stated simply, a node executes only when data arrives at all its input terminals; the node supplies data to all of its output terminals when it finishes executing; and the data passes immediately from source to destination terminals. Dataflow contrasts strikingly with the control flow method of executing a text-based program, in which instructions are executed in the sequence in which they are written. This difference may take some getting used to. While traditional execution flow is instruction driven, dataflow execution is data driven or *data dependent*.

The Icon and the Connector

When your VI operates as a *subVI*, its controls and indicators receive data from and return data to the VI that calls it. A VI's *icon* represents it as a subVI in the block diagram of another VI. An icon can include a pictorial representation or a small textual description of the VI, or a combination of both.

The VI's *connector* functions much like the parameter list of a C or Pascal function call; the connector terminals act like little graphical parameters to pass data to and from the subVI. Each terminal corresponds to its very own control or indicator on the front panel. During the subVI call, the input parameter terminals are copied to the connected controls, and the subVI executes. At completion, the indicator values are copied to the output parameter terminals.

Icon   Connector

An icon and its underlying connector

Every VI has a default icon, which is displayed in the icon pane in the upper right corner of the panel and diagram windows. The default icon is depicted in the following illustration.

Icon pane

A VI's connector is hidden under the icon; access it by choosing **Show Connector** from the front panel icon pane pop-up menu (we'll talk more about pop-up menus later). When you show the connector for the first time, LabVIEW helpfully suggests a connector pattern that has one terminal for each control and indicator currently on the front panel. You can select a different pattern if you desire, and you can assign up to twenty-eight terminals before you run out of real estate on the connector.

Activity 3-1
Getting
Started

Okay, you've read enough for now. It's time to get some hands-on experience. Go ahead and launch LabVIEW. You will step through the creation of a simple LabVIEW VI that generates a random number and plots its value on a waveform chart. You'll learn more in the next chapter about the steps you'll be taking; for now, just get a feel for the environment.

If you are using the full version of LabVIEW, just launch it and you'll be ready to start building your first VI.

If you are using the LabVIEW sample software, start the application and select the <u>Explore LabVIEW for Your Own Applications</u> button in the lower right corner of the main menu. Click OK on the first dialog box that pops up. If you are using Windows or Unix, select New VI on the next dialog box to start building a fresh VI. On Macintosh machines, select **New** from the **File** menu to create a new VI or **Open** to bring up one that's already built.

If you're not comfortable working through the activities in this chapter without more background information or you have trouble getting them to work, read Chapter 4 and then come back and try again.

1. You should have an "Untitled 1" front panel on your screen.

Go to the floating **Controls** palette and click on the **Graph** button to access the **Graph** subpalette. If the **Controls** palette isn't visible, select **Show Controls Palette** from the **Windows** menu. Also make sure the front panel window is active, or you will see the **Functions** palette instead of the **Controls** palette. While continuing to hold down the mouse button, drag the mouse over to the **Graph** subpalette and select **Waveform Chart** by releasing the mouse button. You will notice that, as you run the cursor over the icons in the **Controls** palette and subpalettes, the selected button or icon's name appears at the top of the palette, as shown in the following illustration.

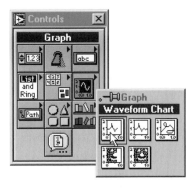

Positioning Tool

You will see the outline of a chart with the cursor "holding" it. Position the cursor in a desirable spot on your front panel and click. The chart magically appears exactly where you placed it. If you want to move it, select the **Positioning** tool from the **Tools** palette, then drag the chart to its new home. If the **Tools** palette isn't visible, select **Show Tools Palette** from the **Windows** menu.

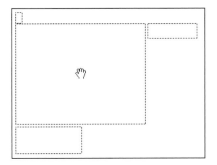

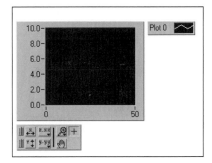

2. Go back to the floating **Controls** palette and select **Vertical Toggle Switch** from the **Boolean** subpalette.

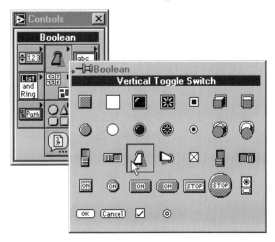

Place it next to the chart as shown in the following illustration.

Operating Tool

3. Select the Operating tool from the floating **Tools** palette.

Enter Button

Now change the scale on the chart. Highlight the number "10" by click-dragging or by double-clicking on it with the Operating tool. Now type in 1.0 and click on the enter button that appears in the Toolbar at the top of the window.

4. Switch to the block diagram by selecting **Show Diagram** from the **Windows** menu. You should see two terminals already there.

5. Now you will put the terminals inside a While Loop to repeat execution of a segment of your program. Go to the **Structures** subpalette of the floating **Functions** palette and select the **While Loop**. Make sure the block diagram window is active, or you will see the **Controls** palette instead of the **Functions** palette.

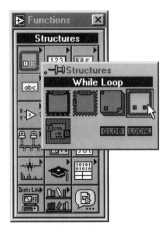

Your cursor will change to a little loop icon. Now enclose the DBL and TF terminals: click and hold down the mouse button while you drag the cursor from the upper left to the lower right corners of the objects you wish to enclose.

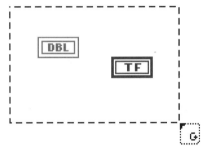

When you release the mouse button, the dashed line that is drawn as you drag will change into the While Loop border. Make sure to leave some extra room inside the loop.

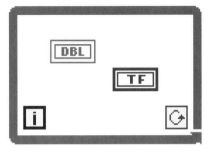

6. Go to the **Functions** palette and select **Random Number (0-1)** from the **Numeric** subpalette. Place it inside the While Loop.

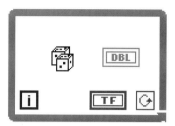

The While Loop is a special LabVIEW structure that repeats the code inside its borders until it reads a FALSE value. It is the equivalent of a Do-While Loop in a more conventional language. You'll learn more about loops in Chapter 6, *Controlling Your Program with Structures.*

Positioning Tool

7. Select the Positioning tool from the floating **Tools** palette and arrange your diagram objects so that they look like the previously shown block diagram.

Wiring Tool

8. Now select the Wiring tool from the **Tools** palette. Click once on the **Random Number (0–1)** icon, drag the mouse over to the DBL terminal, and click again.

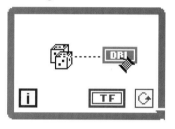

You should now have a solid orange wire connecting the two icons. If you mess up, you can select the wire or wire fragment with the Positioning tool and then hit the <delete> key to get rid of it. Now wire the Boolean TF terminal to the conditional terminal of the While Loop. The loop will execute while the switch on the front panel is TRUE (in the "up" position) and stop when the switch becomes FALSE.

Operating Tool

Run Button

9. You should be about ready to run your VI. First, switch back to the front panel by selecting **Show Panel** from the **Windows** menu. Using the Operating tool, flip the switch to the "up" position. Now click on the run button to run your VI. You will see a series of random numbers plotted continuously across the chart. When you want to stop, click on the switch to flip it to the down position.

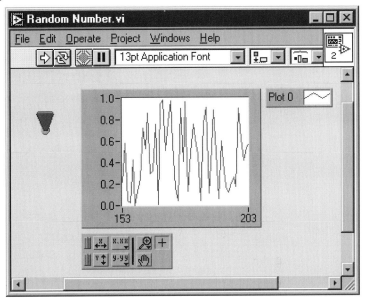

10. If you are using the full version of LabVIEW, create a directory or folder called MYWORK in your LabVIEW directory. Save your VI in your MYWORK directory or folder by selecting **Save** from the **File** menu and pointing out the proper location to save to. Name it **Random Number.vi**.

Save all of your subsequent activities in MYWORK *so you can find them easily!*

If you are using Windows 3.1 and the full version of LabVIEW, you must make MYWORK *a VI library instead of a directory. VI libraries store VIs and allow you to overcome the operating system limitation of eight-character filenames. We'll talk more about them in Chapter 5, Yet More Foundations. For now, select* **Save** *from the* **File** *menu. Make sure you are in the* LabVIEW *directory, then create a VI library by clicking the New VI Library button in the Save dialog box. Name the library* MYWORK.LLB *and click the VI Library button. The library is created and you can save this activity in it.*

If you are using the sample version, save all of your work in the existing MYWORK.LLB. Your VI will be saved until you quit the session, and then it will be deleted.

Remember, if you get stuck or just want to compare your work, the solutions to every activity in this book can be found in the Everyone *directory or folder on the accompanying CD (which you should have installed on your machine—see the Preface for instructions). You can view them in the sample version or the full version of LabVIEW. Keep in mind that activities in the Advanced Topics section may not work in the sample software.*

Congratulate yourself—you've just written your first Lab-VIEW program! Don't worry that it doesn't actually DO much—your programs will be more powerful and have more of a purpose soon enough!

Pull-Down Menus

Keep in mind that LabVIEW's capabilities are many and varied. This book by no means provides an exhaustive list of all of LabVIEW's ins and outs (it would be several thousand pages long if that were the case); instead, we try to get you up to speed comfortably and give you an overview of what you can do. If you want to know everything there is to know about a subject, we'd recommend looking it up in one of LabVIEW's many manuals, attending a seminar, trying out National Instruments' faxback system, or maybe even surfing the web. See Appendix A for an exhaustive list of other resources. Feel free to skim through this section and some of the subsequent ones, but remember that they're here if you need a reference.

LabVIEW has two main types of menus: pull-down and popup. You used some of them in the last activity, and you will use both extensively in all of your program development henceforth. Now you will learn more about what they can do. We'll cover pull-down menu items very briefly in this section. You might find it helpful to look through the menus on your computer as we explain them, and maybe experiment a little.

The menu bar at the top of a VI window contains several pull-down menus. When you click on a menu bar item, a menu appears below the bar. The pull-down menus contain items common to many applications, such as **Open**, **Save**, **Copy**, and **Paste**, and many other functions particular to LabVIEW. We'll discuss some basic pull-down menu functions here. You'll learn about more the advanced capabilities later.

Many menus also list shortcut keyboard combinations for you to use if you choose. To use keyboard shortcuts, press the appropriate key in conjunction with the <control> key on PCs, the <command> key on Macintoshes, the <meta> key on Suns, and the <alt> key on HP machines.

Many of the menu items show keyboard shortcuts to the right of their corresponding commands. You may want to use the shortcuts instead of the menus.

■ File Menu

Pull down the **File** menu, which contains commands common to many applications such as **Save** and **Print**. You can also create new VIs or open existing ones from the **File** menu.

■ Edit Menu

Take a look at the **Edit** menu. It has some universal commands, like **Cut**, **Copy**, and **Paste**, that let you edit your window. You can also change the layering of objects (for example, if two are in the same place, which one is "on top"?) and remove bad wires from the **Edit** menu, as well as customize your palettes and access the myriad of LabVIEW **Preferences**.

■ Operate Menu

You can run or stop your program from the **Operate** menu

(although you'll usually use Toolbar buttons). You can also change a VI's default values, control 'print and log at completion' features, and switch between run mode and edit mode.

■ Project Menu

The **Project** menu contains features to simplify navigation among large sets of VIs. You can see VI hierarchy, determine all of a VI's subVIs, find specific objects or text, and view the Profile window to optimize VI performance.

■ Windows Menu

Pull down the **Windows** menu. Here you can toggle between the panel and diagram windows, show the error list and the clipboard, "tile" both windows so you can see them at the same time, and switch between open VIs. You can also bring up floating palettes if you've closed them. In addition, you can show VI information and development history from this menu.

■ Help Menu

You can show, hide, or lock the contents of the Help window using the **Help** menu. You can also access LabVIEW's online reference information and view the About LabVIEW information window.

Floating Palettes

LabVIEW has three often-used floating palettes that you can place in a convenient spot on your screen: the **Tools** palette, the **Controls** palette, and the **Functions** palette. You can move them around by clicking on their title bar and dragging. Close them just like you would close any window in your operating system. If you decide you want them back, use the **Show... Palette** function in the **Windows** menu.

■ Controls and Functions Palettes

You will be using the **Controls** palette a lot, since that's where you select the controls and indicators that you want on your front panel. You will probably use the **Functions** palette even more often, since it contains the functions and structures used to build a VI.

The **Controls** and **Functions** palettes are unique in several ways. *Most importantly, the **Controls** palette is only visible when the front panel window is active, and the **Functions** palette is only visible when the block diagram window is active.* Both palettes have *subpalettes* containing the objects you need to access. As you pass the cursor over each subpalette button in the **Controls** and **Functions** palettes, you will notice that the subpalette's name appears at the top of the window.

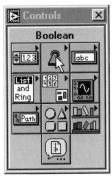

If you click on a button and continue holding down the mouse button, the associated subpalette appears. To select an object in the subpalette, release the mouse button over the object, then click on the front panel or block diagram to place it where you want it. Like palette button names, subpalette object names appear when you run the cursor over them.

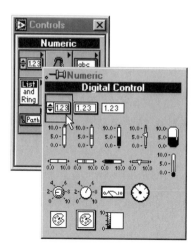

Note that some subpalettes have subpalettes containing more objects; these are denoted by a little triangle in the upper right corner of the icon and a raised appearance. We'll discuss specific subpalettes and their objects in the next chapter.

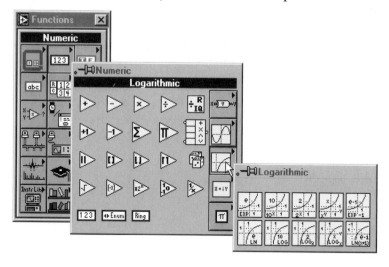

The **Controls** and **Functions** palettes can also be accessed by popping up in an empty area of the front panel or block diagram. "Popping up" is defined as right-mouse-clicking on the PC, Sun, and HP machines, and <command>-clicking on the Mac. You can also pop up by clicking with the soon-to-be-discussed Pop-up Tool.

■ The Thumbtack

If you use a subpalette frequently, you may want to 'tear it off' by releasing the mouse button over the *thumbtack* located at the upper left of the palette. You now have a stand-alone window that you can position anywhere and then close when you're done with it. LabVIEW uses the thumbtack instead of a standard tear-off palette to avoid accidental breakaways when you're buried several levels deep in subpalettes.

▦ Customizable Palettes

If LabVIEW's default organization of the **Controls** and **Functions** palettes doesn't fit your needs, you can customize them according to your whim. Access the menu editor by selecting **Edit Control and Function Palettes** from the **Edit** menu. From here, you can create your own palettes and customize existing views by adding new subpalettes, hiding items, or moving them from one palette to another. For example, if you create a VI using trigonometric functions, you can place it in the existing **Trigonometric** subpalette for easy access. Editing the palettes is handy for placing your most frequently used functions at the top level for easy access and burying those pesky functions you never want to see again at the bottom of a subpalette. You'll learn more about how to customize palettes in Chapter 4. You can also use the built-in **daq_view** or **t&m_view** (short for Test and Measurement) palettes if those configurations are more convenient for you; choose **Select Palette Set▶** from the **Edit** menu to access them.

▉ Tools Palette

A *tool* is a special operating mode of the mouse cursor. You use tools to perform specific editing and operation functions, similar to how you would use them in a standard paint program.

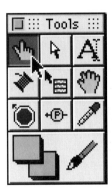

Like the **Controls** and **Functions** palettes, the **Tools** palette window can be relocated or closed. To select a tool, click the appropriate button on the **Tools** palette and your mouse cursor will change accordingly. If you're not sure which tool is which, hold your cursor over the button until a *tip strip* appears describing the tool.

Operating Tool

The Operating tool lets you change values of front panel controls and indicators. You can operate knobs, switches, and other objects with the Operating tool—hence the name. It is the only front panel tool available when your VI is running or in run mode (described shortly).

Positioning Tool

The Positioning tool selects, moves, and resizes objects.

Labeling Tool

The Labeling tool creates and edits text labels.

Wiring Tool

The Wiring tool wires objects together on the block diagram. It is also used to assign controls and indicators on the front panel to terminals on the VI's connector.

Color Tool

The Color tool brightens objects and backgrounds by allowing you to choose from a multitude of hues. You can set both foreground and background colors by clicking on the appropriate color area in the **Tools** palette. If you pop up on an object with the Color tool, you can choose a hue from the color palette that appears.

Pop-up Tool

The Pop-up tool opens an object's pop-up menu when you click on the object with it. You can use it to access pop-up menus instead of the standard method for popping up (right-clicking under Windows and UNIX, and <command>-clicking on Macintosh).

Scroll Tool

The Scroll tool lets you scroll in the active window.

Breakpoint Tool

The Breakpoint tool sets breakpoints on VI diagrams to help you debug your code. It causes execution to suspend so you can see what is going on and change input values if you need to.

Probe Tool

The Probe tool creates probes on wires so you can view the data traveling through them while your VI is running.

Color Copy Tool

Use the Color Copy tool to pick up a color from an existing object, then use the Color tool to paste that color onto other objects. This technique is very useful if you need to duplicate an exact shade but can't remember which one it was. You can also access the Color Copy tool when the Color tool is active by holding down the <control> key on Windows, <option> on Macintosh, <meta> on Sun, and <alt> on HP-UX.

*You can use the <tab> key to tab through the **Tools** palette instead of clicking on the appropriate tool button to access a particular tool. Or press the space bar to toggle between the Operating tool and the Positioning tool when the panel window is active, and between the Wiring tool and the Positioning tool when the diagram window is active. The <tab> and space bar shortcuts cycle through the most frequently used tools for your convenience—try using them, and see if they don't save you time!*

*You can also access a temporary copy of the **Tools** palette by pop-up clicking (<shift>-right click for Windows and UNIX, and <command-shift>-click on Macintosh).*

The Toolbar

The *Toolbar,* located at the top of LabVIEW windows, contains buttons you will use to control the execution of your VI, as well as text configuration options and commands to control the alignment and distribution of objects. You'll notice that the Toolbar has a few more options in the block diagram than in the front panel, and that a few editing-related options disappear when you run your VI. If you're not sure what a button does, hold the cursor over it until a tip strip appears describing its function.

Run Button

Run Button (active)

The Run button, which looks like an arrow, starts VI execution when you click on it. It changes appearance when a VI is actually running. When a VI won't compile, the run button is broken.

Run Button (broken)

The Continuous Run button causes the VI to execute over and over until you hit the stop button. It's kind of like a GOTO statement (sort of a "programming no-no"), so use it sparingly.

Continuous Run Button

Abort Button

The Abort button, easily recognizable because it looks like a tiny stop sign, becomes active when a VI begins to execute; otherwise the Abort button is grayed out. You can click on this button to halt the VI.

Using the Abort button is like hitting the <break> key. Your program will stop immediately rather than coming to a graceful end, and data integrity can be lost this way. You should always code a more appropriate stopping mechanism into your program, as we will demonstrate later.

Pause Button

The Pause button pauses the VI so that you can use single-step debugging options such as step into, step over, and step out. Hit the pause button again to continue execution.

Step Into Button

Step Over Button

Step Out Button

The single-step buttons, Step Into, Step Over, and Step Out, force your VI to execute one step at a time so you can troubleshoot. We'll talk more about how to use them in Chapter 5, *Yet More Foundations*.

Execution Highlighting Button

The Execution Highlighting button causes the VI to highlight the flow of data as it passes through the diagram. When execution highlighting is on, you can see intermediate data values in your block diagram that would not otherwise appear.

Warning Button

The Warning button appears if you have configured your VI to show warnings and you have any warnings outstanding. You can list the warnings by clicking on the button. A warning is not an error; it just alerts you that you are doing something you may not have intended (for example, you have a front panel control with nothing wired to it).

You can change the font, size, style, justification, and color of LabVIEW text from the **Font** ring on the Toolbar.

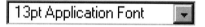

13pt Application Font

LabVIEW has an automatic alignment mechanism to help you line up and evenly space your icons. Select the objects you want to align by dragging around them with the Positioning tool, then go to the **Alignment** ring on the Toolbar and choose how you want to align them (top edges flush, left edges flush, vertical centers, etc.) If you want to set uniform spacing between objects, use the **Distribution** ring in a similar fashion.

Alignment Ring

Distribution Ring

■ Run Mode and Edit Mode

When you open a VI, it opens in *edit mode* so you can make changes to it. When you run a VI, it automatically goes into *run mode* and you can no longer edit. Only the Operating tool is available on the front panel when the VI is in run mode. When your VI completes execution, your VI reverts to edit mode (unless you manually switched it to run mode before you ran it—

then it stays in run mode). You can switch to run mode by selecting **Change to Run Mode** from the **Operate** menu; switch to
edit mode by choosing **Change to Edit Mode**. To draw a parallel with text-based languages, if a VI is in run mode, it has been
successfully compiled and awaits your command to execute.
Most of the time, you will not need to concern yourself with run
and edit modes. But if you accidentally find that you suddenly
have only the Operating tool, and you can't make any changes,
at least now you'll know why.

*If you prefer to open VIs in run mode (perhaps so uninvited users can't make changes), select **Preferences...***
*from the **Edit** menu. Go to the **Miscellaneous** options and choose **Open VIs in Run Mode**.*

Pop-Up
Menus

As if pull-down menus didn't give you enough to learn about,
we will now discuss the other type of LabVIEW menu, the pop-
up menu. You will probably use pop-up menus more often than
any other LabVIEW menu. To pop up, position the cursor over
the object whose menu you you desire; then click the right
mouse button on Windows and UNIX machines, or hold down
the <command> key and click on the Mac. You can also click on
the object with the Pop-up tool. A pop-up menu will appear.

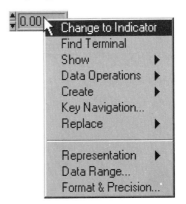

Virtually every LabVIEW object has a pop-up menu of options and commands. Options available in this pop-up menu
depend on the kind of object, and they are different when the VI
is in edit mode or run mode. For example, a numeric control
will have a very different pop-up menu than a graph indicator.
If you pop up on empty space in a front panel or block diagram,
you will get the **Controls** or **Functions** palette, respectively.

You will find that instructions throughout this book guide you to select a command or option from an object pop-up menu, so try popping up now!

How to Pop Up
Windows and UNIX: right mouse click on the object
Mac: <command>-click on the object
All Platforms: Click on the object with the Pop-up tool
Pop-up menus are ever-present in LabVIEW. They contain most configuration options for an object. So remember, when in doubt about how to do something, try POPPING UP!

If the Color tool is active, you will see a color palette when you pop up instead of the pop-up menu that appears when other tools are active.

■ Pop-Up Menu Things to Keep in Mind

Many pop-up menu items expand into submenus called hierarchical menus, denoted by a right arrowhead.

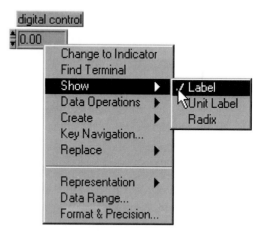

Hierarchical menus sometimes have a selection of mutually exclusive options. The currently selected option is denoted by a check mark for text-displayed options, or surrounded by a box for graphical options.

Some menu items pop up dialog boxes containing options for you to configure. Menu items leading to dialog boxes are denoted by ellipses (...).

Menu items without right arrowheads or ellipses are usually commands that execute immediately upon selection. A command usually appears in verb form, such as **Change to Indicator**. When selected, some commands are replaced in the menu by their inverse commands. For example, after you choose **Change to Indicator**, the menu selection becomes **Change to Control**.

*Sometimes different parts of an object have different pop-up menus. For example, if you pop up on an object's label, the menu contains only a **Size to Text** option. Popping up elsewhere on the object gives you a full menu of options. So if you pop up and don't see the menu you want, try popping up elsewhere on the object.*

■ Pop-Up Features Described

Pop-up menus allow you to specify many traits of an object. The following options appear in numerous pop-up menus, and we thought they were important enough to describe them individually. We'll let you figure out the other options, since we would put you to sleep detailing them all. Feel free to skim over this section and refer back when necessary.

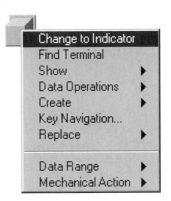

■ Change to Control and Change to Indicator

By selecting **Change to Control**, you can turn an existing control (an input object) into an indicator (an output object), or vice versa if you select **Change to Indicator**. When an object is a control, its pop-up menu contains the option to **Change to Indicator**. When it is an indicator, the pop-up menu reads **Change to Control**.

*Since **Change to Control/Indicator** is the first option in the pop-up menu, it is easy to accidentally select it without realizing what you've done. Controls and indicators are not functionally interchangeable in a block diagram, so the resulting errors may befuddle you.*

A control terminal in the block diagram has a thicker border than an indicator terminal. Always pay attention to whether your objects are controls or indicators to avoid confusion!

■ Find Terminal and Find Control/Indicator

If you select **Find Terminal** from a front panel pop-up menu, LabVIEW will locate and highlight its corresponding block diagram terminal. If you select **Find Control/Indicator** from a block diagram pop-up menu, LabVIEW will show you its corresponding front panel object.

■ Show

Many items have **Show** menus with which you can show or hide certain cosmetic features like labels, scrollbars, or wiring terminals. If you select **Show**, you will get another menu off to the side, listing options of what can be shown (this list varies depending on the object). If an option has a check next to it, that option is currently visible; if it has no check, it is hidden. Release the mouse on an option to toggle its status.

■ Data Operations

The **Data Operations** pop-up menu has several handy options to let you manipulate the data in a control or indicator. Note that these are the only pop-up items available in run mode.

Reinitialize to Default returns an object to its default value, while **Make Current Value Default** sets the default value to whatever data is currently there.

Use **Cut Data**, **Copy Data**, and **Paste Data** to take data out or put data into a control or indicator.

Last but not least, **Description...** brings up a dialog box where you can enter or read text documenting that particular object's use. When you are in run mode, you can view a description but not edit it.

▧ Show or Hide Control/Indicator

You can choose to show or hide a block diagram terminal's corresponding front panel object using this option, which comes in handy when you don't want the user to see the front panel object but still need it in the diagram.

▧ Create...

The **Create...** option is an easy way for you to create an attribute node or local variable for a given object (these topics will be covered in detail in Chapter 12.).

▧ Key Navigation

Use **Key Navigation...** to associate a keyboard key combination with a front panel object. When a user enters that key combination while a VI is running, LabVIEW acts as if the user had clicked on that object, and the object becomes the key focus (key focus means the cursor is active in that field).

▧ Online Help

This one's pretty obvious—if you select **Online Help** from a function's pop-up menu, LabVIEW brings up the corresponding documentation describing that function.

▧ Replace

The **Replace** option is extremely useful. It gives you access to the **Controls** or **Functions** palette (depending on whether you're in the front panel or block diagram) and allows you to replace the object you popped up on with one of your choice. Where possible, wires will remain intact.

Don't worry about memorizing all of these features right now—you'll come across them as you work with LabVIEW and they'll make a lot more sense!!

The same object will have a different pop-up menu in run mode than it will in edit mode. If you cannot find a certain pop-up option, it is either not present for that object, you need to switch modes, or you should pop up elsewhere on the object.

Help!

■ The Help Window

The LabVIEW *Help window* offers indispensable help information for functions, constants, subVIs, and controls and indicators. To display the window, choose **Show Help** from the **Help** menu or use the keyboard shortcut: <control-H> under Windows, <command-H> on the Mac, <meta-H> on the Sun, and <alt-H> on HP-UX. If your keyboard has a <help> key, you can press that instead. You can resize the Help window and move it anywhere on your screen to keep it out of the way.

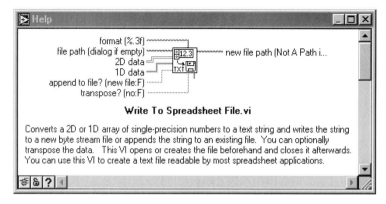

Help Window

When you hold the cursor over a function, a subVI node, or a VI icon (including the icon of the VI you have open, at the top right corner of the VI window), the Help window shows the icon for the function or subVI with wires of the appropriate data type attached to each terminal. *Input wires point to the left, and output wires point to the right.* Terminal names appear beside each wire. If the VI has a description associated with it, this description is also displayed.

For some subVIs or functions, the Help window will show the names of required inputs in bold, with default values shown in parentheses. In some cases, the default value can be used and you do not need to wire an input at all. You can lock the Help window so that its contents do not change when you move the mouse by selecting **Lock Help** from the **Help** menu or by pressing the Lock button in the Help window.

Lock Button

If you position the Wiring tool over a specific node on a function or subVI, the Help window will flash the labeled corresponding node so that you can make sure you are wiring to the right place. Sometimes you may need to use the scrollbar to see all of the text in the Help window.

For VIs and functions with large numbers of inputs and outputs, the Help window can be overwhelming, so LabVIEW gives you the choice between simple or detailed views. You can use the simple view to emphasize the important connections and de-emphasize less commonly-used connections.

Simple/Detailed
Help Button

Switch between views by pressing the Simple/Detailed Diagram Help button on the lower left corner of the Help window or by toggling **Simple Help** on and off in the **Help** menu. In simple help view, required connections appear in bold text; recommended connections appear in plain text; and optional connections are not shown. Wire stubs appear in the place of inputs and outputs that are not displayed, to inform you that additional connections exist (and you can see them in the detailed help view).

In detailed help view, required connections appear in bold text; recommended connections appear in plain text; and optional connections appear as disabled text.

If a function input does not need to be wired, the default value often appears in parentheses next to the input name. If the function can accept multiple data types, the Help window shows the most common type.

■ Online Help

Online Help Button

LabVIEW's Help window provides a quick reference to functions, VIs, controls, and indicators. However, there are times when you'd prefer to look at a more detailed, indexed description for information on using a VI or function. LabVIEW has extensive online help that you can access by selecting **Online Reference...** from the **Help** menu or by pressing the Online Help button in the Help window.

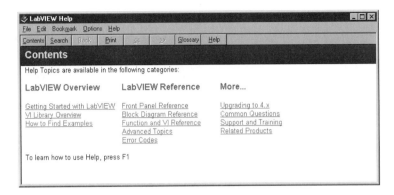

You can type in a keyword to search for, view an extensive keyword index, or choose from a variety of topics to browse through. You can also set your own links to online help documents, which we'll talk about in Chapter 15.

Currently, not all LabVIEW VIs link to online help; if this is the case, the online help menu item and online help button will be grayed out.

A Word about SubVIs

If you want to take full advantage of LabVIEW's abilities, you must understand and use the hierarchical nature of the VI. A *subVI* is simply a stand-alone program that is used by another program. After you create a VI, you can use it as a subVI in the block diagram of a higher-level VI as long as you give it an icon and define its connector. A LabVIEW subVI is analogous to a subroutine in C or another text-based language. Just as there is no limit to the number of subroutines you can use in a C program, there is no limit to the number of subVIs you can use in a LabVIEW program (memory permitting, of course).

If a block diagram has a large number of icons, you can group them into a subVI to maintain the simplicity of the block diagram. You can also use one subVI to accomplish a function common to several different top-level VIs. This modular approach makes applications easy to debug, understand, and modify. We'll talk more about how to build subVIs later, but it's such an important part of the LabVIEW programming environment that we want you to keep it in mind as you're learning the basics.

Activity 3-2
Front Panel and Block Diagram Basics

In this activity, you will practice some simple exercises to get a feel for the LabVIEW environment. Try to do the following basic things on your own. If you have any trouble, glance back through the chapter for clues.

1. Open a new VI and toggle between the front panel and block diagram.

Use the keyboard shortcuts listed in the pull-down menus!

2. Resize the windows so that both front panel and block diagram are visible simultaneously. You may need to move them around.

Do this using the standard resizing technique for your platform.
*Another hint: Try the **Tile** function!*

3. Drop a digital control, a string control, and a Boolean indicator on the front panel by selecting them from the **Controls** palette.

To get the digital control, click on the **Numeric** subpalette button in the **Controls** palette and select **Digital Control** from the subpalette that appears.

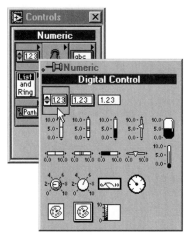

Now click your mouse on the front panel in the location where you want your digital control to appear. Voilà—there it is! Now create the string control and Boolean indicator in the same fashion.

Notice how LabVIEW creates corresponding terminals on the block diagram when you create a front panel object. Also notice that floating-point numeric terminals are orange (integer numerics will be blue), strings are pink, and Booleans are green. This color coding makes it easier for you to distinguish between data types.

4. Now pop up on the digital control (by right mouse-clicking on Windows and UNIX platforms or <command>-clicking on Mac) and select **Change to Indicator** from the pop-up menu. Notice how the appearance of the numeric's front panel changes (the little arrows go away). Also notice how the terminal on the block diagram changes (the border is much thinner for indicators). Switch the object back and forth between control and indicator until you can easily recognize the differences on both front panel and block diagram. Note that for some objects (like a few Booleans), front panel indicators and controls can look the same, but their block diagram terminals will always be different.

Positioning Tool

5. Choose the Positioning tool from the floating **Tools** palette, then select an object on the front panel. Hit the <delete> key to remove it. Delete all front panel objects so you have an empty front panel and block diagram.

Enter Button

6. Drop another digital control from the **Numeric** subpalette of the **Controls** palette onto the front panel. If you don't click on anything first, you should see a little box above the control. Type Number 1, and you will see this text appear in the box. Click the Enter button on the Toolbar to enter the text. You have just created a label. Now create another digital control labeled Number 2, a digital indicator labeled N1+N2, and a digital indicator labeled N1-N2.

Operating Tool

Use the Operating tool to click on the increment arrow of <u>Number 1</u> until it contains the value "4.00." Give <u>Number 2</u> a value of "3.00."

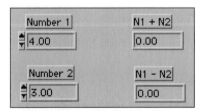

7. Switch back to the diagram. Drop an **Add** function from the **Numeric** subpalette of the **Functions** palette in the block diagram (this works just like creating front panel objects). Now repeat the process and drop a **Subtract** function.

8. Pop up on the **Add** function and select the **Show▶Terminals** option (you'll notice that before you select it, the option is not checked, indicating that terminals are not currently shown). Once you show them, observe how the input and output terminals are arranged; then redisplay the standard icon by again selecting **Show▶Terminals** (this time the option appears with a checkmark next to it, indicating that terminals are currently shown).

9. Bring up the Help window by using either the keyboard shortcut or the **Show Help** command from the **Help** menu. Position the cursor over the **Add** function. The Help window provides valuable information about the function's use and wiring pattern. Now move the cursor over the **Subtract** function and watch the Help window change.

Wiring Tool

10. You may have to use the Positioning tool to reposition some of the terminals as shown in the following illustration. Then use the Wiring tool to wire the terminals together. First select it from the **Tools** palette, then click once on the DBL terminal and once on the appropriate terminal on the **Add** function to draw a wire. A solid orange line should appear. If you mess up and get a dashed black line instead of a solid orange one, select the wire fragment with the Positioning tool and hit the <delete> key, then try again. Click once and release to start the wire, click any time you want to tack a new segment (which turns a corner), and click on a destination to finish the wire.

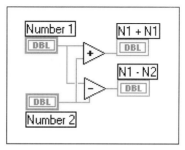

Notice that when you pass the Wiring tool over the **Add** and **Subtract** functions, little wire stubs appear showing where the terminals are located. In addition, as you pass the cursor over a terminal, its name appears in a tip strip. Like learning to type, wiring can be kind of tricky until you get the hang of it, so don't worry if it feels a little awkward right now!

11. Switch back to the front panel and pop up on the icon pane (the little window in the upper right corner). Select **Show Connector** from the menu. Observe the connector that appears. If you can't get the pop-up menu to appear, you are probably trying to pop up in the icon pane of the block diagram.

Now pop up again on the connector and look at its menu to see the configuration options you have. The connector defines the input and output parameters of a VI so that you can use it as a subVI and pass data to it. You can choose different patterns for your connectors depending on how many parameters you need to pass. Show the icon again by selecting **Show Icon**. Remember, the icon is just the pictorial representation of a VI; when you use a VI as a subVI, you will wire to this icon in the block diagram of the top-level VI just like you wired to the **Add** function.

Run Button

12. Run the VI by clicking on the Run button. The N1+N2 indicator should display a value of "7.00" and N1-N2 should be "1.00." Feel free to change the input values and run it over and over.

13. If you have the full version of LabVIEW, save the VI by selecting **Save** from the **File** menu. Call it **Add.vi** and place it in your MYWORK directory or VI library. Remember that any work you save with the sample software will be lost when you quit the session.

Congratulations! You have now mastered several important basic LabVIEW skills!

Wrap It Up!

The LabVIEW environment has three main parts: the *front panel*, *block diagram*, and the *icon/connector*. The front panel is the user interface of the program—you can input data through *controls* and observe output data through *indicators*. When you place an object on the front panel using the **Controls** menu, a corresponding terminal appears in the block diagram, making the front panel data available for use by the program. Wires carry data between *nodes*, which are LabVIEW program execution elements. A node will execute only when all input data is available to it, a principle called *dataflow*.

A VI should also have an *icon* and a *connector*. When you use a VI as a subVI, its icon represents it in the block diagram of the VI you use it in. Its connector, usually hidden under the icon, defines the input and output parameters of the subVI.

LabVIEW has two types of menus: pull-down and pop-up. *Pull-down* menus are located in the usual menu spot at the top of your window or screen, while *pop-up* menus can be accessed by "popping up" on an object. To pop up, right-mouse click on Windows and UNIX machines and <command>-click on the Mac, or click with the Pop-up tool. Pull-down menus tend to have more universal commands, while pop-up menu commands affect only the object you pop up on. Remember, when in doubt about how to do something, pop up to see its menu options!

The **Tools** palette gives you access to the special operating modes of the mouse cursor. You use these tools to perform specific editing and operation functions, similar to how you would use them in a standard paint program. You will find front panel control and indicator graphics located in the **Controls** palette, and block diagram constants, functions, and structures in the **Functions** palette. These palettes often have objects nestled several layers down in *subpalettes*, so make sure you don't give up your search for an object too soon.

The Help window provides priceless information about functions and how to wire them up; you can access it from the **Help** menu. LabVIEW also contains extensive online help that you can call up from the **Help** menu or by pressing the online help button in the Help window. Between these two features, your questions should never go unanswered!

You can easily turn any VI into a subVI by creating its icon and connector and placing it in the block diagram of another VI. Completely stand-alone and modular, subVIs offer many advantages: they facilitate debugging, allow many VIs to call the same function without duplicating code, and offer an alternative to huge messy diagrams.

Don't worry if this seems like a lot to remember. It will all become natural to you as you work your way through the book.

OVERVIEW

Get ready to learn about LabVIEW's basic principles in this chapter. You will learn how to use different data types and how to build, change, wire, and run your own VIs. You will also learn some helpful shortcuts to speed your development. Make sure you understand these fundamentals before you proceed, because they are integral to all developments you will achieve in LabVIEW.

YOUR GOALS

- Become comfortable with LabVIEW's editing techniques
- Learn the different types of controls and indicators, and the special options available for each
- Master the basics of creating a VI, such as wiring and editing
- Create and run a simple VI

KEY TERMS

- preferences
- numeric
- string
- Boolean
- path
- ring control
- format and precision
- numeric representation
- label

LabVIEW
Foundations

4

Creating VIs—It's Your Turn Now!

We've gone over a few basics of the LabVIEW environment, and now we're going to show you exactly how to build your own VIs. Since people remember things better if they actually *do* them, you might step through these instructions on your computer as you read them so that you can learn the techniques more quickly.

■ Placing Items on the Front Panel

You will usually want to start your programs by "dropping" controls and indicators on a front panel to define your user inputs and program outputs. You've done this once or twice before in activities, but we'll mention it here for reference (and as a good way to start this interactive section). As you run the cursor over the **Controls** palette, you will see the names of subpalettes appear at the top of the palette. Click and hold down the mouse button on one of these buttons to access the corresponding sub-palette. Choose the desired object from the subpalette by releasing the mouse; again, you will notice that objects' names appear at the top of the subpalette when the mouse is over them.

Now click on the front panel, in the place you want your object to appear... and there it is!

 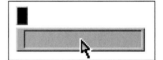

*You can also access the **Controls** palette by popping up in an empty area of the front panel.*

Now create a new VI and drop a digital control on your front panel.

Remember, when you drop an item on the front panel, its corresponding terminal appears on the block diagram. You might find it helpful to select **Tile Left and Right** from the **Windows** menu so that you can see the front panel and block diagram windows at the same time.

■ Labeling Items

Labels are blocks of text that annotate components of front panels and block diagrams. An object first appears in the front panel window with a little rectangle representing a label. If you want to retain the label at this time, enter text from the keyboard. If you click somewhere else with the mouse first, the label disappears. After you enter text into a label, any one of the following actions completes the entry:

◆ *Press <enter> on the numeric keypad.*

◆ *Click on the enter button in the **Tools** palette.*

◆ *Click somewhere outside the label on the front panel or block diagram.*

◆ *Press <shift-enter> on Windows and HP, <shift-return> on Mac and Sun, from the alphanumeric keyboard.*

The label appears on the corresponding block diagram terminal as well as the front panel object.

LabVIEW has two kinds of labels: owned labels and free labels. Owned labels belong to and move with a particular object; they annotate that object only. When you create a control or indicator on the front panel, a blank owned label accompanies it, awaiting input. A front panel object and corresponding block diagram terminal will have the same owned label. A free label is not associated with any particular object and can be created and deleted at will.

You can select **Show►Label** from the owning object's pop-up menu to create or change a label that isn't currently visible. You can hide owned labels, but you cannot copy or delete them independently of their owners. Structures and functions come with a default label that is hidden until you show it. You may want to edit this label to reflect the object's function in your program. You can also show labels of subVIs (which are really just their names), but you cannot edit them.

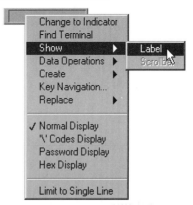

Now label the digital control you just created Number 1. You may have to **Show►Label** if you've already clicked somewhere after creating the control.

■ Creating Free Labels

Free labels are not attached to any object, and you can create, move, or dispose of them independently. Use them to annotate your panels and diagrams. Use the Labeling tool to create free labels and to edit virtually any visible text.

Labeling Tool

To create a free label, select the Labeling tool from the **Tools** palette and click anywhere in empty space. A small, bordered box appears with a text cursor at the left margin ready to accept typed input. Type the text you want to appear in the label and enter it in one of the four ways previously described. If you do not type any text in the label, the label disappears as soon as you click somewhere else.

Create a free label on the front panel that says `hippopotamus`.

■ Changing Font, Style, Size, and Color of Text

You can change text attributes in LabVIEW using the options in the **Font** ring on the Toolbar. Select objects with the Positioning tool or highlight text with the Labeling or Operating tools, then make a selection from the **Font** ring. The changes apply to everything selected or highlighted. If nothing is selected, the changes apply to the default font and will affect future instances of text.

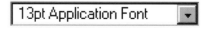

Change your `hippopotamus` label so that it uses eighteen-point font.

If you select **Font Dialog...** from the menu, a dialog box appears; you can change multiple font attributes at the same time using this dialog box.

LabVIEW uses System, Application, and Dialog fonts for specific portions of its interface. These fonts are predefined by LabVIEW, and changes to them affect all controls that use them.

♦ *The Application font is the default font, used for the **Controls** palette, the **Functions** palette, and text in new controls.*

♦ *The System font is used for menus.*

♦ *LabVIEW uses the Dialog font for text in dialog boxes.*

■ Placing Items on the Block Diagram

A user interface isn't much good if there's no program to support it. You create the actual program by placing functions, sub-VIs, and structures on the block diagram. To do this, access the **Functions** palette just like you did the **Controls** palette. Then select the item you want from a subpalette, and click on the diagram to place it.

Drop an **Add** function from the **Numeric** subpalette of the **Functions** palette onto the block diagram.

■ Editing Techniques

Once you have objects in your windows, you will want to be able to move them around, copy them, delete them, etc. Read on to learn how.

■ Selecting Objects

Positioning Tool

You must select an item before you can move it. To select something, click the mouse button while the Positioning tool is on the object. When you select an object, LabVIEW surrounds it with a moving dotted outline called a marquee, shown in the following illustration.

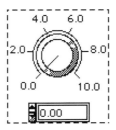

To select more than one object, <shift>-click on each additional object. You can also deselect a selected object by <shift>-clicking on it.

Another way to select single or multiple objects is to drag a selection rectangle around them. To do this, click in an open area with the Positioning tool and drag diagonally until all the objects you want to select lie within or are touched by the selection rectangle that appears. When you release the mouse button, the selection rectangle disappears and a marquee surrounds each selected object. The marquee is sometimes referred to as "marching ants," for obvious reasons. Once you have selected the desired objects, you can move, copy, or delete them at will.

You cannot select a front panel object and a block diagram object at the same time. However, you can select multiple objects on the same front panel or block diagram.

Clicking on an unselected object or clicking in an open area deselects everything currently selected. <shift>-clicking on an object selects or deselects it without affecting other selected objects.

Now select the digital control you created earlier.

▓ Moving Objects

You can move an object by selecting and dragging it to the desired location. If you hold down the <shift> key and then drag an object, LabVIEW restricts the direction of movement horizontally or vertically (depending on which direction you first move the object). You can also move selected objects in small, precise increments by pressing the appropriate arrow key; hold down the <shift> key at the same time to make the arrow keys move objects by a larger amount.

If you change your mind about moving an object while you are dragging it, drag the cursor outside all open windows and the dotted outline will disappear. Then release the mouse button, and the object will remain in its original location. If the object lingers at the window edge, the window will auto-scroll.

Move your digital control to the other side of the screen.

▓ Duplicating Objects

Positioning Tool

You can duplicate LabVIEW objects after you have selected them. From the **Edit** menu, select the **Copy** option, click the cursor where you want the new object, and then select the **Paste** option. You can also clone an object by using the Positioning tool to <control>-click on the object if you use Windows, <option>-click if you use a Mac, <meta>-click on a Sun, and <alt>-click on HP machines. Then drag the cursor away while still holding down the mouse. You will drag away the new copy, displayed as a dotted line, while the original stays in place. You can also duplicate front panel and block diagram objects from one VI to another. For example, if you select a block of code on one diagram and drag it to another, the appropriate wires will stay in place and any necessary front panel objects will be created.

You cannot duplicate control and indicator terminals on the block diagram—you must copy the items on the front panel.

Copy your digital control using both methods. You should now have three digital controls labeled Number 1, Number 1 copy, and Number 1 copy 2 on your front panel and three corresponding terminals on the block diagram. To find out which one belongs to which, pop up on a control or on a terminal and select **Find Terminal** or **Find Control**. LabVIEW will find and highlight the object's counterpart.

■ Deleting Objects

To delete an object, select it and then choose **Clear** from the **Edit** menu or press <delete> (Windows and HP-UX) or <backspace> (Mac and Sun).

You can only delete controls and indicators from the front panel. If you try to delete their terminals on the block diagram, the deletes are ignored.

Although you can delete most objects, you cannot delete control or indicator components such as labels and digital displays. Instead, you can hide these components by selecting **Show▶** from the pop-up menu and then deselecting the appropriate option.

Delete one of your digital controls.

■ Resizing Objects

Positioning Tool

You can change the size of most objects. When you move the Positioning tool over a resizable object, resizing handles appear at the corners of the object, as in the following illustration.

When you pass the Positioning tool over a resizing handle, the cursor changes to the Resizing tool. Click and drag this cursor until the dotted border outlines the size you want.

To cancel a resizing operation, continue dragging the frame corner outside the window until the dotted frame disappears. Then release the mouse button. The object maintains its original size.

If you hold down the <shift> key while you resize, the object will change size only horizontally, vertically, or in the same proportions in both directions, depending on the object and in which direction you drag first.

Resize one of your digital controls.

Some controls and indicators don't allow certain sizing operations. For example, digital controls can only grow horizontally (however, you can use a larger font to make the object bigger horizontally and vertically).

■ Moving Object to Front, to Back, Forward, and Backward

Objects can sit on top of and often hide other objects, either because you placed there or through some wicked twist of fate. LabVIEW has several commands in the **Edit** menu that move them relative to each other. You may find these commands very useful for finding "lost" objects in your programs. If you see an object surrounded by a shadow, chances are it's sitting on top of something. In this picture, the string control is not actually inside the loop, it is sitting on it.

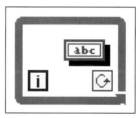

Move To Front moves the selected object to the top of a stack of objects.

Move Forward moves the selected object one position higher in the stack.

Move To Back and **Move Backward** work similarly to **Move To Front** and **Move Forward** except that they move items down the stack rather than up.

■ Coloring Objects

LabVIEW appears on the screen in black and white, shades of gray, or vivid color, depending on the capability and settings of your monitor. You can change the color of many LabVIEW objects, but not all of them. For example, block diagram terminals of front panel objects and wires use color codes for the type of data they carry, so you cannot change them. You also cannot change colors in black and white mode.

Color Tool

To change the color of an object or the background window, pop up on it with the Color tool. The following palette appears in color (assuming you have coloring ability on your machine).

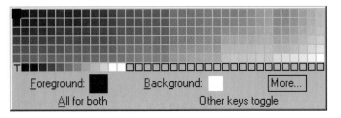

As you move through the palette while depressing the mouse button, the object or background you are coloring redraws with the color currently touched by the cursor. This gives you a preview of the object in the new color. If you release the mouse button on a color, the object retains the selected color. To cancel the coloring operation, move the cursor out of the color palette before releasing the mouse button. Selecting the **More** option from the color palette calls up a dialog box for picking custom colors.

Some objects have both a foreground and a background that you can color separately. For example, the foreground color of a knob is the main dial area, and the background color is the base color of the raised edge. The display at the bottom of the color palette indicates if you are coloring foreground, background, or both. You can press <f> to specify foreground or <b> for background; pressing <a> for "all" selects both foreground and background. Pressing any other key also toggles the selection between foreground and background.

You can also set colors on the **Tools** palette by clicking on the foreground or background color square and choosing a color from the palette that appears. When you subsequently *click* on objects (not pop up and use the color palette, like previously described), they will adopt the specified color scheme.

Color one of your digital controls by popping up and selecting the color. Then color another control using the **Tools** palette method.

■ Matching Colors

Sometimes it's hard to match a shade you've used, so you can also duplicate the color of one object and transfer it to a second object without going through the color palette. You can use the Color Copy tool on the **Tools** palette, which looks like an eye dropper (we call it the "sucker" tool), to set the active colors. Simply click with it on an object displaying the colors you want to pick up, then switch to the Color tool to color other things.

Color Copy Tool

You can also access the Color Copy tool by <control>-clicking under Windows, <option>-clicking on the Mac, <meta>-clicking on Sun, and <alt>-clicking on HP machines with the Color tool on the object whose color you want to duplicate. Then you can release the keystroke and click on another object with the Color tool; that object assumes the color you chose.

■ Transparency

If you select the box with a "T" in it from the color palette and color an item, LabVIEW makes the object transparent. You can use this feature to layer objects. For instance, you can place invisible controls on top of indicators, or you can create numeric controls without the standard three-dimensional container. Transparency affects only the appearance of an object. The object responds to mouse and key operations as usual.

■ Object Alignment and Distribution

Sometimes you want to make your VIs look just perfect, and you need a way to evenly line up and space your objects. LabVIEW's alignment and distribution functions make this easy. To align objects, select them with the Positioning tool (it's usually easiest just to drag a rectangle around them all, rather than <shift>-clicking on each one individually), then go to the **Align** ring, located in the Toolbar right next to the **Font** ring, and choose how you want them lined up. The **Distribute** ring works similarly to space objects evenly.

Align Ring   Distribute Ring

Be careful when you use these functions, because sometimes you'll end up with all of your objects on top of each other and you'll wonder what happened. For example, if you have three buttons in a row, and you align by Left Edges, all left edges will be flush, and all the objects will be stacked on top of each other. If this happens, use the Positioning tool to pick them off one by one.

Activity 4-1
Editing
Practice

In this activity, you will practice some of the editing techniques you've just learned. Remember, just as the **Controls** palette is visible only when the front panel window is active, the **Functions** palette can only be seen when the block diagram window is up.

1. Open **Editing Exercise.vi**, found in CH4.LLB in the EVERYONE directory or folder. The front panel of the **Editing Exercise** VI contains a number of LabVIEW objects. Your objective is to change the front panel of the VI shown.

Positioning Tool

2. First, you will reposition the digital control. Choose the Positioning tool from the **Tools** palette. Click on the digital control and drag it to another location. Notice that the label follows the control—the control *owns* the label. Now, click on a blank space on the panel to deselect the control; then click on the label and drag it to another location. Notice that the control does not follow. An owned label can be positioned anywhere relative to the control, but when the control moves, the label will follow.

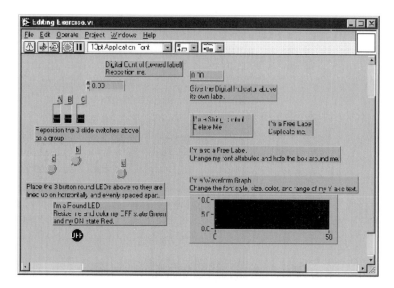

3. Reposition the three slide switches as a group. Using the Positioning tool, click in an open area near the three switches, hold down the mouse button, and drag until all the switches lie within the selection rectangle. Click on the selected switches and drag them to a different location.

4. Delete the string control by selecting it with the Positioning tool, then pressing <delete> or selecting **Clear** from the **Edit** menu.

5. Duplicate the free label. Hold down the <control> key on a computer running Windows, the <option> key on the Mac, the <meta> key on the Sun, or the <alt> key under HP-UX, then click on the free label and drag the duplicate to a new location.

Labeling Tool

6. Change the font style of the free label. Select the text by using the Labeling tool. You can double-click on the text, or click and drag the cursor across the text to select it. Modify the selected text using the options from the **Font** ring. Then hide the box around the label by popping up on the box with the Color tool and selecting the **T** (for transparent) from the color palette.

Color Tool

Remember, use the right mouse button on Windows, Sun, and HP, and <command>-click on Macintosh, to pop up on an item. Or you can use the Pop-up tool from the **Tools** palette and simply click on the object to access its pop-up menu.

Pop-up Tool

7. Now use the **Font** ring again to change the font style, size, and color of the Y-axis text on the Waveform Graph.

8. Create an owned label for the digital indicator. Pop up on the

digital indicator by clicking the right mouse button under Windows, Sun, and HP, or by clicking the mouse button while holding down the <command> key on the Mac, then choose **Show▶ Label** from the pop-up menu. Type `Digital Indicator` inside the bordered box. Press <enter> on the numeric keypad, click the enter button on the Toolbar, or click the mouse button outside the label to enter the text.

Enter Button

9. Resize the round LED. Place the Positioning tool over a corner of the LED until the tool becomes the resizing cursor. Click and drag the cursor outward to enlarge the LED. If you want to maintain the current ratio of horizontal to vertical size of the LED, hold down the <shift> key while you resize.

Positioning Tool

10. Change the color of the round LED. Using the Color tool, pop up on the LED. While continuing to depress the mouse button, choose a color from the selection palette. When you release the mouse button, the object assumes the last color you selected. Now click with the Operating tool on the LED to change its state to ON, then color the new state.

Color Tool

11. Place the three LED indicators so they are aligned horizontally and evenly spaced. Using the Positioning tool, click in an open area near the LEDs and drag a rectangle around them. Align them horizontally by choosing the **Vertical Centers** axis from the **Align** ring in the Toolbar. Then space the LEDs evenly by choosing **Horizontal Centers** axis from the **Distribute** ring.

Vertical Centers Axis

Horizontal Centers Axis

12. Your panel should now look something like the one shown.

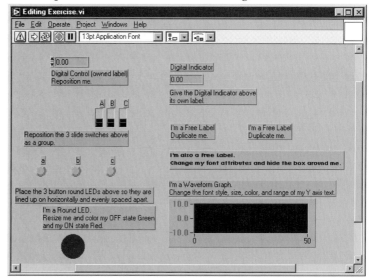

13. Close the VI by selecting **Close** from the **File** menu. Do not save any changes. Pat yourself on the back—you've mastered LabVIEW's editing techniques!

Basic Controls and Indicators and the Fun Stuff They Do

We're now going to talk a bit about the goodies contained in subpalettes of the **Controls** palette. LabVIEW has four types of simple controls and indicators: *numeric*, *Boolean*, *string*, and the less frequently used *path*. You will also encounter a few more complex data types such as arrays, clusters, tables, charts, and graphs, that we will expand on later.

*The **Controls** palette is visible only when the front panel window is active, NOT when the block diagram is up. You can drive yourself crazy looking for it if you don't keep this in mind.*

Enter Button

When you need to enter numeric or text values into any controls or indicators, you can use the Operating or Labeling tool. New or changed text is not registered until you press the <enter> key on the numeric keypad, click the enter button on the Toolbar, or click outside the object to terminate the editing session.

Hitting the <enter> key (Windows and HP-UX) or <return> key (Macintosh and Sun) on the alphanumeric keyboard (NOT the one on the numeric keypad) enters a carriage return and does NOT register your change (unless you've configured your system otherwise).

You must use the <enter> key on the numeric keypad to enter text to LabVIEW. If you must use the alphanumeric keyboard, hit <shift-enter> or <shift-return> to enter text.

■ Numeric Controls and Indicators

Numeric controls allow you to enter numeric values into your VIs; numeric indicators display numeric values you wish to see. LabVIEW has many types of numeric objects: knobs, slides, tanks, thermometers, and, of course, the simple digital display. To use numerics, select them from the **Numeric** palette of the **Controls** subpalette. All numerics can be either controls or indicators, although each type defaults to one or the other. For example, a thermometer defaults to an indicator because you will most likely use it as one. By contrast, a knob appears on the front panel as a control because knobs are usually input devices.

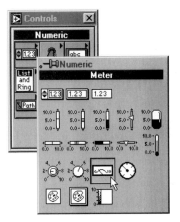

■ Representation

The appearance of numeric terminals on the block diagram depends on the *representation* of the data. The different representations provide alternative methods of storing data, to help use memory more effectively. Different numeric representations may use a different number of bytes of memory to store data, or may view data as *signed* (having the capacity for negative values) or *unsigned* (having only zero or positive values). Block diagram terminals are blue for integer data and orange for floating-point data (integer data has no digits to the right of the decimal point). The terminals contain a few letters describing the data type, such as "DBL" for double-precision floating-point data.

The numeric data representations available in LabVIEW are shown in the following table, along with their size in bytes and a picture of a digital control terminal with that representation.

Representation	Abbreviation	Terminal	Size (bytes)
byte	I8	`I8`	1
unsigned byte	U8	`U8`	1
word	I16	`I16`	2
unsigned word	U16	`U16`	2
long	I32	`I32`	4

Data Type	Abbreviation	Representation	Size (bytes)
unsigned long	U32	U32	4
single precision	SGL	SGL	4
double precision	DBL	DBL	8
extended precision	EXT	EXT	10(a)/12(b)/16(c)
complex single	CSG	CSG	8
complex double	CDB	CDB	16
complex extended	CXT	CXT	20(a)/24(b)/32(c)

(a) Windows
(b) Macintosh
(c) Sun and HP-UX

You can change the representation of numeric constants, controls, and indicators by popping up on the object and selecting Representation➤. Remember, you pop up on a numeric control or indicator by right-mouse-clicking on machines running Windows or UNIX or <command>-clicking on MacOS. You can then choose from the following palette.

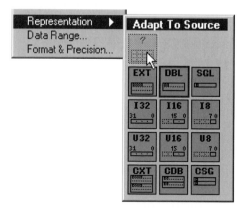

If you are concerned about memory requirements, you will want to use the smallest representation that will hold your data without losing information, especially if you are using larger structures like arrays. **Adapt To Source** automatically assigns the representation of the source data to your indicator—a good habit to get into. LabVIEW also contains functions that convert one data type to another, which will be covered in detail in Chapter 9, *Exploring Strings and File I/O*, and in Chapter 12, *Advanced LabVIEW Functions & Structures*.

▓ Format & Precision

LabVIEW lets you select whether your digital displays are formatted for numeric values or for time and date. If numeric, you can choose whether the notation is floating point, scientific, engineering, or relative time in seconds; you can also choose the *precision* of the display, which refers to the number of digits to the right of the decimal point, from 0 through 20. The precision affects only the display of the value; the internal accuracy still depends on the representation.

You can specify the format and precision by selecting **Format & Precision...** from an object's pop-up menu. The following dialog box appears.

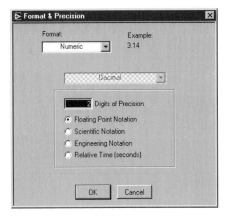

If you'd rather show time and date, choose **Time & Date** from the **Format** ring and your dialog box will change accordingly.

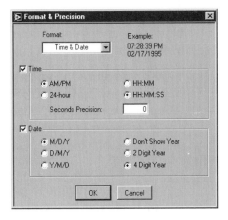

Sometimes it can be hard to discern exact values from graphical controls and indicators like graphs and thermometers. Use the pop-up option to **Show▶Digital Display** to bring up a digital window next to the object and display the precise numeric value. This digital display is part of the object itself and will not have a block diagram terminal.

Numeric Range Checking

LabVIEW gives you the option to enforce a certain valid range of numeric values and increments of data. For example, you might only want an input between zero and 100, in increments of two. You can set range checking by popping up on the appropriate numeric value and selecting **Data Range...**

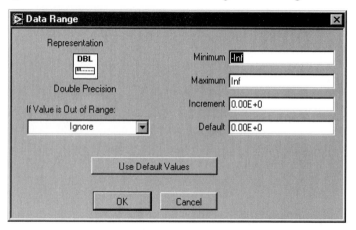

From the dialog box that appears, you can change numeric representation, input maximum and minimum acceptable values, set the increments you want, change the default value for that object, and select a course of action to follow if values are out of range.

◆ If you choose to **Ignore** out-of-range values, LabVIEW does not change or flag them. Clicking on the increment or decrement arrows of a control will change the value by the increment you set, up to the maximum values (or down to the minimum). However, you can still type in or pass a parameter out of the limits.

◆ If you choose to **Coerce** your data, LabVIEW will set all values below the minimum to equal the minimum and all values above the maximum to the maximum. Values in improper increments will be rounded.

◆ If you choose **Suspend**, LabVIEW will not begin to execute or will halt execution of a VI if any data is not valid. The invalid control(s) will be ringed in red and the run button will hide under a red circle with a slash through it. After you set all controls to valid inputs, the red rings disappear.

Rings

Rings are special numeric objects that associate unsigned 16-bit integers with strings, pictures, or both. You can find them in the **List & Ring** subpalette of the **Controls** palette. They are particularly useful for selecting mutually exclusive options such as modes of operation, calculator function, etc.

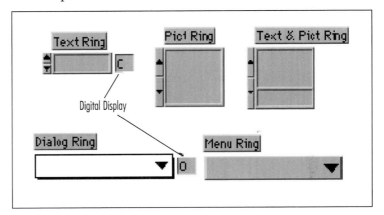

When you create a ring, you enter text or paste a picture into the ring that becomes associated with a certain number (zero for the first text message, one for the next, and so on). You can see this number (shown in the previous illustration) by selecting **Show▶Digital Display** from the ring's pop-up menu.

A new ring contains one item with a value of zero and an empty display. If you want to add another number and corresponding message, select **Add Item After** or **Add Item Before** from the pop-up menu and a blank entry window will appear. You can then type in text with the Labeling tool or import a picture.

If you click on a ring with the Operating tool, you will see a list of all possible messages or pictures, with the current one checked. Rings are useful if you want a user to select an option that will then correspond to a numeric value in the block diagram. Try dropping a ring on the front panel; then show the digital display and add a few items.

■ Booleans

Booleans are named for George Boole, an English logician and mathematician whose work forms the basis for Boolean algebra. For our purposes, you can think of Boolean as just a fancy word for "on or off." Boolean data can have one of two states: true or false. LabVIEW provides a myriad of switches, LEDs, and buttons for your Boolean controls and indicators, all accessible from the **Boolean** subpalette of the **Controls** palette. You can change the state of a Boolean by clicking on it with the Operating tool. Like numeric controls and indicators, each type of Boolean has a default type based on its probable use (i.e., switches appear as controls, LEDs as indicators).

Boolean Terminal

Boolean terminals appear green on the block diagram and contain the letters "TF."

Control terminals have THICK borders, while indicator terminal borders are THIN. It is VERY important to distinguish between the two since they are NOT functionally equivalent (Control=Input=data source, Indicator=Output=data sink, so they are NOT interchangeable).

The dialog button, checkmark, and radio button look different on each platform so that you can create VIs with the standard appearance for that platform.

■ Labeled Buttons

LabVIEW has two buttons with text messages built into them—the labeled square button and the labeled round button.

This text is merely informative for the user. Each labeled button can contain two text messages: one for the TRUE state and one for the FALSE state. When you first drop the button, the TRUE state says "ON" and the FALSE state says "OFF." You can then use the Labeling tool to change each message.

All Booleans also have a **Show▶Boolean Text** option that will display the word "ON" or "OFF," depending on their states.

■ Mechanical Action

A Boolean control has a handy pop-up option called **Mechanical Action**, which lets you determine how the Boolean behaves when you click on it (e.g., whether the value switches when you press the mouse button, switches when you release it, or changes just long enough for one value to be read and then returns to its original state). Mechanical Action is covered in more detail in Chapter 8, *Charts and Graphs*.

■ Data Range for Booleans

If you want to be able to flag an incorrect Boolean value, you can select **Suspend if True** or **Suspend if False** from the **Data Range▶** submenu of the Boolean's pop-up menu. If the unexpected value occurs, the VI will stop execution as described for the **Suspend** option of numeric data range.

■ Customizing Your Boolean With Imported Pictures

You can design your own Boolean style by importing pictures for the TRUE and FALSE state of any of the Boolean controls or indicators. You can learn more about how to do this in Chapter 15, *The Art of LabVIEW Programming*.

■ Strings

Simply put, string controls and indicators display text data. Strings most often contain data in ASCII format, the standard way to store alphanumeric characters. String terminals and wires carrying string data appear pink on the diagram. The terminals contain the letters "abc." You can find strings in the **String & Table** subpalette of the **Controls** palette.

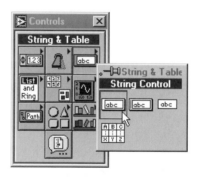

 Although string controls and indicators can contain numeric characters, they do NOT contain numeric data. You cannot do any numerical processing on string data; that is, you can no more add an ASCII "9" character than you can an "A." If you need to use numeric information that is stored in string format (to perform arithmetic, for example), you must first convert it to numeric format using the appropriate functions (see Chapter 9, Exploring Strings and File I/O).

String controls and indicators are fairly simple. Their pop-up menus contain few special options. We'll talk more about strings and their more complex cousin, the table, in Chapter 9, *Exploring Strings and File I/O*.

■ Paths

You use *path* controls and indicators to display paths to files, folders, or directories. If a function that is supposed to return a path fails, it will return <Not A Path> in the path indicator. Paths are a separate, platform-independent data type especially for file paths, and their terminals and wires appear bluish-green on the block diagram. A path is specified by `drivename` followed by `directory` or `folder` names and then finally the `filename` itself. On a computer running Windows, directory and file names are separated by a backslash (\); on a Mac, folder and file names are separated by a colon (:); on UNIX machines, a forward slash (/) separates files and directories. You'll learn more about paths in Chapter 14, *Communications and Advanced File I/O*.

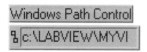

Windows   Mac

■ Decorations

Just for fun, you can use LabVIEW's special **Decorations** sub-palette of the **Controls** palette to enhance your front panel's appearance. These decorations have a solely aesthetic function—they are the only objects from the **Controls** palette that do not have corresponding block diagram terminals.

■ Custom Controls and Indicators

To make programming even more fun, LabVIEW lets you create your own custom controls and indicators. So if LabVIEW doesn't provide exactly the one you want, make your own! You'll learn how to do that in Chapter 15, *The Art of LabVIEW Programming*.

■ Summary of Basic Controls and Indicators

Just to make sure you get your data types straight, we'll recap the four types of simple controls and indicators:

Numerics contain standard numeric values.

Booleans can have one of two states: on or off (true or false, one or zero).

Strings contain text data. Although they can contain numeric characters (such as zero to nine), you must convert string data to numeric data before you can perform any arithmetic on it.

Paths give you a platform-independent data type especially for file paths.

Wiring Up

Your neatly arranged front panel full of sharp-looking controls and indicators won't do you much good if you don't connect the wires in your diagram to create some action in your program. The following sections detail everything you need to know about wiring techniques.

Wiring Tool

You use the Wiring tool to connect terminals. The cursor point or "hot spot" of the tool is the tip of the unwound wire segment, as shown.

Wiring Cursor Hot Spot

To wire from one terminal to another, click the Wiring tool on the first terminal, move the tool to the second terminal, and then click on the second terminal. It does not matter which terminal you click on first. The terminal area blinks when the hot spot of the Wiring tool is correctly positioned on the terminal. Clicking connects a wire to that terminal.

Once you have made the first connection, LabVIEW draws a wire as you move the cursor across the diagram, as if the wire were reeling off the spool. You do not need to hold down the mouse button.

To wire from an existing wire, perform the operation described above, starting or ending the operation on the existing wire. The wire blinks when the Wiring tool is correctly positioned to fasten a new wire to the existing wire.

You can wire directly from a terminal outside a structure to a terminal within the structure using the basic wiring operation (you'll learn more about structures in Chapter 6). LabVIEW creates a tunnel where the wire crosses the structure boundary, as shown in the following illustration. The first picture shows what the tunnel looks like as you are drawing the wire; the second picture depicts a finished tunnel.

■ Wiring Complicated Objects

When you are wiring a complicated built-in node or subVI, it helps to pay attention to the wire "whiskers" and tip strips that appear as the Wiring tool approaches the icon. Wire whiskers, the truncated wires shown around the pictured VI icon, indicate the data type needed at that terminal by their style, thickness, and color. Dots at the end of stubs indicate inputs, while outputs have no dots. Whiskers are drawn in the suggested direction to use if you want to wire clean diagrams.

You may also want to take advantage of the Help window feature that highlights each connector pane terminal. When you pass the Wiring tool over a terminal, the corresponding Help window terminal will blink so that you can be sure you are wiring to the right spot. You can also use the Help window to determine which connections are recommended, required, or optional.

▨ Bad Wires

When you make a wiring mistake, a broken wire—a black dotted line—appears instead of the usual colored wire pattern. Until all bad wires have been vanquished, your run button will appear broken and the VI won't compile. You can remove a bad wire by selecting and deleting it. A better method is to obliterate all bad wires at once by selecting **Remove Bad Wires** from the **Edit** menu or by using the keyboard shortcut, <control-B> under Windows, <command-B> on the Mac, <meta-B> on the Sun, and <alt-B> on HP machines.

*Sometimes bad wires are mere fragments, hidden under something or so small you can't even see them. In some cases, all you need to do to fix a broken run arrow is **Remove Bad Wires.***

If you don't know why a wire is broken, click on the broken run button or pop up on the broken wire and choose **List Errors**. A dialog box will appear describing your problem(s).

▨ Wiring Tips

The following tips may make wiring a little easier for you:

◆ *You can make a 90-degree turn in your wire, or "elbow" it, only once without clicking.*

◆ *Click the mouse to tack the wire and change direction.*

◆ *Change the direction from which the wire leaves a tack point by pressing the space bar.*

◆ *Double-click with the wiring tool to begin or terminate a wire in an open area.*

◆ *When wires cross, a small gap appears in the first wire drawn, as if it were underneath the second wire, illustrated below. You can also* **Edit➤Preferences** *and go to the* **Block Diagram** *menu, then check the box to* **Show dots at wire junctions***.*

◆ *Right-mouse-click to delete a wire while you're wiring, or <command>-click on the Macintosh.*

◆ *Use the Help window for more information about an object and to help you wire to the right terminal.*

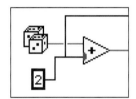

Create two numeric controls on a front panel and wire them to the inputs of the **Add** function. Don't wire the output just yet.

■ Wire Stretching

You can move wired objects individually or in groups by dragging the selected objects to the new location using the Positioning tool. Wires connected to the selected objects stretch automatically. If you duplicate the selected objects or move them from one diagram or subdiagram into another (for example, from the block diagram into a structure subdiagram such as a While Loop), LabVIEW leaves behind the connecting wires, unless you select them as well.

Wire stretching occasionally creates wire stubs or loose ends. You must remove these by using the **Remove Bad Wires** command from the **Edit** menu (or the keyboard shortcut) before the VI will execute.

Now move the **Add** function with the Positioning tool and watch how the attached wires adjust.

■ Selecting and Deleting Wires

Positioning Tool

A wire segment is a single horizontal or vertical piece of wire. The point at which three or four wire segments join is a *junction*. A *bend* in a wire is where two segments join. A wire branch contains all the wire segments from junction to junction, terminal to junction, or terminal to terminal if there are no junctions in between. One mouse click with the Positioning tool on a wire selects a segment. A double-click selects a branch. A triple-click selects an entire wire. Press the <delete> or <backspace> key to remove the selected portion of wire.

Select and delete one of your wires; then rewire it.

■ Moving Wires

Positioning Tool

You can reposition one or more segments by selecting and dragging them with the Positioning tool. For fine tuning, you can also move selected segments one pixel at a time by pressing the arrow keys on the keyboard. LabVIEW stretches adjacent, unselected segments to accommodate the change. You can select and drag multiple wire segments, even discontinuous segments, simultaneously. When you move a tunnel, LabVIEW normally maintains a wire connection between the tunnel and the wired node.

Move a wire segment first using the Positioning tool, then the arrow keys.

■ Wiring to Off-Screen Areas

If a block diagram is too large to fit on the screen, you can use the scroll bars to move to an off-screen area and drag whatever objects you need to that area. Dragging the Wiring tool slightly past the edge of the diagram window while you are wiring automatically scrolls the diagram. You can also click in empty space with the Positioning tool and drag outside the block diagram, and more space will be created.

■ Adding Constants, Controls, and Indicators Automatically

Instead of creating a constant, control, or indicator by selecting it from a palette and then wiring it manually to a terminal, you can pop up on the terminal and choose **Create Constant, Create Control**, or **Create Indicator** to automatically create an object with an appropriate data type for that terminal. The new object will be automatically wired for you, assuming that makes sense. Remember this feature as you develop your programs, because it's amazingly convenient!

Create an indicator to display the results of your **Add** by popping up on the function and selecting **Create Indicator**. Lab-VIEW will create an indicator terminal wired to the **Add** output on the block diagram as well as a corresponding front panel indicator, saving you the effort.

Running Your VI

Run Button

Run Button (active)

Run Button (subVI)

Continuous Run Button

Abort Button

Pause Button

You can run a VI using **Run** command from the **Operate** menu, the associated keyboard shortcut, or by clicking on the Run button. While the VI is executing, the Run button changes appearance.

The VI is currently running at its top level if the Run button is black and looks like it's "moving."

The VI is executing as a subVI, called by another VI, if the Run button has a tiny arrow inside the larger arrow.

If you want to run a VI continuously, press the Continuous Run button, but be careful—this is not a good programming habit to get into. You can accidentally catch your program in an endless loop and have to reboot to get out. If you do get stuck, try hitting the keyboard shortcut for the Abort command: <control-.> under Windows, <command-.> on Macs, <meta-.> on Suns, and <alt-.> under HP-UX.

Press the Abort button to abort execution of the *top level* VI. If a VI is used by more than one running top level VI, it is grayed out. Using the Abort button causes an immediate halt of execution and is not good programming practice, as your data may be invalid. You should code a "soft halt" into your programs that gracefully wraps up execution. You will learn how very soon.

The Pause button pauses execution when you press it, then resumes execution when you press it again.

You can run multiple VIs at the same time. After you start the first one, switch to the panel or diagram window of the next one and start it as previously described. Notice that if you run a sub-VI as a top-level VI, all VIs that call it as a subVI are broken until the subVI completes. You cannot run a subVI as a top-level VI and as a subVI at the same time.

Activity 4-2
Building a
Thermometer

Now you're going to put together a VI that actually does something! This program will take a voltage reading from a channel on your data acquisition (DAQ) board if you happen to have one, or from a simulation if you don't, and display it in a thermometer on the front panel. You should have channel 0 of your board connected to a temperature sensor or similar voltage (preferably between 0 and 1 volt). If you do have a DAQ board, read the instructions that accompany it for information on how to set it up. You will also learn more about DAQ setup in Chapters 10 and 11.

Make sure you save this activity if you are not using the sample version, because you will be adding to it later. If you are using the demo or didn't save it, you can find our version of **Thermometer.vi** *in* EVERYONE\ CH4.LLB *when you need it.*

1. Open a new front panel.

2. Drop a thermometer on the panel by selecting it from the **Numeric** palette of the **Controls** menu. Label it `Temperature` by typing inside the owned label box as soon as the thermometer appears on the panel. Rescale the thermometer by dragging over the "10.0" with the Operating tool or the Labeling tool and entering `100`.

Operating Tool

Labeling Tool

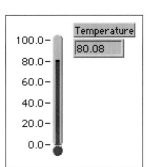

3. Build one of the block diagrams shown in the following illustration. You might find it helpful to select **Tile Left and Right**

from the **Windows** menu so that you can see both the front panel and the block diagram at the same time. Build diagram A if you are using a DAQ board or diagram B if you need to use simulated data. The sample software cannot acquire data, so if you are using it, build diagram B. *You do not need to build both unless you really want to practice!!*

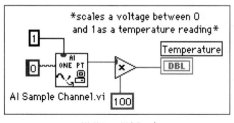

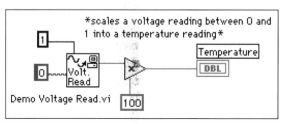

(A) Using a DAQ Board   (B) Using Simulated Data

You will find **AI Sample Channel.vi**, used in (A), under **Functions►Data Acquisition►Analog Input**. If you do not have a board, use the **Demo Voltage Read**, located under **Functions►Tutorial**, shown in (B). Remember to use the Help window to assist you in wiring to the correct terminals! Notice how, when you place the wiring tool over a node of a function, the corresponding terminal in the Help window will blink to let you see that you are wiring to the right place.

You can create a constant by popping up on the appropriate terminal of **Demo Voltage Read** or **AI Sample Channel** and selecting **Create Constant**. Just make sure you pop up on the terminal you want to wire to, or you create a constant for the wrong terminal.

Other block diagram components are described in the following list:

You will need a string constant, located in the **String** subpalette of the **Functions** palette, to specify the analog input channel you want to use, which may or may not be channel 0 as shown. In the **Demo Voltage Read**, this value is ignored, but we left it there to simulate the real thing. *Even though the string constant contains a numeric character, you must use a constant of string data type rather than numeric data type, or you will not be able to wire it up.*

Enter Button

When you drop a string constant on the block diagram, it contains no data. You can type a string immediately (if you don't click anywhere else first), then enter the data either by hitting <enter> on the numeric keypad, by clicking elsewhere in the diagram, or clicking the Enter button. You can change the value of a constant at any time using the Operating tool or the Labeling tool.

This numeric constant, located in the **Numeric** subpalette of the **Functions** palette, specifies the device number of your board, which may or may not be 1. In the **Demo Voltage Read**, this value is ignored.

When you drop a numeric constant on the block diagram, it contains the highlighted value "0." You can type and enter a new number immediately (if you don't click anywhere else first).

Under Windows, the device number is assigned in the NI-DAQ Configuration Utility (it is probably 1 unless you have multiple boards). On UNIX machines, the device number is assigned through a configuration file. On a Mac, the device number is the slot number of your DAQ board; you can see this by opening up NI-DAQ in the control panels. For more information on device numbers, see the manuals accompanying your DAQ board, or look in Chapter 10.

Diagram A assumes you have a temperature sensor wired to channel 0 of your DAQ board (another voltage source is fine, but will probably not provide a very accurate "temperature" unless it reads around 0.8 volts).

This numeric constant simply scales your voltage (somewhat artificially) into a "valid" temperature. If your voltage input is not between 0 and 1.0, you might want to change the value of this constant to make your output "temperature" more appropriate.

4. Use the Help window, found under the **Help** menu, to show you the wiring pattern of the function terminals. Pay special attention to color coding to avoid broken wires. Remember, numeric data types are blue or orange, strings are pink, and Booleans are green.

Run Button

5. Run the VI by clicking on the Run button. You will see the thermometer display the voltage brought in from your board or from the simulation function. If you can't get your VI to compile, read the next chapter, *Yet More Foundations*, which explains debugging techniques. Then try again. If you're having trouble getting the data acquisition to work, don't worry for now and just try the VI with the simulation function (hint: use the **Replace** feature to drop in **Demo Voltage Read.vi**). We just want to give you a good sample of LabVIEW programming, not drag you through a data acquisition trouble shooting session.

6. If you have the full version of LabVIEW, save the VI in your MYWORK directory by selecting **Save** from the **File** menu. Name it **Thermometer.vi**. You will be using this VI as a subVI later on in the book.

Useful Tips

As you do more and more programming in LabVIEW, you'll find some of these shortcuts very useful for putting together your VI more quickly. Look over all of these, and refer back to them later as reminders; you're sure to find some that will make you say, "I wish I'd known that!"

■ Keyboard Shortcuts

Many LabVIEW menu options have keyboard shortcuts. For example, to create a new front panel window, you can select the **New** option from the **File** menu or press the keyboard equivalent <control-N> (for Windows) or <command-N> (for MacOS). The following table gives frequently used shortcuts for Windows and MacOS.

In general, the keyboard auxiliary keys, <control> on Windows or <command> on MacOS, have the equivalent of <meta> key for the Sun and the <alt> key for HP-UX.

Windows	MacOS	Action
<control-B>	<command-B>	Remove all bad wires from the diagram
<control-E>	<command-E>	Flip between the Panel and Diagram windows
<control-F>	<command-F>	Find a LabVIEW object or text
<control-G>	<command-G>	Finds the next occurrence of an object or text
<control-H>	<command-H>	Show/Hide the Help window
<control-N>	<command-N>	Create a new VI
<control-Q>	<command-Q>	Quit the current session of LabVIEW
<control-R>	<command-R>	Run the current VI
<control-W>	<command-W>	Close the current VI
<control-.>	<command-.>	Stop the current VI

■ Examples

Glance through the examples that ship in LabVIEW's EXAMPLES directory. You can use these programs as is or modify them to suit your application. Open and play with **readme.vi**, also found in the EXAMPLES directory, to give you a good idea of what each example does.

■ Changing Tools

When LabVIEW is in edit mode, pressing <tab> toggles through the tools. If the front panel is active, LabVIEW rotates from the Operating tool to the Positioning tool to the Labeling tool to the Color tool. If the block diagram is active, LabVIEW toggles through the tools in the same order, except that it selects the Wiring tool instead of the Color tool.

You can also press the space bar to alternate between the Operating and Positioning tools in the front panel, and the Wiring and Positioning tools in the block diagram.

■ Changing the Direction of a Wire

Pressing the space bar while you wire changes the direction that the current wire branch leaves the last tack point. Thus, if you accidentally move horizontally from a tack point but want the wire to move down initially, pressing the space bar changes the initial orientation from horizontal to vertical.

■ Canceling a Wiring Operation

To delete a wire as you are wiring under Windows and UNIX, click the right mouse button. On a Mac, wire off the screen and click.

■ Removing the Last Tack Point

Clicking while wiring tacks a wire down. <control>-clicking (Windows) or <command>-clicking (MacOS) while wiring removes the last tack point, and <control>-clicking or <command>-clicking again removes the next-to-the-last tack point. If the last tack point is the terminal, <control>-clicking or <command>-clicking removes the wire.

■ Inserting an Object into Existing Wires

You can insert an object, such as an arithmetic or logic function, into an existing wiring scheme without breaking the wire and rewiring the objects. Pop up on the wire where you wish to insert the object, and choose **Insert➤**; then go ahead and choose the object you want to insert from the **Functions** palette that appears.

■ Moving an Object Precisely

You can move selected objects very small distances by pressing the arrow keys on the keyboard once for each pixel you want the objects to move. Hold down the arrow keys to repeat the action. To move the object in larger increments, hold down the <shift> key while pressing the arrow key.

■ Incrementing Digital Controls More Quickly

If you press the <shift> key while clicking on the increment or decrement buttons of a digital control, the display increments or decrements very quickly. The size of the increment increases by successively higher orders of magnitude; for example, by ones, then by tens, then by hundreds, and so on. As the range limit approaches, the increment decreases by orders of magnitude, slowing down to normal as the value reaches the limit.

■ Entering Items in a Ring Control

To add items quickly to ring controls, press <shift-enter> or <shift-return> after typing the item name to accept the item and position the cursor to add the next item.

■ Cloning an Object

To clone objects, select the objects to be copied, hold down the <control> key (Windows) or the <option> key (MacOS), and drag the duplicates to the new position. The original objects remain where they are. You can clone objects into another VI window as well.

■ Moving an Object in Only One Direction

If you hold down the <shift> key while moving or cloning objects, LabVIEW restricts the direction of movement horizontally or vertically, depending on which direction you move the mouse first.

▓ Matching the Color

To pick a color from an object, click on the object with the Color Copy tool. Then color other objects by clicking on them using the Color tool.

▓ Replacing Objects

You can easily replace a front panel or block diagram object by popping up on it and selecting **Replace➤**. A **Controls** or **Functions** palette will appear (depending on which window you're in), and you can choose a new object or function. The new one will replace the old, and any wires that are still legal will remain intact.

▓ Making Space

To add more working space to your panel or diagram window, drag out a region with the Positioning tool that goes beyond the window border . The front panel or block diagram will scroll and you will see a rectangle marked by a dotted line, which defines your new space.

▓ Designing Custom Palettes

If you're using a certain front panel object or function very often in your application, take advantage of the "sticky" palettes and keep the palette that contains your object open. Just tear off the palette by releasing the mouse button on its thumbtack. LabVIEW also gives you the option of adding VIs or custom controls you've created to the standard palettes you're using for quick access. To create a custom palette, follow these steps:

1. Select the **Edit Control & Function Palettes...** from the **Edit** menu.

2. The **Controls** and **Functions** top-level palettes will appear. Open the palette you wish to edit or add an object to.

3. Pop up on the palette and select the desired option from the **Insert➤** menu.

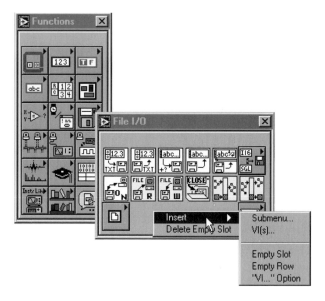

4. A dialog box will let you choose your VI or custom control.

5. When you're finished, save the new palettes by giving them a name. Later you can switch between your custom palettes, the default palettes, and others by choosing **Select Palette Set►** from the **Edit** menu.

■ Configuring Your Preferences

LabVIEW has many preferences you can configure to suit your taste and convenience by selecting **Preferences...** from the **Edit** menu. You can select which preference menu to view from the menu ring at the top of the Preferences dialog window.

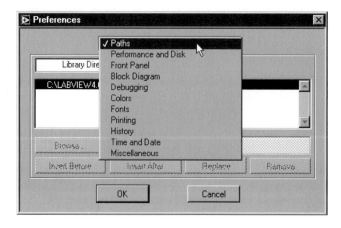

Select **Preferences** from the **Edit** menu and browse through the different options available to you. If you want to know more about Preference options, look in the LabVIEW manuals or on-line help.

Wrap It Up!

LabVIEW has special editing tools and techniques fitting to its graphical environment. The Operating tool changes an object's value. The Positioning tool selects, deletes, and moves objects. The Wiring tool creates the wires that connect diagram objects. The Labeling tool creates and changes owned and free labels. Owned labels belong to a particular object and cannot be deleted or moved independently, while free labels have no such restrictions.

LabVIEW has four types of simple controls and indicators: *numeric*, *Boolean*, *string*, and *path*. Each holds a separate data type and has special pop-up options. Control and indicator terminals and wires on the block diagram are color coded according to data type: floating-point numbers are orange, integer numbers are blue, Booleans are green, strings are pink, and paths are bluish-green.

You place objects on the front panel or block diagram using the **Controls** or **Functions** palette, respectively. You can also access these palettes by popping up in an empty section of the panel or diagram.

To run your VI, click on the run button or select **Run** from the **Operate** menu. If your run button is broken, it means something is wrong in your VI. Read the next chapter to learn good debugging techniques.

Additional Activities

■ Activity 4-3 **Comparison Practice**

Build a VI that compares two input numbers. If they are equal, an LED on the front panel turns on. Name it **Comparison Practice.vi**.

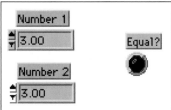

■ Activity 4-4 **Very Simple Calculator**

Build a VI that adds, subtracts, multiplies, and divides two in-put numbers and displays the result on the front panel. Use this front panel to get started. Call it **Very Simple Calculator.vi.**

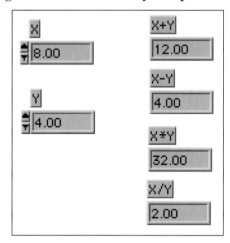

If you get stuck, you can find the answers in EVERYONE\CH4.LLB.

OVERVIEW

In this chapter, you will learn more fundamentals of the LabVIEW environment. We'll talk about load and save basics, LabVIEW's special library files, debugging features, subVI implementation, and documentation procedures.

YOUR GOALS

- Be able to load and save your VIs (and then save regularly!)
- Learn how to make LabVIEW's powerful debugging features work for you
- Create your very first subVI and understand how it's used
- Document your achievements for all the world to see

KEY TERMS

- VI library
- broken VI
- single-step mode
- node
- execution highlighting
- probe
- breakpoint
- subVI
- Icon Editor
- required, recommended, and optional inputs

Yet More Foundations

5

Loading and Saving VIs

Obviously, you will be loading and saving VIs quite a bit during your development. LabVIEW has many features that can accommodate your file storage needs, and this section discusses how you can make them work for you.

You can load a VI by selecting **Open** from the **File** menu, then choosing the VI from the dialog box that appears. As the VI loads, you will see a status window that describes the VIs that are currently being loaded and allows you to cancel the loading process. You can load a specific VI and launch LabVIEW at the same time by double-clicking on the VI's icon, or, on Windows and Macintosh operating systems, by dragging the icon on top of the LabVIEW icon.

Save VIs by selecting **Save** (or a similar option) from the **File** menu. LabVIEW then pops up a file dialog box so you can choose where you want to save. If you save VIs as individual files, they must conform to the file naming restrictions of your operating system (such as eight character filenames under Windows 3.1). To avoid these restrictions, you can save VIs in a compressed form in a special LabVIEW file called a *VI library*, described later in this chapter.

Keep in mind that LabVIEW references VIs by name. You cannot have two VIs with the same name in memory at one time. When searching for a VI of a given name, LabVIEW will load the first VI it finds, which may not be the one you intended.

Note that an asterisk (*) marks the titles of VIs that you've modified but not yet saved. We're sure you already know to save your work constantly, but we'd like to stress the importance of saving frequently and backing up everything you do on a computer—you never know when lightning will strike (literally)!

Do not ever save your VIs in the vi.lib directory. This directory is updated by National Instruments during new version releases of LabVIEW, and if you put anything in there, you may lose your work.

■ Save Options

You can save VIs with one of four save options in the **File** menu.

Select the **Save** option to save a new VI and then specify a name for the VI and its destination in the disk hierarchy; or use this option to save changes to an existing VI in a previously specified location.

Select the **Save As...** option to rename the VI in memory and to save a copy of the VI to disk under the new name. If you enter a new name for the VI, LabVIEW does not overwrite the disk version of the original VI. In addition, all VIs currently in memory that call the old VI now point to the new VI. If you do not change the name of the VI in the dialog box, LabVIEW prompts you to verify that you want to overwrite the original file.

When you select the **Save a Copy As...** option, LabVIEW saves a copy of the VI in memory to disk under a different name, which you enter in the dialog box. Unlike **Save As...**, this option does not affect the name or linkage of the VI in memory.

Save with Options... brings up a dialog box in which you can choose to save the VI for application or development distribution, save the entire VI hierarchy, or do a custom save. You also have the option to save VIs without block diagrams, but make sure you keep an extra copy somewhere that retains the block diagram, in case you ever need to modify it! To save a specified VI or VIs to a single new location without being interrupted by multiple prompts, save **To new location—single prompt**.

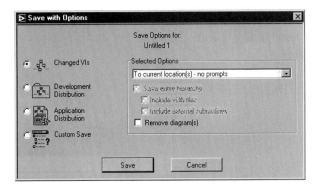

You cannot edit a VI after you save it without a block diagram. Always make a copy of the original VI before saving it without a block diagram.

■ Revert

You can use the **Revert...** option in the **File** menu to return to the last saved version of the VI you are working on. A dialog box will appear to confirm whether you want to discard any changes in the VI.

■ Save and Load Dialogs

LabVIEW supports the File dialog box format that your system uses. When you open or save a VI, your system File dialog box appears. If you click on a VI library (a special LabVIEW file storage structure), LabVIEW replaces the system dialog box with its own File dialog box so that you can select files within the library.

Because the interface for selecting and saving into a VI library is somewhat awkward with the system dialog box, you may prefer to set the LabVIEW preferences to use the standard LabVIEW dialog box if you commonly use VI libraries. Select **Preferences...** from the **Edit** menu, then go to the **Miscellaneous** menu and uncheck the **Use native file dialogs** box.

The MacOS Save dialog box disables files from VI libraries. You will need to click the **Use LLBs** button in order to save into a VI library, or use the standard LabVIEW dialog box instead of the native one.

■ Filter Rings

At the bottom of your Save or Load dialog box, you will see a filter ring that allows you to **View All,** view only **VIs & Controls, VIs,** or **Controls** files, or just see those with a **Custom Pattern** that you specify.

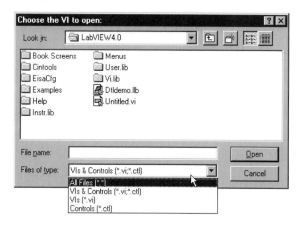

*The **Custom Pattern** option is not available on some native-style dialog boxes.*

If you choose **Custom Pattern**, another box appears in which you can specify a pattern. Only files matching that pattern will show up in the dialog box. Notice that an asterisk automatically appears inside the box; this is the "wild card character" and is considered a match with any character or characters.

VI Libraries

VI Libraries are special LabVIEW files that have the same load, save, and open capabilities as directories and folders within the LabVIEW environment. You can group several VIs together and save them as a VI library. VI libraries offer several advantages and disadvantages; for example, they can contain only compressed versions of VIs, not data or other files. In addition, your operating system sees VI libraries as single files, and you can access their contents only from LabVIEW.

Read on to determine if VI libraries or individual files best meet your storage needs.

■ Reasons for Using VI Libraries

◆ If you are using Windows 3.1 or plan to transfer your files to it, VI libraries allow you to use up to 255 characters to name your VIs, instead of just the eight characters allowed by the operating system.

◆ If you will be transferring VIs to other platforms, you may find VI libraries easier to move than multiple individual files.

◆ VI libraries compress their contents and reduce disk space requirements.

■ Reasons to Save VIs as Individual Files

◆ You can use the file system on your computer to manage individual files (e.g. copy, move, rename, back up) without having to go through LabVIEW.

◆ You cannot have hierarchy within VI libraries—they cannot contain subdirectories or folders.

◆ Loading and saving files occurs faster from the file system than from VI libraries. Less disk space is required for temporary files during the load and save processes.

◆ Storing VIs and controls in individual files is more robust than storing your entire project in the same file.

Note that many of the VIs shipped with LabVIEW are kept in VI libraries so that they are stored in consistent locations on all platforms. For the activities in this book, we've asked you to save your work in a MYWORK directory if you are on all platforms except Windows 3.1, so that you can access individual files more easily. Since Windows 3.1 has the eight-character file name limitation, we recommend you make a `MYWORK.LLB` VI library, as you probably did in Activity 3-1, so you can use longer VI names.

■ How to Use VI Libraries

Create a VI library from the **Save, Save As...**, or **Save a Copy As...** dialog box by clicking on the New VI Library button under Windows or the New... button on MacOS. If you're on a Mac configured to use native dialog boxes, you will have to click on the Use LLBs button from the save dialog box, then select New... from the dialog box that appears.

Enter the name of the new library in the dialog box that appears, shown in the next figure, and append a `.llb` extension. Then click on the VI Library button and the library is created. If you do not include the `.llb` extension, LabVIEW adds it.

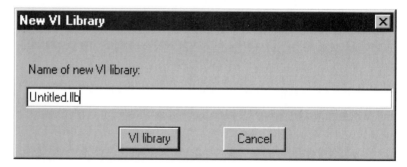

Usually you will create a VI library when you are saving a VI, so after the library is created, a dialog box appears to let you name your VI and save it in your new library.

Once you've created a VI library, you can save VIs in it and access them through LabVIEW much like a directory or folder, but you cannot see the individual VIs from your operating system. Remember that on Macs configured to use native dialogs, you'll have to select Use LLBs from the Save dialog box in order to access them.

■ The Edit VI Library Dialog

Since you can't do it through your operating system, you must use the Edit VI Library dialog box to edit the contents of a VI library. The Edit VI Library dialog box, shown in the following illustration, initially displays a list of the files in the VI library. As you move through the list, the creation and last modification dates for the selected file are shown at the bottom of the dialog box.

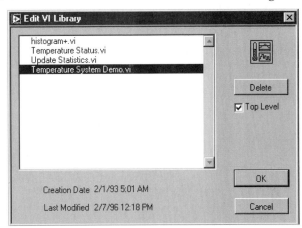

If you mark a VI **Top Level**, it will load automatically when you open the VI library. You can have more than one top-level VI in a library. Top-level VI names will also appear in a separate section at the top of the Load dialog, so it's easier for you to determine which VIs are main VIs and which are subVIs.

Debugging Techniques

Have you *ever* written a program with no errors in it? Lab-VIEW has many built-in debugging features to help you develop your VIs. This section explains how to use these conveniences to your best advantage.

■ Fixing a Broken VI

Run Button (broken)

A *broken VI* is a VI that cannot compile or run. The Run button appears as a broken arrow to indicate that the VI has a problem. It's perfectly normal for a VI to be broken while you are creating or editing it, until you finish wiring all the icons in the diagram. Sometimes you may need to **Remove Bad Wires** (found in the **Edit** menu) to clean up loose wires, but be careful not to delete wires you want!

To find out why a VI is broken, click on the broken Run button or select **Show Error List** from the **Windows** menu. An information box titled "Error List" appears listing all errors for the VI. You can choose to see the error list for other open VIs using a menu ring at the top of the window. To find out more about a particular error, click on it. The Error List window will display more information. To locate a particular error in your VI, double-click on the error in the list or highlight it and press the Find button. LabVIEW brings the relevant window to the front and highlights the object causing the error.

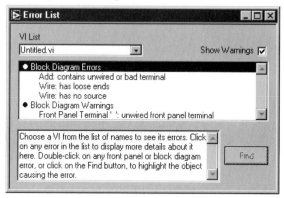

■ Warnings

If you want extra debugging help, you can choose to **Show Warnings** in the Error List window by clicking in the appropriate box. A warning is something that's not illegal and won't cause a broken run arrow, but does not make sense to LabVIEW, such as a control terminal that is not wired to anything. If you have **Show Warnings** checked and have any outstanding warnings, you will see the Warning button on the Toolbar. You can click on the Warning button to see the Error List window, which will describe the warning.

Warning Button

You can also configure LabVIEW's preferences to show warnings by default. Go to the **Debugging** menu in the **Preferences** dialog box (accessed by selecting **Edit➤Preferences...**) and check the **Show warnings in error box by default** box.

■ Most Common Mistakes

Certain mistakes are made more frequently than others, so we thought we'd list them to make your life easier. If your run button is broken, one of these might describe your problem. Please see Appendix B for a more exhaustive troubleshooting list.

◆   A function terminal requiring an input is unwired. You cannot leave unwired functions on the diagram while you run a VI to try out different algorithms.

◆   The block diagram contains a bad wire due to a data-type mismatch or a loose, unconnected end, which may be hidden under something or so tiny that you can't see it. The **Remove Bad Wires** command from the **Edit** menu eliminates the bad wires, but you might have to look a little harder to find a data-type conflict.

◆   A subVI is broken, or you edited its connector after placing its icon on the diagram. Use the **Replace** or **Relink to SubVI** pop-up option to re-link to the subVI.

◆   You have a problem with an object that is disabled, invisible, or altered using an attribute node (which we'll talk more about in Chapter 13).

◆   You have unwittingly wired two controls together, or wired two controls to the same indicator. The Error List window will bear the message, "Signal: has multiple sources," for this problem. You can often solve it by changing one of those controls to an indicator.

■ Single-Stepping Through a VI

Pause Button

For debugging purposes, you may want to execute a block diagram node by node. *Nodes* include subVIs, functions, structures, Code Interface Nodes (CINs), Formula Nodes, and attribute nodes. To begin single-stepping, you can start a VI by clicking on one of the single-step buttons (instead of the Run button), pause a VI by setting a breakpoint, or click on the Pause button. To resume normal execution, hit the Pause button again.

You may want to use execution highlighting (described next) as you single-step through your VI, so you can visually follow data as it flows through the nodes.

While in *single-step mode*, press any of the three step buttons that are active to proceed to the next step. The step button you press determines how the next step will be executed.

Step Into Button

Press the *Step Into* button to execute the first step of a subVI or structure and then pause at the next step of the subVI or structure. Or use the keyboard shortcut: down arrow key in conjunction with <control> under Windows and <command> on Macs (<meta> on Suns and <alt> on HP workstations).

Step Over Button

Press the *Step Over* button to execute a structure (sequence, loop, etc.) or a subVI and then pause at the next node. Or use the keyboard shortcut: right arrow key in conjunction with <control> under Windows and <command> on Macs.

Step Out Button

Press the *Step Out* button to finish executing the current block diagram, structure, or VI and then pause. Or use the keyboard shortcut: up arrow key in conjunction with <control> under Windows and <command> on Macs.

■ Execution Highlighting

Execution Highlighting
Button

Sometimes it's nice to see exactly where your data is and what's happening to it. In LabVIEW, you can view an animation of VI block diagram execution. To enable this mode, click on the *Execution Highlighting* button in the Toolbar.

As data passes from one node to another, the movement of data is marked by bubbles moving along the wires. You will notice that highlighting greatly reduces the performance of a VI. Click again on the Execution Highlighting button to resume normal execution. The following illustration shows a VI running with execution highlighting enabled.

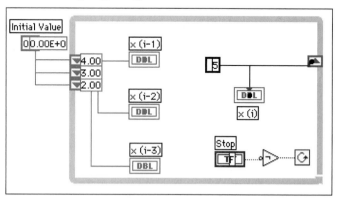

Node values are automatically shown during execution highlighting, as in the previous illustration, if you select **Auto probe during execution highlighting** from the **Debugging** menu of the **Preferences** dialog.

You commonly use execution highlighting in conjunction with single-step mode to gain an understanding of how data flows through nodes. When these two modes are used together, execution glyphs on the subVI's icon indicate which VIs are running and which are waiting to run.

■ Using the Probe

Probe Tool

Use the *probe* to check intermediate values in a VI that executes but produces questionable or unexpected results. For instance, assume you have a diagram with a series of operations, any one of which may be the cause of incorrect output data. To fix it, you could create an indicator to display the intermediate results on a wire, or you can leave the VI running and simply use a probe. To access the probe, select the Probe tool from the **Tools** palette and click its cursor on a wire, or pop up on the wire and select **Probe**. The probe display, which is a floating window, first appears empty if your VI is not running. When you run the VI, the probe display shows the value carried by its associated wire.

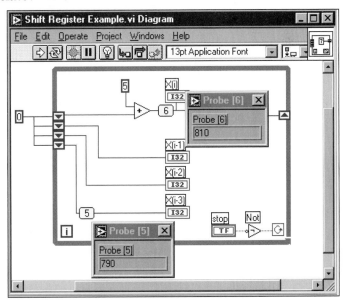

You can use the probe with execution highlighting and single-step mode to view values more easily. Each probe and the wire it references are automatically numbered uniquely by LabVIEW to help you keep track of them. A probe's number will not be visible if the name of the object probed is longer than the Probe window itself; if you lose track of which probe goes with which wire, you can pop up on a probe or a wire and select **Find Wire** or **Find Probe**, respectively, to highlight the corresponding object.

You cannot change data with the probe.

You can also probe using any conventional indicator by selecting **Custom Probe►** from the wire's pop-up menu, then choosing the desired indicator with which to probe. For example, you could use a chart to show the progress of a variable in a loop, since it displays past values as well as current ones. LabVIEW won't let you select an indicator of a different data type than the wire.

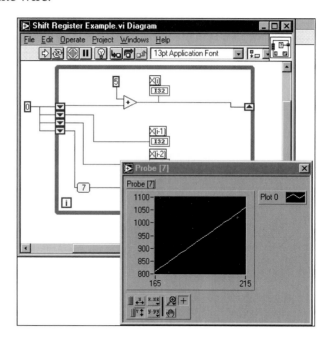

■ Setting Breakpoints

Don't panic—*breakpoints* do not "break" a VI; they only suspend its execution so that you can debug it. Breakpoints are handy if you want to inspect the inputs to a VI, node, or wire during execution. When the diagram reaches a breakpoint, it activates the pause button; you can single-step through execution, probe wires to see their data, change values of front panel objects, or simply continue running by pressing the Pause button or the Run button.

Breakpoint Tool

To set a breakpoint, click on a block diagram object with the Breakpoint tool from the **Tools** palette. Click again on the object to clear the breakpoint. The appearance of the breakpoint cursor indicates whether a breakpoint will be set or cleared.

Set Breakpoint Cursor   Clear Breakpoint Cursor

Depending on where they are placed, breakpoints behave differently.

◆ If the breakpoint is set on a *block diagram*, a red border appears around the diagram and the pause will occur when the block diagram completes.

◆ If the breakpoint is set on a *node*, a red border frames the node and execution pauses just before the node executes.

◆ If the breakpoint is set on a *wire*, a red bullet appears on the wire and any attached probe will be surrounded by a red border. The pause will occur after data has passed through the wire.

When a VI pauses because of a breakpoint, the block diagram comes to the front, and the object causing the break is highlighted with a marquee.

Breakpoints are saved with a VI but only become active during execution.

▓ Suspending Execution

You can also enable and disable breakpoints with the **Suspend when Called** option, found in the **Execution** menu of **VI Setup...** (which you access from the icon pane pop-up menu in a VI's front panel). **Suspend when Called** causes the breakpoint to occur at all calls to the VI on which it's set. If a subVI is called from two locations in a block diagram, the subVI breakpoint suspends execution at both calls.

If you want a breakpoint to suspend execution only at a particular call to the subVI, set the breakpoint using the **SubVI Node Setup...** option. Pop up on the subVI icon (in the block diagram of the calling VI) to access this option. You will learn more about VI setup options in Chapter 13, *Advanced LabVIEW Features*.

Activity 5-1
Debugging
Challenge

In this activity, you will troubleshoot and fix a broken VI. Then you will practice using other debugging features, including execution highlighting, single-step mode, and the probe.

1. Open up the VI called **Debug Exercise.vi**, located in EVERY-ONE\CH5.LLB.

2. Switch to the block diagram. Notice that the run arrow is broken. You must find out why and rectify the situation so the VI will run.

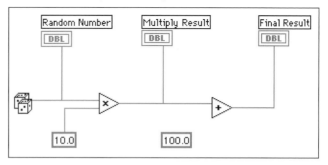

3. Click on the broken run arrow. An Error List dialog box appears describing the errors in the VI.

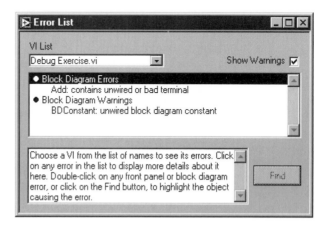

4. Click on the "Add: contains unwired or bad terminal" error. The Error List window will give you a more detailed description of the error. Now double-click on the error, or click the Find button. LabVIEW will highlight the offending function in the block diagram, to help you locate the mistake.

5. Draw in the missing wire. The Run button should appear solid. If it doesn't, try to **Remove Bad Wires**.

*If you can't find the missing wire, think about how many inputs the **Add** function should have.*

6. Switch back to the front panel and run the VI a few times.

Execution Highlighting Button

Step Into Button

Step Over Button

Step Out Button

7. Tile the front panel and block diagram (using the **Tile** command under the **Windows** menu) so you can see both at the same time. Enable execution highlighting and run the VI in single-step mode by pressing the appropriate buttons on the Toolbar in the block diagram.

8. Click the Step Over button each time you wish to execute a node (or click the Step Out button to finish the block diagram). Notice that the data appears on the front panel as you step through the program. First, the VI generates a random number and then multiplies it by 10.0. Finally, the VI adds 100.0 to the multiplication result. Notice how each of these front panel indicators is updated as new data reaches its block diagram terminals, a perfect example of dataflow programming. Remember, you can exit single-step mode and complete the VI by pressing the pause button. Also notice that the tip strips describing the single-step buttons change text to give you an exact description of what they will do when you click them, given the context of where you are.

9. Now enable the probe by popping up on any wire segment and selecting **Probe**.

10. Step through the VI again and note how the probe displays the data carried by its corresponding wire.

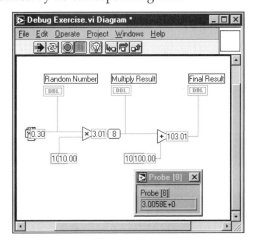

11. Turn off execution highlighting by clicking its button. You're almost done....

12. If you have the full version of LabVIEW, save your finished VI in your MYWORK directory (or in MYWORK.LLB if you are using Windows 3.1) by selecting the **Save As...** option from the **File** menu so you won't overwrite the original. Name it **Debugged Exercise.vi**. If you're feeling adventuresome, or you think VI libraries will suit your needs, try creating a VI library and saving your work again in it for practice.

13. Close the VI by selecting **Close** from the **File** menu. Good job!

Creating SubVIs

Much of LabVIEW's power and convenience stems from its modularity. You can build the parts of your program one complete module at a time by creating subVIs. A subVI is simply a VI used in (or called by) another VI. A subVI node (comprised of icon/connector in a calling VI block diagram) is analogous to a subroutine call in a main program. A block diagram can contain several identical subVI nodes that call the same subVI several times.

You can use any VI as a subVI in the block diagram of another VI, provided its icon has been created and its connector assigned. Drop existing VIs on a block diagram to use as subVIs with the **Select a VI...** button in the **Functions** palette. Choosing this option produces a file dialog box from which you can select any VI in the system; its icon will appear on your diagram.

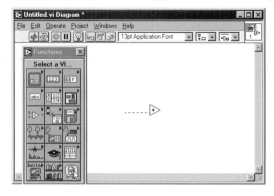

A VI can't call itself directly or indirectly. This is a capability of recursive execution, which LabVIEW can't do. You can't put a VI in its own block diagram or in the diagram of any of its subVIs (even buried several layers down in subVIs).

■ Creating a SubVI from a VI

Before you use a VI as a subVI, you must supply a way for the VI to receive data from and pass data to the calling VI. To do this, you need to assign the VI controls and indicators to terminals on its connector pane, and you must create an icon to represent the VI.

■ Designing the Icon

Every subVI must have an icon to represent it in the block diagram of a calling VI; the icon is its graphical symbol. You can create the icon by selecting **Edit Icon** from the pop-up menu of the icon pane in the upper right-hand corner of the front panel. You must be in edit mode to get this menu.

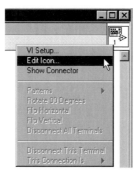

You can also access the *Icon Editor* by double clicking on the icon in the icon pane. The Icon Editor window, shown in the following illustration, will appear. Use its tools to design the icon of your choice.

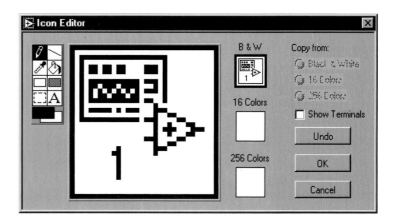

	Pencil	Draws and erases pixel by pixel.
	Line	Draws straight lines. Press <shift> to restrict drawing to horizontal, vertical, and diagonal lines.
	Dropper	Copies the foreground color from an element in the icon. Use the <shift> key to select the background color with the dropper.
	Fill bucket	Fills an outlined area with the foreground color.
	Rectangle	Draws a rectangle in the foreground color. Double-click on this tool to frame the icon in the foreground color. Use the <shift> key to constrain the rectangle to a square shape.
	Filled rectangle	Draws a rectangle bordered with the foreground color and filled with the background color. Double-click to frame the icon in the foreground color and fill it with the background color.
	Select	Selects an area of the icon for moving, cloning, or other changes.
	Text	Enters text into the icon design. Double-click on this tool to change the font attributes.
	Foreground/ background	Displays the current foreground and background colors. Click on each to get a palette from which you can choose new colors.

The buttons at the right of the editing screen perform the following functions:

◆ *Undo*   *Cancels the last icon edit operation you performed*

◆ *OK*   *Saves your drawing as the VI icon and returns to the front panel window*

◆ *Cancel*   *Returns to the front panel window without saving any changes*

Depending on your monitor, you can design a separate icon for display in monochrome, 16-color, and 256-color mode. You can design and save each icon version separately; you can also copy an icon from color to black and white (or vice versa) using the **Copy from...** buttons. Your VIs should always have at least a black and white icon, because color icons do not show up in a palette menu, and they don't show up on black and white screens. If no black and white icon exists, LabVIEW will show a blank icon.

■ Assigning the Connector

Before you can use a VI as a subVI, you will need to assign connector terminals. The connector is LabVIEW's way of passing data into and out of a subVI, just like you must define parameters for a subroutine in a conventional language. The connector of a VI assigns the VI's control and indicators to input and output terminals. To define your connector, pop up in the icon pane and select **Show Connector** (when you want to see the icon again, pop up on the connector and select **Show Icon**). LabVIEW chooses a default connector based on the number of controls and indicators on the front panel. If you want a different one, choose it from the **Patterns** menu, obtained by popping up on the connector. You can also rotate and flip your connector if it doesn't have a convenient orientation, using commands in the connector pop-up menu.

Follow these steps to assign a terminal to a control or indicator:

1. Click on a terminal in the connector. The cursor automatically changes to the Wiring tool, and the terminal turns black as shown.

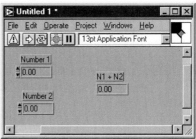

2. Click on the control or indicator you want that terminal to represent. A moving dotted line frames the control or indicator.

3. Click in an open area on the front panel. The dotted line disappears and the selected terminal dims as pictured, indicating that you have assigned the control or indicator to that terminal.

If the terminal is white or black, you have not made the connection correctly. Repeat the previous steps if necessary. You can have up to twenty-eight connector terminals for each VI.

You can reverse the order of the first two steps.

If you make a mistake, you can **Disconnect** a particular terminal or **Disconnect All** by selecting the appropriate action from the connector's pop-up menu.

■ SubVIs from Selection

Sometimes you won't realize you should have used a subVI for a certain section of code until you've already built it into the main program. Fortunately, you can also create subVIs by converting a part of the code in an existing VI. Use the Positioning tool to select the section of the VI that you want to replace with a subVI, choose **SubVI From Selection** from the **Edit** menu, and watch LabVIEW replace that section with a subVI, complete with correct wiring and an icon. You can double click on your new subVI to view its front panel, edit its icon, look at its connector, and save it under its new name. Use **SubVIs from Selection** with caution, as you may cause some unexpected results.You will learn more about this handy feature in Chapter 13, where we'll describe the restrictions that apply as well as common pitfalls.

■ SubVI Help: Recommended, Required, and Optional Inputs

If you bring up the Help window on a subVI node in a block diagram, its description and wiring pattern will appear. Input labels appear on the left, while outputs appear on the right. You can also bring up the Help window on the current VI's icon pane to see its parameters and description. You will learn how to edit the description in the next section.

Built-in LabVIEW functions automatically detect if you have not wired a required input and break the VI until you do. You can also configure your subVIs to have the same types of required, recommended, and optional inputs as functions. When an input is *required*, you cannot run the VI as a subVI without wiring that input correctly. When an input or output is *recommended*, you can run the VI, but the Error List window will list a warning (if you have warnings enabled) that tells you a recommended input or output is unwired. When an input is *optional*, no restrictions are enforced, and the connection is often considered advanced.

To mark a connection as required, recommended, or optional (or see what state it's in currently), pop up on the assigned terminal in the connector pane and take a look at the **This Connection Is▶** pullout menu. A checkmark next to **Required**, **Recommended**, or **Optional** indicates its current state.

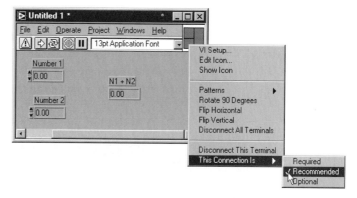

In the Help window, required connections appear bold, recommended connections are in plain text, and optional connections are grayed if you are using detailed help view. If the Help window is set to simple help view, optional connections are hidden.

Documenting Your Work

It's important to document your VIs so that others can understand them and so you don't forget why you did something or how it works. This section discusses a few ways you can document your work in LabVIEW.

■ Creating Descriptions for Individual Objects

If you want to enter a description of a LabVIEW object, such as a control, indicator, or function, choose **Description...** from the **Data Operations** submenu of the object's pop-up menu. Enter the description in the resulting Description dialog box, shown in the next illustration, and click OK to save it. LabVIEW displays this description whenever you subsequently choose **Description...** from the object's pop-up menu; this text also appears in the Help window whenever you pass the cursor over a front panel control or indicator. The **Description...** option is particularly useful on functions and subVIs to document how they are used in your block diagram. However, documentation entered in the Description window of functions and subVIs does not appear in the Help window—you have to open the window to see it.

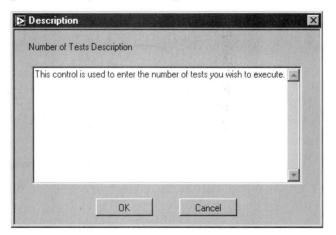

The best way to set up online help for your VIs is to enter descriptions for all of their controls, indicators, and functions.

■ Documenting VIs with the Show VI Info... Option

LabVIEW also gives you an easy way to document an entire VI. Selecting **Show VI Info...** from the **Windows** menu displays the VI Information dialog box for the current VI.

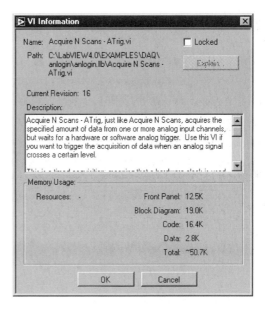

You can use the VI Information dialog box to perform the following functions:

◆   Enter a description of the VI. The description area has a scrollbar so that you can edit or view lengthy descriptions. When you use the VI as a subVI, the Help window will display this description when the cursor is over its block diagram icon.

◆   See a list of changes made to the VI since you last saved it by pressing the Explain button.

◆   View the path of the VI (i.e., where it is stored).

◆   See how much memory the VI uses. The memory usage portion of the information box displays the disk and system memory used by the VI. The figure applies only to the amount of memory the VI is using and does not reflect the memory used by any of its subVIs.

*Choosing **Description**... from an object's **Data Operations** menu documents the individual object, while entering text in the **Show VI Info**... dialog box of a VI documents the entire VI.*

■ A Little About Printing

LabVIEW has three kinds of printing you can use when you want to make a hard copy of your work.

◆ You can use the **Print Window** option from the **File** menu to make a quick printout of the contents of the current window.

◆ You can make a comprehensive printout of a VI, including information about the front panel, block diagram, subVIs, controls, VI history, and so on, by selecting the **Print Documentation** option from the **File** menu. Choosing **Print Documentation** pops up a dialog box where you can specify the format you want. We will talk more about it in Chapter 14, *Communications and Advanced File I/O.*

◆ You can use the LabVIEW programmatic printing features to make VI front panels print under the control of your application. Select **Print at Completion** from the **Operate** menu to enable programmatic printing. LabVIEW will then print the contents of the front panel any time the VI finishes executing. If the VI is a subVI, LabVIEW prints when that subVI finishes, before returning to the caller. We'll also talk more about this option in Chapter 14.

Activity 5-2
Creating SubVIs— Practice Makes Perfect

Okay, time to go back to the computer again. You will turn the **Thermometer** VI you created in the last chapter into a subVI so that you can use it in the diagram of another VI.

1. Open the **Thermometer.vi** that you created in Activity 4-2. If you saved it in your MYWORK directory (or VI library) like we told you to, it should be easy to find. If you are using the sample software or can't find it, use the **Thermometer.vi** found in EVERYONE\CH4.LLB.

2. Create an icon for the VI. Pop up on the icon pane in the front panel (it doesn't work in the diagram) and select **Edit Icon...** from the menu to open the Icon Editor. Use the tools described earlier in this chapter (in the *Designing the Icon* section) to create the icon, then click the OK button to return to the main VI. Your icon should appear in the icon pane as shown.

3. Create the connector by popping up in the icon pane and selecting **Show Connector**. Since you have only one indicator on

your front panel, your connector should have only one terminal and should appear as a white box, as shown.

Wiring Tool

4. Assign the terminal to the thermometer indicator. Using the Wiring tool (which is probably what the cursor will be automatically), click on the terminal in the connector. The terminal will turn black. Then click on the thermometer indicator. A moving dotted line will frame the indicator. Finally, click in an open area on the panel. The dotted line will disappear and the selected terminal will turn from black to gray, indicating that you have assigned the indicator to that terminal. Pop up and choose **Show Icon** to return to the icon.

5. Document the Temperature indicator by selecting **Description...** from its **Data Operations▶** pop-up menu. Type in the description shown and click OK when you're finished.

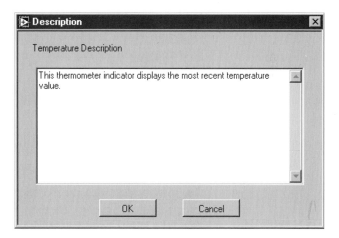

6. Document **Thermometer.vi** by selecting **Show VI Info...** from the **Windows** menu and typing in a description of what it does, as shown on the following page. Click OK to return to the main VI.

7. Now bring up the Help window by choosing **Show Help** from the **Help** menu. When you place the cursor over the icon pane, you will see the VI's description and wiring pattern in the Help window. If your Temperature indicator is not labeled in the VI, it won't have a label in the Help window either.

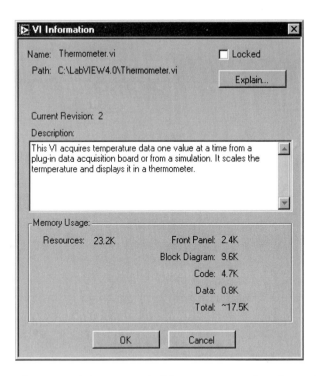

8. If you have a printer connected to your computer, choose **Print Window** from the **File** menu to print the active window. You can decide whether you want to print the front panel or block diagram.

9. Save the changes by selecting **Save** from the **File** menu unless you are using the sample software. If you are, you will find an updated **Thermometer.vi** in CH5.LLB for you to build on in later activities. Excellent work! You will use this VI as a subVI in the next chapter, so make sure to put it in MYWORK so you can find it!

10. Just for fun, use the Positioning tool to select portion of the block diagram shown, then choose **SubVI From Selection** from the **Edit** menu to automatically turn it into a subVI. Notice that the <u>Temperature</u> indicator remains part of the caller VI. Double-click on the new subVI to see its front panel.

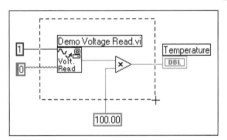

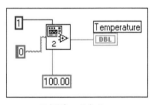

Selected Portion of Block Diagram   SubVI from Selection

11. Close both the new subVI and **Thermometer.vi**. *This time, do not save any changes!*

Frequently, people create an icon for a VI and forget about the connector. If you haven't assigned the connector, you will be unable to wire inputs and outputs to your VI when you try to use it as a subVI, and you may find the source of your wiring inability very difficult to locate!

Wrap It Up!

LabVIEW offers several ways to save your VIs. You may want to save them in *VI libraries*, which are special LabVIEW files containing groups of VIs. Your operating system sees VI libraries as single files; only LabVIEW can access the individual VIs inside. However, if you have Windows 3.1, which only allows filenames to contain eight characters, VI libraries helpfully allow you to have filenames of up to 255 characters. You should review the pros and cons of using VI libraries before you decide how you want to save your work. Regardless of how you save, be sure to do it frequently!!

You can take advantage of LabVIEW's many useful debugging features if your VIs do not work right away. You can *single-step* through your diagram node by node, animate the diagram using *execution highlighting*, and suspend subVIs when they are called so that you can look at input and output values by setting a *breakpoint*. You can also use the *probe* to display the value a wire is carrying at any time. Each of these features allows you to take a closer look at your VI and its data. If you still can't find your problem, Appendix B provides tips on avoiding frequent mistakes, describes common errors in more detail, and answers some common questions.

SubVIs are the LabVIEW equivalent of subroutines. All subVIs must have an icon and a connector. You can create subVIs from existing VIs or from a selected part of a block diagram. To prevent wiring mistakes with subVIs, use their online help and specify their inputs as *required*, *recommended*, or *optional*. Import an existing subVI into a calling VI's block diagram by choosing **Select a VI...** from the **Functions** palette, then selecting the subVI you want from the dialog box. SubVIs represent one of LabVIEW's most powerful features. You will find it very easy to develop and debug reusable, low-level subVIs and then call them from higher-level VIs.

As with all programming, it is a good idea to document your work in LabVIEW. You can document an entire VI by entering a description under **Show VI Info...** from the **Windows** menu. This description is also visible in the Help window if you pass the cursor over the VI's icon. You can document individual front panel objects and block diagram functions by selecting **Description...** from the **Data Operations** menu and then entering your text in the resulting dialog box. Descriptions for front panel objects will also appear in the Help window when you pass the cursor over the object.

LabVIEW offers several options for printing VIs; you can print the active window, specify what parts of a VI you want to print (such as front panel, block diagram, or subVI information), or set the VI to print programmatically.

Congratulations! You've now covered LabVIEW's fundamental operations. Now you're ready to learn about some of LabVIEW's powerful structures and functions, and how to write cool programs with them.

Additional Activities

Here's some more practice for you. If you get really stuck, look in EVERYONE\CH5.LLB for the solutions.

■ Activity 5-3 **Find the Average**

Create a subVI that averages three input numbers and outputs the result. Remember to create both the icon and connector for the subVI. Save the VI as **Find the Average.vi**.

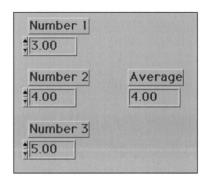

■ Activity 5-4 **Divide by Zero**

Build a VI that generates a random number between zero and ten, then divides it by an input number and displays the result on the front panel. If the input number is zero, the VI lights an LED to flag a "divide by zero" error. Save the VI as **Divide by Zero.vi**.

*Use the **Equal?** function found in the **Comparison** subpalette of the **Functions** palette.*

Next-Generation Gas Delivery System For Semiconductor Manufacturers

By Jeffrey Travis,
VI Technology, Inc.

The Challenge
Developing a PC-based automated gas delivery system.

The Solution
Using LabVIEW, SCXI (signal conditioning), and DAQ for data acquisition and control.

Introduction

Tylan General Intelligent Gas Panel

When engineers at Tylan General's Austin facility needed to manufacture a new gas delivery system, they chose to partner with VI Technology, Inc., a member of the National Instruments Alliance Program, to develop a PC-based process control system with LabVIEW and data acquisition (DAQ) products. As a result, they created a new benchmark for precision control and flexibility.

Semiconductor manufacturers use gas delivery systems to provide processed gas mixtures at critical flows, pressures, and temperatures in their clean rooms. The manufacturing process requires a gas supply that is accurate, repeatable, and contaminant-free to ensure the highest possible yield in chip fabrication processes such as etching. Earlier gas delivery systems that were not centrally controlled required manual observations and adjustments at several locations in the plant. The key to developing a successful new product was to provide a more automated system that could deliver higher levels of cleanliness, reduce costs by reducing downtime, and provide an easy-to-use interface for operators.

> Thanks to this system, with the power of LabVIEW and the flexibility of SCXI/DAQ, IC manufacturers such as Atmel can now move from a reactive to a predictive mode in their fabrication processes.

Intelligent Gas Panel Design and Benefits

Tylan's next-generation gas delivery system, called the Intelligent Gas Panel®, employs a 486 PC that contains an AT-MIO-16 DAQ board connected to several SCXI signal conditioning modules—SCXI-1100 analog input, SCXI-1124 optically isolated analog output, and SCXI-1162/1163 optically isolated digital I/O modules. The I/O capabilities of the RS-232 port of the PC are also used in controlling the gas panel. The LabVIEW software program used was written by VI Technology Inc. Engineers at Tylan chose LabVIEW for the gas panel because of its efficient programming environment and its appealing graphical user interface (GUI).

One of the key features of the Intelligent Gas Panel is that it provides a way to predict problems with gas delivery in progress, automatically adjusting flow and pressure if necessary. With previous gas control systems, manufacturers could not determine if an incorrect or contaminated mixture of gases

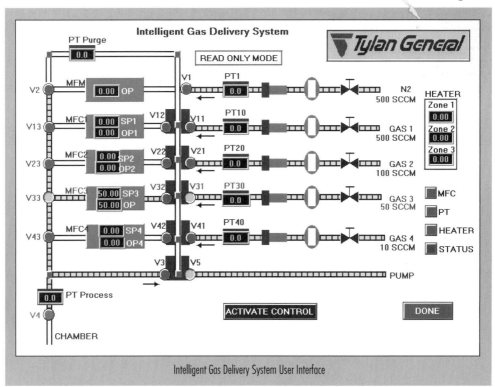

Intelligent Gas Delivery System User Interface

was affecting a batch of silicon wafers until processing was completed. Now, however, the software can acquire data in real time to check the status of the valves, pressure, temperature, and flow. After statistical process control (SPC) analysis is performed with LabVIEW, the gas delivery is automatically adjusted, if necessary. Another benefit of the Intelligent Gas Panel is its ability to use only one *digital* mass flow controller (MFC) to control multiple gases, rather than resorting to several *analog* MFCs. The single digital MFC facilitates communication with the software in addition to being more economical than using several analog units. Tylan's gas panel also has the capability to perform automatic gas purging. Using LabVIEW to control the automated gas purge makes it easy to implement safeguards to avoid human errors, such as selecting multiple or incorrect gas lines. Previously, operators had to manually purge the gas.

The goal of the Intelligent Gas Panel software portion was to create an easy-to-use turnkey system. Today, the built-in diagnostic capabilities of the resulting system substantially reduce or eliminate the downtime traditionally associated with gas panel diagnostics and repair. The Intelligent Gas Panel has already been incorporated into an Applied Materials Precision Etch 8330 metal etcher in Atmel Semiconductor's Colorado Springs Lab. Its onboard calibration and diagnostic capabilities were key factors in Atmel's decision to purchase the system. Thanks to this system, with the power of LabVIEW and the flexibility of SCXI/DAQ, IC manufacturers such as Atmel can now move from a reactive to a predictive mode in their fabrication processes. More importantly, future expansion is assured because VI Technology has built in the ability to easily accommodate additional I/O ports and process control technologies.

For More Information Contact

VI Technology
3616 Far West Blvd, Suites 101–212
Austin, TX 78731
tel (512) 327-3348
fax (512) 327-3341

OVERVIEW

Structures, an important type of node, govern execution flow in a VI, just as control structures do in a standard programming language. This chapter introduces you to the four main structures in LabVIEW: the While Loop, the For Loop, the Case Structure, and the Sequence Structure. You will also learn how to implement lengthy formulas using the Formula Node, how to pop up a dialog box containing your very own message, and a few basics on how to control the timing of your programs. You might want to take a look at some examples of structures in EXAMPLES\GENERAL\ STRUCTS.LLB *if you have the full version of LabVIEW.*

YOUR GOALS

- Know the uses of the While Loop and the For Loop and understand the differences between them
- Recognize the necessity of shift registers in graphical programming
- Understand the two types of Case Structures—numeric and Boolean
- Learn how to regulate execution order using Sequence Structures
- Use the formula node to implement long mathematical formulas
- Make LabVIEW pop up a dialog box that says anything you tell it to
- Understand how to use some of LabVIEW's simple timing functions

KEY TERMS

- For Loop
- While Loop
- iteration terminal
- conditional terminal
- count terminal
- tunnel
- coercion dot

- shift register
- Case Structure
- selector terminal
- dialog box
- Sequence Structure
- sequence local
- Formula Node

Controlling Program Execution with Structures

6

Two Loops

If you've ever programmed in any language, you've probably wanted to repeat a section of code. LabVIEW offers two loop structures to make this easy. You can use the *For Loop* and *While Loop* to control repetitive operations in a VI. A For Loop executes a specified number of times; a While Loop executes until a specified condition is no longer true. You can find both loops under the **Structures** subpalette of the **Functions** palette.

■ The For Loop

A *For Loop* executes the code inside its borders, called its *subdiagram*, for a total of *count* times, where the *count* equals the value contained in the *count terminal*. You can set the count by wiring a value from outside the loop to the count terminal. If you wire 0 to the count terminal, the loop does not execute.

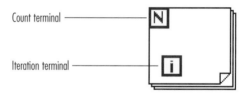

For Loop

The *iteration terminal* contains the current number of completed loop iterations; 0 during the first iteration, 1 during the second, and so on, up to N–1 (where N is the number of times you want the loop to execute).

The For Loop is equivalent to the following pseudocode:

```
for i = 0 to N-1
    Execute subdiagram
```

■ The While Loop

The *While Loop* executes the subdiagram inside its borders until the Boolean value wired to its *conditional terminal* is FALSE. LabVIEW checks the conditional terminal value at the *end* of each iteration. If the value is TRUE, another iteration occurs. The default value of the conditional terminal is FALSE, so if you leave it unwired, the loop iterates only once.

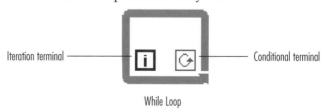

While Loop

The While Loop's *iteration terminal* behaves exactly like the one in the For Loop.

The While Loop is equivalent to the following pseudocode:

```
Do
    Execute subdiagram (which sets condition)
While condition is TRUE
```

■ Placing Objects inside Structures

For Loop Cursor

When you first select a structure from the **Structures** subpalette of the **Functions** palette, the cursor appears as a miniature of the structure you've selected; for example, the For Loop or the While Loop. You can then click where you want one corner of your structure to be, and drag to define the borders of your structure. When you release the mouse button, the structure will appear containing all objects you captured in the borders.

Once you have the structure on the diagram, you can place other objects inside either by dragging them in, or by placing them inside when you select them from the **Functions** palette. To make it clear to you that you are dragging something *into* a structure, the *structure's* border will highlight as the object moves inside. When you drag an object *out* of a structure, the *block diagram's* border (or that of an outer structure) will highlight as the object moves outside.

You can resize an existing structure by grabbing and dragging a corner with the Positioning tool.

If you move an existing structure so that it overlaps another object, the overlapped object will be visible above the edge of the structure. If you drag an existing structure completely over another object, that object will display a thick shadow to warn you that the object is *over* or *under* rather than *inside* the structure. Both of these situations are shown in the following illustration.

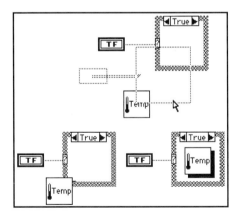

■ Terminals Inside Loops and Other Behavioral Issues

Data passes into and out of a loop through a little box on the loop border called a *tunnel*. Since LabVIEW operates according to dataflow principles, inputs to a loop must pass their data in before the loop executes. *Loop outputs pass data out only after the loop completes all iterations.*

As well, according to dataflow, *you must place a terminal inside a loop if you want that terminal checked or updated on each loop iteration.* For example, the left While Loop in the following illustration checks its Boolean control each time it loops. When the loop reads a FALSE value, it terminates.

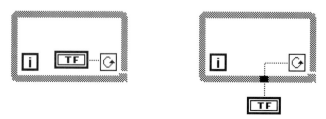

If you place the terminal of the Boolean control outside the While Loop, as pictured in the right loop, you create an infinite loop or a loop that executes only once, depending on the Boolean's initial value. True to dataflow, LabVIEW reads the value of the Boolean *before* it enters the loop, not within the loop or after completion.

Similarly, the <u>Digital Indicator</u> in the left loop will update during each loop iteration. The <u>Digital Indicator</u> in the right loop will update only once, after the loop completes. It will contain the random number value from the last loop iteration.

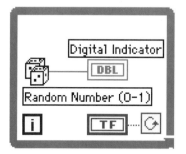
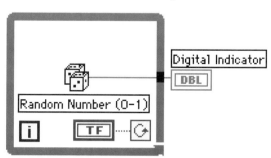

If you want to remove a loop without deleting its contents, pop up on its border and select **Remove While Loop** or **Remove For Loop**, respectively. If you simply highlight the loop with the Positioning tool and delete it, all of the objects inside will be deleted too.

You can generate arrays of data in a loop and store them on loop boundaries using LabVIEW's *auto-indexing* capability. We'll talk more about arrays and auto-indexing in the next chapter.

Remember, the first time through a For Loop or a While Loop, the iteration count is zero! If you want to show how many times the loop has actually executed, you must add one to the count!

Activity 6-1
Counting with Loops

In this activity, you get to build a For Loop that displays its count in a chart on the front panel. You will choose the <u>Number of Iterations</u>, and the loop will count from zero up to that number minus one (since everything is zero-based). You will then build a While Loop that counts until you stop it with a Boolean switch. Just for fun (and also to illustrate an important point), you will observe the effect of putting controls and indicators outside the While Loop.

1. Create a new panel by selecting **New** from the **File** menu or by clicking the New VI button in the "There are no VIs open" dialog box.

2. Build the front panel and block diagram shown. The For Loop is located in the **Structures** subpalette of the **Functions** palette. You might use the **Tile Left and Right** command from the **Windows** menu so that you can see both the front panel and the block diagram at the same time.

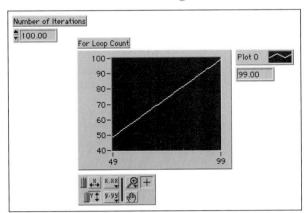

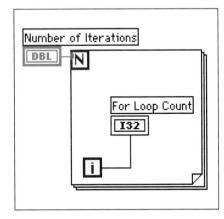

Drop a **Waveform Chart** from the **Graph** subpalette of the **Controls** palette onto your front panel. Label it `For Loop Count`. We'll talk more about charts and graphs in Chapter 8. Use a digital control from the **Numeric** subpalette for your <u>Number of Iterations</u> control.

3. Pop up on the Waveform Chart and select **AutoScale Y** from the **Y Scale** pull-out menu so that your chart will scale to fit the

For Loop count. Then pop up on the chart and **Show▶Digital Display**. Input a number to your <u>Number of Iterations</u> control and run the VI. Notice that the digital indicator counts from 0 to N-1, NOT 1 to N (where N is the number you specified)! Each time the loop executes, the chart plots the For Loop count on the Y axis against time on the X axis. In this case, each unit of time represents one loop iteration.

4. Notice the little grey dot present at the junction of the count terminal and the <u>Number of Iterations</u> wire. It's called a coercion dot, and we'll talk about it after this exercise. Pop up on the <u>Number of Iterations</u> control and choose **I32 Long** from the **Representation** subpalette to make it go away.

5. You can save the VI if you want, but we won't be using it again. Open up another new window so you can try out the While Loop.

6. Build the VI shown in the following illustration. Remember, Booleans appear on the front panel in their default FALSE position.

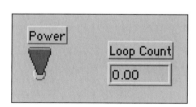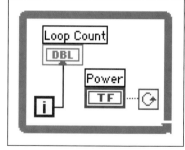

Operating Tool

7. Flip the switch up to its TRUE position by clicking on it with the Operating tool and run the VI. When you want to stop, click on the switch to flip it down to FALSE. <u>Loop Count</u> will update during each loop iteration.

8. With the switch still in the FALSE position, run the VI again. Notice that the While Loop executes once, but only once. Remember, the loop checks the conditional terminal at the *end* of an iteration, so it always executes at least once, even if nothing is wired to it.

9. Now go to the block diagram and move the <u>Loop Count</u> indicator outside the loop as shown. You will have to re-wire the indicator; the tunnel is created automatically as the wire leaves the loop.

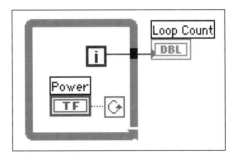

10. Make sure the switch is TRUE and run the VI. Notice that the indicator updates only after the loop has finished executing; it contains the final value of the iteration count, which is passed out after the loop completes. You will learn more about passing data out of loops in Chapter 7, *LabVIEW's Composite Data: Arrays and Clusters. Until then, do not try to pass scalar data out of a For Loop like you just did in a While Loop, or you will get bad wires and you won't understand why.* It can easily be done, but you will have to learn a little about auto-indexing first.

11. Save the VI. Place it in your MYWORK directory and call it **Loop Count.vi**.

12. Now, just to demonstrate what *not* to do, drag the switch out of the loop (but leave it wired). Make sure the switch is TRUE, run the VI, and then hit the switch to stop it. It won't stop, will it? Once LabVIEW enters the loop, it will not check the value of controls outside of the loop (just like it didn't update the <u>Loop Count</u> indicator until the loop completed). Go ahead and hit the Abort button on the Toolbar to halt execution. If your switch had been FALSE when you started the loop, the loop would have only executed once instead of forever. Close the VI and do not save changes.

Abort Button

■ The Coercion Dot

Remember the little gray dot present at the junction of the For Loop's count terminal and the <u>Number of Iterations</u> wire in the last activity? It's the *coercion dot*, so named because LabVIEW is coercing one numeric representation to fit another. If you wire two terminals of different numeric representations together, LabVIEW converts one to the same representation as the other. In the previous exercise, the count terminal has a 32-bit integer representation, while the <u>Number of Iterations</u> control is by default a double-precision floating-point number until you change it. In this case, LabVIEW converts the double-precision floating-point number to a long integer. In doing so, LabVIEW makes a new copy of the number in memory, in the proper representation. This copy takes up space. While the extra space is negligible for scalar numbers (single-valued data types), it can add up quickly if you are using arrays (which store multiple values). Try to minimize the appearance of the coercion dot on large arrays by changing the representation of your controls and indicators to exactly match the representation of data they will carry.

When a VI converts floating-point numbers to integers, it rounds to the nearest integer. A number with a decimal value of ".5" is rounded to the nearest even integer.

An easy way to create a count terminal input with the correct data type and representation is to pop up on the count terminal and select **Create Constant** *(for a block diagram constant) or* **Create Control** *(for a front panel control).*

Shift Registers

Shift registers, available for While Loops and For Loops, are a special type of variable used to transfer values from one iteration of a loop to the next. They are unique to and necessary for LabVIEW's graphical structure; we'll talk more about their uses in a little while. You create a shift register by popping up on the left or right loop border and selecting **Add Shift Register** from the pop-up menu.

Right Terminal

Left Terminal

A shift register is comprised of a pair of terminals directly opposite each other on the vertical sides of the loop border. The *right* terminal stores the data upon the completion of an iteration. That data is "shifted" at the end of the iteration and appears in the *left* terminal at the beginning of the next iteration, as shown in the following illustration. A shift register can hold any data type—numeric, Boolean, string, array, and so on. The shift register automatically adapts to the data type of the first object that you wire to it. It appears black when you first create it, but the shift register assumes the color of the data type wired to it.

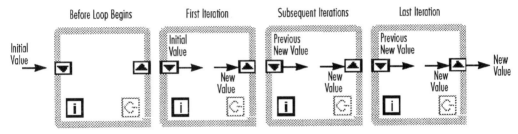

Shift Registers

You can configure the shift register to remember values from several previous iterations, as shown in the following illustration, a useful feature when you are averaging data values acquired in different iterations. To access values from previous iterations, create additional terminals by popping up on the *left* terminal and choosing **Add Element** from the pop-up menu.

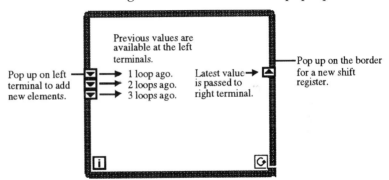

You can have many different shift registers storing many different variables on the same loop. Just pop up on the loop border and add them until you have as many pairs as you need. The left terminal will always stay parallel with its right terminal; if you move one, they both move. So if you have a lot of shift registers on your loop and can't tell exactly which ones are parallel, just select one and its partner will be automatically selected, or move one terminal a little and watch its mate follow.

Don't make the common mistake of confusing *multiple* variables stored in multiple shift registers with a *single* variable stored from multiple previous iterations in one shift register. The following illustration shows the difference.

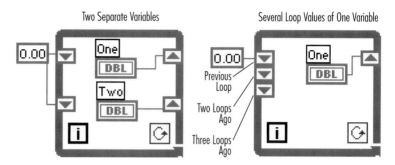

Two Separate Variables   Several Loop Values of One Variable

If you're still a little confused, don't worry. Shift registers are a completely new and different concept, unlike anything you may have encountered in a traditional programming language. Stepping through the next exercise should demonstrate them more clearly for you.

Make sure to wire directly to the shift register terminals so that you don't accidentally create an unrelated tunnel into or out of the loop.

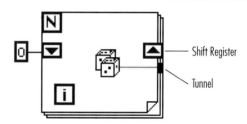

Activity 6-2
Shift Register
Example

To give you an idea of how shift registers work, you will observe their use to access values from previous iterations of a loop. In this VI, you will be retrieving count values from previous loops.

1. Open **Shift Register Example.vi,** located in EVERYONE\
CH6.LLB.

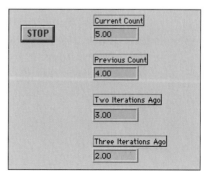

The front panel has four digital indicators. The <u>Current Count</u> indicator will display the current value of the loop count (it is wired to the iteration terminal). The <u>Previous Count</u> indicator will display the value of the loop count one iteration ago. The <u>Two Iterations Ago Count</u> indicator will display the value from two iterations ago, and so on.

2. Open the block diagram window by choosing **Show Diagram** from the **Windows** menu.

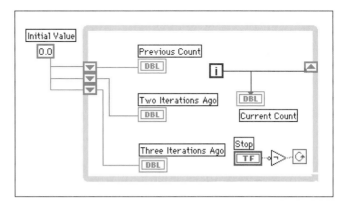

The zero wired to the left shift register terminals initializes the elements of the shift register to zero. At the beginning of the next iteration, the old <u>Current Count</u> value will shift to the top left terminal to become <u>Previous Count</u>. <u>Previous Count</u> shifts down into <u>Two Iterations Ago Count</u>, and so on.

3. After examining the block diagram show both the panel and the diagram at the same time by choosing **Tile Left and Right** from the **Windows** menu.

Execution Highlighting
Button

4. Enable the execution highlighting by clicking on the Execution Highlighting button.

Step Into Button

5. Run the VI and carefully watch the bubbles. If the bubbles are moving too fast, stop the VI and click on the Step Into button to put the VI in single-step mode. Click on the button again to execute each step of the VI. Watch how the front panel indicator values change.

Notice that in each iteration of the While Loop, the VI "funnels" the previous values through the left terminals of the shift register using a first in, first out (FIFO) algorithm. Each iteration of the loop increments the count terminal wired to the right shift register terminal, <u>Current Count</u>, of the shift register. This value shifts to the left terminal, <u>Previous Count</u>, at the beginning of the next iteration. The rest of the shift register values at the left terminal funnel downward through the terminals. In this example, the VI retains only the last three values. To retain more values, add more elements to the left terminal of the shift register by popping up on it and selecting **Add Element**.

Stop the VI by pressing the STOP button on the front panel. If you are in single-step mode, keep pressing the step button until it completes.

6. Close the VI. Do not save any changes. Another job well done!

■ Why You Need Shift Registers

Observe the following example. In loop (A), you are creating a running sum of the iteration count. Each time through the loop, the new sum is saved in the shift register. At the end of the loop, the total sum of 45 is passed out to the numeric indicator. In loop (B), you have no shift registers, so you cannot save values between iterations. Instead, you add zero to the current "i" each time, and only the last value of 9 will be passed out of the loop.

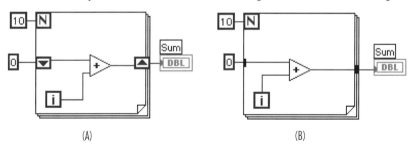

(A)   (B)

Or what about a case where you need to average values from successive loop iterations? Maybe you want to take a temperature reading once per second, then average those values over an hour. Given LabVIEW's graphical nature, how could you wire a value produced in one loop iteration into the next iteration without using a shift register?

■ Initializing Shift Registers

To avoid unforeseen and possibly nasty behavior, you should *always initialize your shift registers* unless you have a specific reason not to and make a conscious decision to that effect. To initialize the shift register with a specific value, wire that value to the left terminal of the shift register from outside the loop, as shown in the left two loops in the following illustration. If you do not initialize it, the initial value will be the default value for the shift register data type the first time you run your program. In subsequent runs, the shift register will contain whatever values are left over from previous runs.

For example, if the shift register data type is Boolean, the initial value will be FALSE for the first run. Similarly, if the shift register data type is numeric, the initial value will be zero. The second time you run your VI, an uninitialized shift register will contain values left over from the first run! Study the following figure to make sure you understand what initialization does. The two loops in the left column show what happens when you run a program that contains an initialized shift register twice. The right column shows what happens if you run a program containing an uninitialized shift register two times. Note the initial values of the shift registers in the two bottom loops.

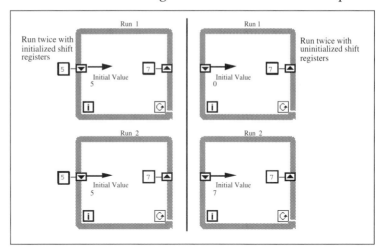

LabVIEW does not discard values stored in the shift register until you close the VI and remove it from memory. In other words, if you run a VI containing uninitialized shift registers, the initial values for the subsequent run will be the ones left over from the previous run. You seldom want this behavior, and the resulting problems can be very difficult to spot!

Case Structures

Enough about loops for now—let's move on to another powerful structure. A *Case Structure* is LabVIEW's method of executing conditional text, sort of like an "if-then-else" statement. You

can find it in the **Structures** subpalette of the **Functions** palette.
The Case Structure, shown in the following illustration, has two
or more subdiagrams, or cases; only one of them executes, depending on the value of the Boolean or numeric value you wire
to the *selector terminal*.

Selector Terminal

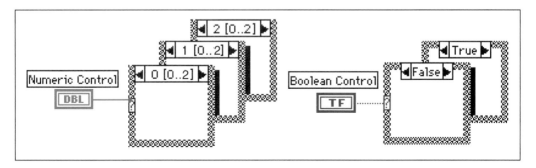

Case Structures

If a Boolean value is wired to the selector terminal, the structure has two cases, FALSE and TRUE. If a numeric data type is
wired to the selector, the structure can have from 0 to $2^{15}-1$ cases.
For numeric Case Structures, the total number of frames is indicated at the top of the structure, along with the current case number. Initially only the 0 and 1 cases are available, but you can easily add more. When you first place it on the panel, the Case
Structure appears in its Boolean form; it assumes numeric values
as soon as you wire a numeric data type to its selector terminal.

Case Structures can have multiple subdiagrams, but *you can
only see one case at a time,* sort of like a stacked deck of cards (unlike it appears in the previous illustration, where we cheated
and took two pictures). Clicking on the decrement (left) or increment (right) arrow at the top of the structure displays the previous or next subdiagram, respectively. You can also click on the
display at the top of the structure for a pull-down menu listing
all cases, then highlight the one you want to go to. Yet another
way to switch cases is to pop up on the structure border and select **Show Case....**

Decrement Arrow

Increment Arrow

If you wire a floating-point number to the selector, LabVIEW
rounds that number to the nearest integer value. LabVIEW co-
erces negative numbers to 0, and reduces any value higher than
the highest-numbered case to equal the number of that case.

You can position the selector terminal anywhere along the left
border. You must always wire something to the selector terminal,

and when you do, the selector automatically assumes that data type. If you change the data type wired to the selector from a numeric to a Boolean, cases 0 and 1 change to FALSE and TRUE. If other cases exist (2 through n), LabVIEW does not discard them, in case the change in data type is accidental. However, you must delete these extra cases before the structure can execute.

■ Wiring Inputs and Outputs

The data at all Case Structure input terminals (tunnels and selector terminal) is available to all cases. Cases are not required to use input data or to supply output data, *but if any one case outputs a value, all must output a value.* When you wire an output from one case, a little white tunnel appears in the same location on all cases. The run arrow will be broken until you wire data to this output tunnel from every case, at which time the tunnel will turn black and the run arrow will be whole again (provided you have no other errors). Make sure you wire *directly* to the existing output tunnel, or you might accidentally create more tunnels.

Why must you always assign outputs for each case, you ask? Because the Case Structure must supply a value to the next node regardless of which case executes. LabVIEW forces you to select the value you want rather than selecting one for you, as a matter of good programming practice.

■ Adding Cases

If you pop up on the Case Structure border, the resulting menu gives you options to **Add Case After** and **Add Case Before** the current case. You can also choose to copy the currently shown case by selecting **Duplicate Case**. You can delete the current case (and everything in it) by selecting **Remove Case**.

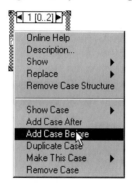

■ Dialog Boxes

Switch gears for a moment; we will digress from structures to tell you about dialog boxes so that you can use them in the next activity. The **One Button Dialog** and **Two Button Dialog** functions, shown in the following illustration, bring up a *dialog box* containing a message of your choice. You can find these functions in the **Time & Dialog** subpalette of the **Functions** palette. The **One Button Dialog** stays open until you click the OK button, while the **Two Button Dialog** box remains until you click either the OK or the Cancel button. You can also rename these buttons by inputting "button name" strings to the functions. These dialog boxes are *modal*; in other words, you can't activate any other LabVIEW window while they are open. They are very useful for delivering messages to or soliciting input from your program's operator.

One Button Dialog

Two Button Dialog

Activity 6-3
Square Roots

This activity will give you some practice with Case Structures and dialog boxes. You will build a VI that returns the square root of a positive input number. If the input number is negative, the VI pops up a dialog box and returns an error.

1. Open a new panel.

2. Build the front panel shown in the following illustration.

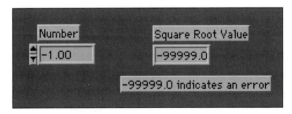

The Number digital control supplies the input number. The Square Root Value indicator will display the square root of the number.

3. Open the block diagram window. You will construct the following code.

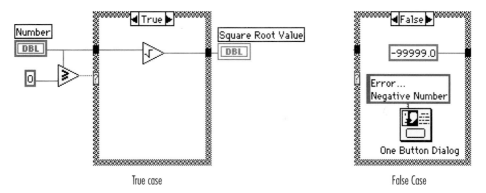

True case   False Case

4. Place the Case Structure (**Structures** subpalette) in the block diagram window. Like you did with the For Loop and While Loop, click with the structure cursor and drag to define the boundaries you want.

The **Greater or Equal?** function returns a Boolean value, so the Case Structure remains in its default Boolean form.

Remember, you can display only one case at a time. To change cases, click on the arrows in the top border of the Case Structure. Note that the previous picture shows two cases from the same structure so you will know what to build. *Do not create two different Case Structures for this activity!*

5. Select the other diagram objects and wire them as shown in the preceding illustration. *Make sure to use the Help window to practice displaying terminal inputs and outputs!*

Greater or Equal?
Function

Greater or Equal? function (**Comparison** subpalette). In this activity, checks whether the number input is negative. The function returns a TRUE if the number input is greater than or equal to zero.

Square Root
Function

Square Root function (**Numeric** subpalette). Returns the square root of the input number.

Numeric Constants

Numeric Constants (**Numeric** subpalette). "-99999.0" supplies the error case output, and "0" provides the basis for determining if the input number is negative.

One Button
Dialog Function

String Constant

One Button Dialog function (**Time & Dialog** menu). In this exercise, displays a dialog box that contains the message "Error . . . Negative Number."

String Constant (**String** subpalette). Enter text inside the box with the Operating or Labeling tool. (You will study strings in detail in Chapter 9, *Exploring Strings and File I/O*.)

In this exercise, the VI will execute either the TRUE case or the FALSE case of the Case Structure. If the input <u>Number</u> is greater than or equal to zero, the VI will execute the TRUE case, which returns the square root of the number. If <u>Number</u> is less than zero, the FALSE case outputs a –99999.00 and displays a dialog box containing the message "Error . . . Negative Number."

Remember that you must define the output tunnel for each case, which is why we bothered with the –99999.00 error case output. When you create an output tunnel in one case, tunnels appear at the same location in the other cases. Unwired tunnels look like white squares. Be sure to wire to the output tunnel for each unwired case, clicking on the tunnel itself each time, or you might accidentally create another tunnel.

6. Return to the front panel and run the VI. Try a number greater than zero and one less than zero.

7. Save and close the VI. Name it **Square Root.vi** and place it in your MYWORK directory or VI library.

■ Square Root VI Logic

```
if (Number >= 0) then
    Square Root Value = SQRT (Number)
    else
    Square Root Value = −99999.00
    Display Message  Error . . . Negative Number
end if
```

■ The Select Function

In simple "if-then-else" cases, you might find it more convenient to use LabVIEW's **Select** function, which works much like a Case Structure.

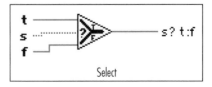

Select

The **Select** function, found in the **Comparison** subpalette of the **Functions** palette, returns a value of **t** if the **s** input value is TRUE, and returns a value of **f** if the **s** input is FALSE. This function could accomplish almost the same thing as the Case Structure in the last activity, with the exception of popping up the dialog box.

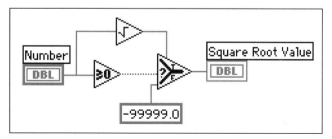

Sequence Structures

Determining the execution order of a program by arranging its elements in a certain sequence is called *control flow*. BASIC, C, and most other programming languages have inherent control flow because statements execute in the order in which they appear in the program. LabVIEW uses the *Sequence Structure* to obtain control flow within a dataflow framework. A Sequence Structure executes frame 0, followed by frame 1, then frame 2, until the last frame executes. Only when the last frame completes does data leave the structure.

The Sequence Structure, shown in the following illustration, looks like a frame of film. It can be found in the **Structures** subpalette of the **Functions** palette. Like the Case Structure, only one frame is visible at a time—you must click the arrows at the top of the structure to see other frames; or you can click on the top display for a listing of existing frames, or pop up on the structure border and choose **Show Frame....** When you first drop a Sequence Structure on the block diagram, it has only one frame; thus, it has no arrows or numbers at the top of the structure to designate which frame is showing. Create new frames by popping up on the structure border and selecting **Add Frame After** or **Add Frame Before**.

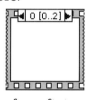

Sequence Structure

You use the Sequence Structure to control the order of execution of nodes that are not data dependent on each other. Within each frame, as in the rest of the block diagram, data dependency determines the execution order of nodes. You will learn about another way to control execution order called artificial data dependency in Chapter 15, *The Art of LabVIEW Programming*.

Output tunnels of Sequence Structures can have only one data source, unlike Case Structures, whose outputs must have one data source per case. The output can originate from any frame, but keep in mind that data is passed out of the structure only when the structure completes execution entirely, not when the individual frames finish. Data at input tunnels is available to all frames.

■ Sequence Locals

To pass data from one frame to any subsequent frame, you must use a terminal called a *sequence local*. To obtain a sequence local, choose **Add Sequence Local** from the *structure border* pop-up menu. This option is not available if you pop up too close to another sequence local or over the subdiagram display window. You can drag the sequence local terminal to any unoccupied location on the border. Use the **Remove** command from the sequence local pop-up menu to remove a terminal, or just select and delete it.

When it first arrives on the diagram, a sequence local terminal is just a small yellow box. The following figure shows the sequence local terminal in its several forms. When you wire source data to the sequence local, an outward-pointing arrow appears in the terminal of the frame containing the data source. The terminals in subsequent frames contain an inward-pointing arrow, indicating that the terminal is a data source for that frame. In frames before the source frame, you cannot use the sequence local (after all, it hasn't been assigned a value), and it appears as a dimmed rectangle.

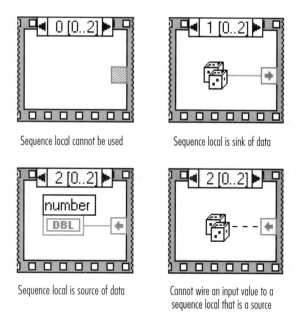

Sequence local cannot be used   Sequence local is sink of data

Sequence local is source of data   Cannot wire an input value to a
                                            sequence local that is a source

Sequence Local Terminals

■ Timing

Sometimes you will find it useful to control or monitor the timing of your VI. **Wait (ms)**, **Tick Count (ms)**, and **Wait Until Next ms Multiple,** located in the **Time & Dialog** subpalette of the **Functions** palette, accomplish these tasks. **Wait (ms)** causes your VI to wait a specified number of milliseconds before it continues execution.

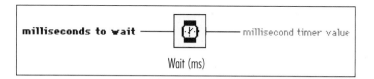

Wait (ms)

Wait Until Next ms Multiple causes LabVIEW to wait until the internal clock equals or has passed a multiple of the **millisecond multiple** input number before continuing VI execution; it is useful for causing loops to execute at specified intervals and synchronizing activities. These two functions are similar, but not identical. For example, **Wait Until Next ms Multiple** will probably wait less than the specified number of milliseconds in the first loop iteration, depending on the value of the clock when it starts (that is, how long it takes until the clock is at the next multiple and the VI proceeds). In addition, if the loop is still executing when the clock passes a millisecond multiple, the VI will wait until the clock reaches the *next* multiple, so the VI may become "out of synch" and slow down. Just make sure you take all possibilities into account when you use these functions.

Wait Until Next ms Multiple

Tick Count (ms) returns the value of your operating system's internal clock in milliseconds; it is commonly used to calculate elapsed time, as in the next activity. Be warned that the internal clock doesn't always have great resolution—one tick of the clock can be up to 55 milliseconds (ms) on Windows 3.1, 10 ms on Windows NT and Windows 95, 17 ms on Macs, and 1 ms on UNIX machines; LabVIEW can't work around this operating system limitation.

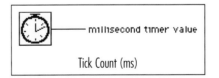

Tick Count (ms)

Activity 6-4
Matching
Numbers

Now you'll have the opportunity to work with the Sequence Structure and one of the timing functions. You will build a VI that computes the time it takes to match an input number with a randomly generated number. Remember this algorithm if you ever need to time a LabVIEW operation.

1. Open a new front panel.

2. Build the front panel shown.

Number to Match	Current Number
4	4

Number of Iterations	Time to Match
16	0.09

3. Change the precision of <u>Number to Match</u>, <u>Current Number</u>, and <u>Number of Iterations</u> to zero by selecting **Format & Precision...** from their pop-up menus. Enter "0" for **Digits of Precision** so that no digits are displayed to the right of the decimal point.

4. Open the diagram window and build the block diagram shown.

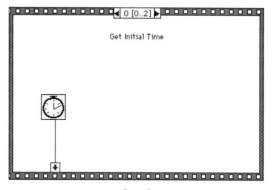

Frame 0

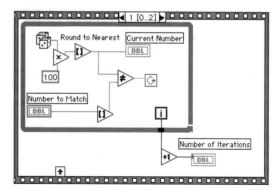

Frame 1

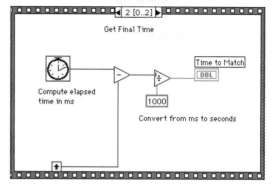

Frame 2

5. Place the Sequence Structure (**Structures** palette) in the diagram window. It works like the For Loop and While Loop; click with the structure cursor and drag to define the boundaries you want.

You will have to build three separate frames of the Sequence Structure. To create a new frame, pop up on the frame border and choose **Add Frame After** from the pop-up menu.

6. Create the sequence local by popping up on the bottom border of Frame 0 and choosing **Add Sequence Local** from the pop-up menu. The sequence local will appear as an empty square. The arrow inside the square will appear automatically when you wire to the sequence local.

7. Build the rest of the diagram. Some new functions are described here. Make sure to use the Help window to display terminal inputs and outputs when you wire!

Tick Count Function

Tick Count (ms) function (**Time & Dialog** palette). Returns the value of the internal clock.

Random Number
(0-1) Function

Random Number (0–1) function (**Numeric** palette). Returns a random number between 0 and 1.

Multiply Function

Multiply function (**Numeric** palette). Multiplies the random number by 100 so that the function returns a random number between 0.0 and 100.0.

Round to Nearest
Function

Round to Nearest function (**Comparison** palette). Rounds the random number between 0 and 100 to the nearest whole number.

Not Equal? Function

Not Equal? function (**Comparison** palette). Compares the random number to the number specified in the front panel and returns a TRUE if the numbers are not equal; otherwise, this function returns a FALSE.

Increment Function

Increment function (**Numeric** palette). Adds one to the loop count to produce the <u>Number of Iterations</u> value (to compensate for zero-based indexing).

In Frame 0, the **Tick Count (ms)** function returns the value of the internal clock in milliseconds. This value is wired to the sequence local, so it will be available in subsequent frames. In Frame 1, the VI executes the While Loop as long as the number specified does not match the number returned by the **Random Number (0–1)** function. In Frame 2, the **Tick Count (ms)** function returns a new time in milliseconds. The VI subtracts the old time (passed from Frame 0 through the sequence local) from the new time to compute the time elapsed, then divides by 1000 to convert from milliseconds to seconds.

Note that a more efficient way to do this would be to use a two-frame Sequence Structure, as shown here. However, we wanted you to practice using sequence locals.

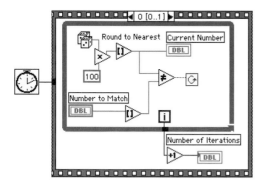
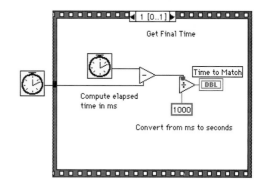

8. Turn on execution highlighting, which slows the VI enough to see the current generated number on the front panel.

9. Enter a number inside the Number to Match control and run the VI. When you want to speed things up, turn off execution highlighting.

10. Use the **Save** command to save the VI in your MYWORK directory or VI library as **Time to Match.vi**, and then close it. Good job!

The Formula Node

Now that you know about LabVIEW's four main control flow structures, we'll introduce a structure that doesn't affect program flow. The *Formula Node* is a resizable box that you use to enter algebraic formulas directly into the block diagram. You will find this feature extremely useful when you have a long formula to solve. For example, consider the fairly simple equation $y = x^2 + x + 1$. Even for this simple formula, if you implement this equation using regular LabVIEW arithmetic functions, the block diagram is a little bit harder to follow than the text equations.

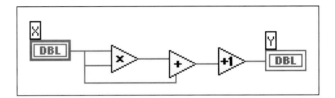

You can implement the same equation using a Formula Node, as shown in the following illustration.

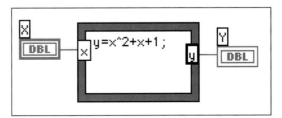

With the Formula Node, you can directly enter a formula or formulas, in lieu of creating complex block diagram subsections. Simply enter the formula inside the box. You create the input and output terminals of the Formula Node by popping up on the border of the node and choosing **Add Input** or **Add Output** from the pop-up menu. Then enter variable names into the input and output boxes. Names are case sensitive, and *each formula statement must terminate with a semicolon (;).*

You will find the Formula Node in the **Structures** subpalette of the **Functions** palette.

These operators and functions are available inside the Formula Node.

```
Formula Node operators, lowest precedence first:
=                         assignment
? :                       conditional
|| &&                     logical
== != > < >= <=           relational
+ - * / ^                 arithmetic
+ - !                     unary

Formula Node functions:
abs   acos  acosh  asin  asinh  atan   atanh   ceil
cos   cosh  cot    csc   exp    expm1  floor  getexp  getman
int   intrz  ln    lnp1  log    log2   max    min    mod    rand
rem   sec   sign   sin   sinc   sinh   sqrt   tan    tanh
```

The following example shows a conditional branching that you could perform inside a Formula Node. Consider the following code fragment, similar to Activity 6-3, that computes the square root of **x** if **x** is positive, and assigns the result to **y**. If **x** is negative, the code assigns value of –99 to **y**.

```
if (x >= 0) then
    y = sqrt(x)
else
    y = -99
end if
```

You can implement the code fragment using a Formula Node, as shown in the next illustration.

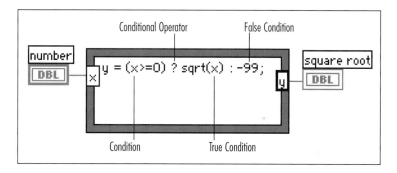

Activity 6-5
Formula Fun

You will build a VI that uses the Formula Node to evaluate the equation `y = sin (x)`, and graph the results.

1. Open a new panel. Select **Waveform Graph** from the **Graph** subpalette of the **Controls** palette. Label it `Graph`. The graph indicator will display the plot of the equation `y = sin (x)`. You'll learn all about graphs in Chapter 8, but this activity would be kind of dull without a nice pictorial representation, so we thought we'd give you a taste of them.

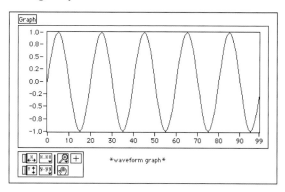

2. Build the block diagram shown.

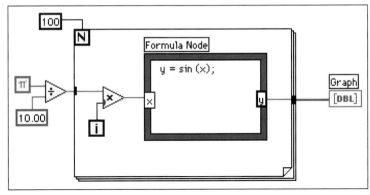

With the **Formula Node** (**Structures** palette), you can directly enter mathematical formulas. Create the input terminal by popping up on the border and choosing **Add Input** from the pop-up menu, then create the output terminal by choosing **Add Output** from the pop-up menu.

When you create an input or output terminal, you must give it a variable name. The variable name must exactly match the one you use in the formula. Remember, variable names are case sensitive.

Notice that a semicolon (;) must terminate the formula statement.

Pi Constant

The π constant is located in the **Functions➤Numeric➤Additional Numeric Constants** palette.

During each iteration, the VI multiplies the iteration terminal value by π/10. The multiplication result is wired to the Formula Node, which computes the sine of the result. The VI then stores the result in an array at the For Loop border (you will learn all about arrays in Chapter 7, *LabVIEW's Composite Data: Arrays and Clusters*. Then you will see why you can wire array data out of a For Loop, while scalar data comes out of a While Loop by default). After the For Loop finishes executing, the VI plots the array.

3. Return to the front panel and run the VI. Note that you could also use the existing **Sine** function (**Functions➤Numeric➤ Trigonometric** palette) to do the same thing as the Formula Node in this activity, but LabVIEW does not have built-in functions for every formula you'll need and we wanted to give you the practice.

4. Save the VI in your MYWORK directory or VI library and name it **Formula Node Exercise.vi**. Close the VI.

▓ VI Logic

```
for i = 0 to 99
  x = i* („/10)
  y = sin (x)
  array [i] = y
next i
Graph (array)
```

Wrap It Up!

LabVIEW has two structures to repeat execution of a subdiagram—the *While Loop* and the *For Loop*. Both structures are resizable boxes; place the subdiagram to be repeated inside the border of the loop structure. The While Loop executes as long as the value at the *conditional terminal* is TRUE. The For Loop executes a specified number of times.

Shift registers, available for While Loops and For Loops, transfer values from the end of one loop iteration to the beginning of the next. You can configure shift registers to access values from many previous iterations. For each iteration you want to recall, you must add a new element to the left terminal of the shift register. You can also have multiple shift registers on a loop to store multiple variables.

LabVIEW has two structures to add control to data flow—the *Case Structure* and the *Sequence Structure*. Only one case or one frame of these structures is visible at a time; you can switch between them using the little arrows at the top of the structure, with the pop-up menu, or by clicking with the Operating tool in the window at the top of the structure.

You use the Case Structure to branch to different subdiagrams depending on the input to its selector terminal, much like an if-then-else statement in conventional languages. Simply place the subdiagrams you want to execute inside the border of each case of the Case Structure and wire an input to the case selector terminal. Case Structures can be either Boolean (with 2 cases) or numeric (with up to $2^{15}-1$ cases)—LabVIEW automatically determines which type when you wire a Boolean or numeric control to the selector terminal.

Sometimes the principles of dataflow do not cause your program to behave the way you want it to, and you need a way to force a certain execution order. The Sequence Structure lets you set a specific order for your diagram functions. The portion of the diagram to be executed first is placed in the first frame (Frame 0) of the Sequence Structure, the subdiagram to be executed second is placed in the second frame, and so on.

You use *sequence locals* to pass values between Sequence Structure frames. The data passed in a sequence local is available only in frames subsequent to the frame in which you created the sequence local, NOT in those frames that precede the frame in which its value is assigned.

With the *Formula Node*, you can directly enter formulas in the block diagram, an extremely useful feature for complex function equations. Remember that variable names are case sensitive and that each formula statement must end with a semicolon (;).

The **Time & Dialog** subpalette of the **Functions** palette provides functions that pop up dialog boxes and control or monitor VI timing. The **One Button Dialog** and **Two Button Dialog** functions pop up a dialog box containing the message of your choice. The **Wait (ms)** function pauses your VI for the specified number of milliseconds. **Wait Until Next ms Multiple** can force loops to execute at a given interval by pausing until the internal clock equals (or has exceeded) a multiple of the millisecond input. These two wait functions are similar but not identical, and you don't really need to worry about the difference right now. **Tick Count (ms)** returns to you the value of the internal clock.

Additional Activities

■ Activity 6-6 **Equations**

Build a VI that uses the Formula Node to calculate the following equations.

$y1 = x3 + x2 + 5$

$y2 = (m * x) + b$

Use only one Formula Node for both equations. (Remember to put a semicolon [;] after each equation in the node.) Name the VI **Equations.vi**.

■ Activity 6-7 **Calculator**

Build a VI that functions like a calculator. The front panel should have digital controls to input two numbers, and a digital indicator to display the result of the operation (add, subtract, multiply, or divide) that the VI performs on the two numbers. Use a slide control to specify the operation to be performed. Name the VI **Calculator.vi**.

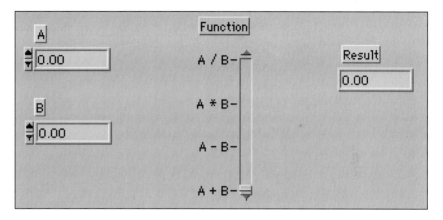

*You will want to use **Text Labels** on the slide control, obtained from the pop-up menu of the slide, to specify the function (add, subtract, multiply, and divide). If you don't see this option in the pop-up menu, make sure you're popping up on the slide itself and not the scale. Slides with text labels behave very much like text ring controls. When you first select **Text Labels**, the slide will have two settings, __max__ and __min__. You can use the Labeling tool to change this text. To add another text marker to the slide, pop up on the text display that appears next to your slide and select **Add Item After** or **Add Item Before** and then type in that marker's text.*

■ Activity 6-8 **Combination For/While Loop Challenge**

Using only a While Loop, build a combination For Loop/While Loop that stops either when it reaches "N" (specified inside a front panel control), or when a user pushes a stop button. Name the VI **Combo For/While Loop.vi**.

*Don't forget that a While Loop only executes while the conditional terminal reads a TRUE value. You might want to use the **And** function (**Boolean** subpalette of the **Functions** palette). Also, remember that while executing a loop, LabVIEW does not update indicators or read controls that are outside of the loop. Your <u>stop</u> button MUST be inside of your loop if you want correct functionality.*

■ Activity 6-9 **Dialog Display**

Write a VI that reads the value of a front panel switch, then pops up a dialog box indicating if the switch is on or off. Name the VI **Dialog Display.vi**. If you've developed the bad habit of using the continuous run button, now is the time to break it, or you will get yourself stuck in an endless loop! If you do get stuck, use the keyboard shortcut to stop your VI: <control-.> under Windows, <command-.> on a Macintosh, <meta-.> on the Sun, and <alt-.> under HP-UX.

OVERVIEW

In this chapter, you will learn about two new, more complex data types—arrays and clusters. These composite data types allow you great flexibility in data storage and manipulation. You will also see some valuable uses for arrays and clusters and learn how you can use built-in functions to manage them.

YOUR GOALS

- Create and use array controls and indicators
- Use auto-indexing to build an array
- Learn about the built-in array manipulation functions
- Grasp the concept of polymorphism
- Learn how to use clusters, and how to bundle and unbundle them
- Understand how clusters differ from arrays

KEY TERMS

- array
- auto-indexing
- polymorphism
- cluster
- bundle
- unbundle

LabVIEW's Composite Data: Arrays and Clusters

7

What Are Arrays?

Until now, we've dealt with scalar numbers only (a scalar is simply a data type that contains a single value, or "non-array"), and now it's time to move onto something more powerful and compound. A LabVIEW *array* is a collection of data elements that are all the same type, just like in traditional programming languages. An array can have one or more dimensions, and up to 2^{31} elements per dimension (memory permitting, of course). An array data element can have any type except another array, a chart, or a graph.

Array elements are accessed by their indices; each element's *index* is in the range 0 to *N–1*, where *N* is the total number of elements in the array. The *one-dimensional* (1D) array shown here illustrates this structure. Notice that the *first* element has index 0, the *second* element has index 1, and so on.

Index		0	1	2	3	4	5	6	7	8	9
10-Element Array		12	32	82	8.0	4.8	5.1	6.0	1.0	2.5	1.7

You will find that waveforms (and many other things) are often stored in arrays, with each point in the waveform comprising an element of the array. Arrays are also useful for storing data generated in loops, where each loop iteration generates one element of the array.

■ Creating Array Controls and Indicators

It takes two steps to make the controls and indicators for compound data types like arrays and clusters. You create the array control or indicator by combining an *array shell* with a *data object*, which can be numeric, Boolean, path, or string (or cluster, but we'll cover that later). You will find the array shell in the **Array & Cluster** subpalette of the **Controls** palette.

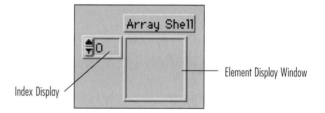

Index Display

Element Display Window

To create an array, drag a data object into the element display window. You can also deposit the object directly by clicking inside the window when you first choose the object from the **Controls** palette. The element display window resizes to accommodate its new data type, as shown in the following illustration, but remains grayed out until you enter data into it. Note that all elements of an array must be either controls or indicators, not a combination.

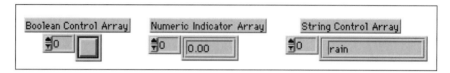

When you first drop an array shell on the front panel, its block diagram terminal is black, characteristic of an undefined data type. The terminal also contains brackets, shown in part

(A) of the next illustration, which are LabVIEW's way of denoting an array structure. When you assign a data type to the array (by placing a control or indicator in its element display window), then the array's block diagram terminal assumes its new type's color and lettering (although it retains its brackets), as in part (B). You will notice that array wires are thicker than wires carrying a single value.

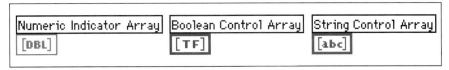

(A)   (B)

You can enter data into your array as soon as you assign a data type to it. Use the Labeling or Operating tool to type in a value, or if your data is numeric, click the arrows of the index display to increment or decrement it.

If you want to resize the object in the display window, use the Positioning tool and make sure it turns into the standard resizing brackets when you place it on the corner of the window (you will probably have to position it slightly inside the box and you may have to move it around a bit to get the correct brackets). If you want to show more elements at the same time, move the Positioning tool around the window corner until you find the grid cursor, then stretch either horizontally or vertically (your data is unchanged by the layout of your array). You will then have multiple elements visible. The element closest to the index display always corresponds to the element number displayed there.

Resizing Brackets Cursor
(Positioning Tool)

Grid Cursor
(Positioning Tool)

You can create array constants on the block diagram just like you can create numeric, Boolean, or string constants. Choosing **Array Constant** from the **Array** subpalette of the **Functions** palette will give you an array shell; then simply place in an appropriate data type (usually another constant) just like you do on the front panel. This feature is useful when you need to initialize shift registers or provide a data type to a file or network function (which you'll learn about later).

If you want to clear an array control, indicator, or constant of data, pop up on the *index display* (NOT the element itself or you'll get the wrong menu) and choose **Data Operations▶ Empty Array**.

■ Using Auto-Indexing

The For Loop and the While Loop can index and accumulate arrays at their boundaries automatically—one new element for each loop iteration. This capability is called *auto-indexing*. One important thing to remember is that *auto-indexing is enabled by default on For Loops but disabled by default on While Loops.* The following figure shows a For Loop auto-indexing an array at its boundary. Each iteration creates the next array element. After the loop completes, the array passes out of the loop to the indicator; none of the array data is available until after the loop finishes. Notice that the wire becomes thicker as it changes to an array wire type at the loop border.

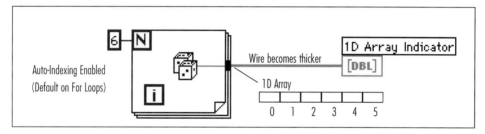

If you need to wire a scalar value out of a For Loop without creating an array, you must disable auto-indexing by popping up on the tunnel (the black square) and choosing **Disable Indexing** from the tunnel's pop-up menu.

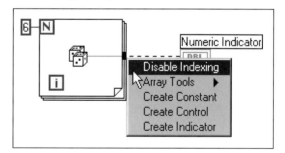

Since auto-indexing is disabled by default on a While Loop, if you need to wire array data out of a While Loop, you must pop up on the tunnel and select **Enable Indexing**.

In the next illustration, auto-indexing is disabled, and only the last value returned from the **Random Number (0–1)** function passes out of the loop. Notice that the wire remains the same size after it leaves the loop. Pay attention to this wire size, because auto-indexing is a common source of problems among beginners. They often create arrays when they don't want to, or don't create them when they need them, and then go crazy trying to figure out why they get a bad wire.

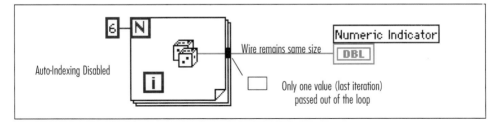

Auto-indexing also applies when you are wiring arrays into loops. If indexing is enabled as in loop (A), the loop will index off one element from the array each time it iterates (note how the wire becomes thinner as it enters the loop). If indexing is disabled as in loop (B), the entire array passes into the loop at once.

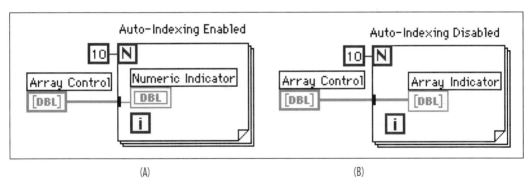

| (A) | (B) |

*Because For Loops are often used to process arrays, LabVIEW enables auto-indexing by default when you wire an array into or out of them. By default, LabVIEW does NOT enable auto-indexing for While Loops. You must pop up on the array tunnel and choose **Enable Indexing** from the pop-up menu if you want your While Loop to auto-index. Pay close attention to the state of your indexing, lest you develop errors that are tricky to spot.*

▨ Using Auto-Indexing to Set the For Loop Count

When you enable auto-indexing on an array *entering* a For Loop, LabVIEW automatically sets the *count* to the array size, thus eliminating the need to wire a value to the count terminal. If you give LabVIEW conflicting counts, e.g. by setting the count explicitly and by auto-indexing (or by auto-indexing two different size arrays), LabVIEW sets the count to the smallest of the choices. In the following figure, the array size, and not the value wired to the count terminal, determines the number of For Loop iterations, because the array size is the smaller of the two.

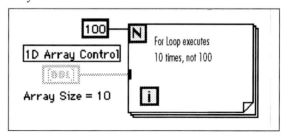

■ Two-Dimensional Arrays

A two-dimensional, or 2D, array stores elements in a grid-like fashion. It requires two indices to locate an element: a column index and a row index, both of which are zero-based like everything else in LabVIEW. The following figure shows how a six-column by four-row array that contains six times four elements is stored.

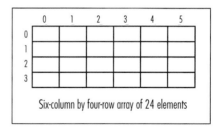

Six-column by four-row array of 24 elements

You can add dimensions to an array control or indicator by popping up on its *index display* (not on the element display) and choosing **Add Dimension** from the pop-up menu. The next illustration shows a 2D array of digital controls. Notice that you now have two indices to specify each element. You can use the grid cursor of the Positioning tool to expand your element display in two dimensions so that you can see more elements.

Grid Cursor
(Positioning Tool)

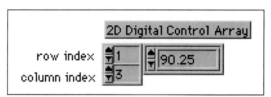

Remove unwanted dimensions by selecting **Remove Dimension** from the index display's pop-up menu.

If you have waveforms from several channels being read from a data acquisition (DAQ) board, they will be stored in a two-dimensional (2D) array, with each column in the 2D array corresponding to one channel's data.

Creating Two-Dimensional Arrays

You can use two For Loops, one inside the other, to create a 2D array if you don't want to type in values on the front panel. The inner For Loop creates a row, and the outer For Loop "stacks" these rows to fill in the columns of the matrix. The following illustration shows two For Loops creating a 2D array of random numbers using auto-indexing.

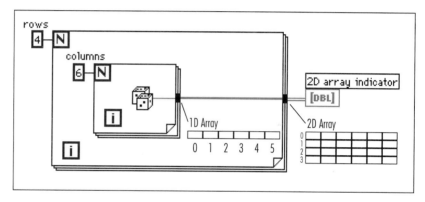

Notice that a 2D array wire is even thicker than a 1D array wire.

Activity 7-1
Building Arrays with Auto-Indexing

Now we'll give you a chance to better understand arrays and auto-indexing by working with them yourself. In this activity, you will open and observe a VI that uses auto-indexing on both a For Loop and a While Loop to create arrays of data.

1. Open the **Building Arrays.vi** example, located in EVERY-ONE\CH7.LLB. This exercise generates two arrays on the front panel, using a For Loop to create a 2D array and a While Loop to create a 1D array. The For Loop executes a set number of times; you must hit the STOP button to halt the While Loop (or it will stop after 101 iterations).

2. Take a look at the front panel, then switch to the block diagram. Notice how the nested For Loops create a 2-D array's rows and columns at their borders, respectively, using auto-indexing. Also

notice how the auto-indexed wires become thicker as they leave the loop boundaries.

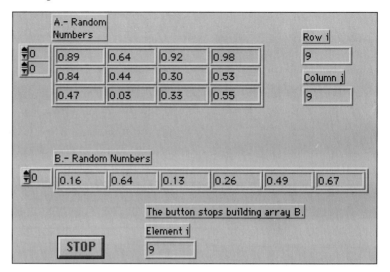

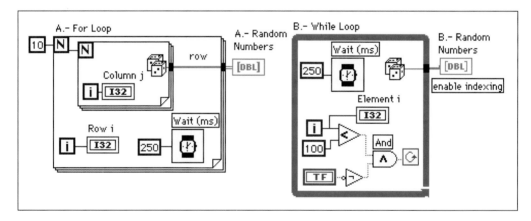

3. Before we could get array data out of the While Loop, we had to pop up on the tunnel containing the random number and select **Enable Indexing**. To see how this works, go ahead and pop up on the tunnel, then select **Disable Indexing**. You will see the wire leaving the loop break. Pop up on the tunnel again and select **Enable Indexing** to fix it.

This loop uses a little logic algorithm to ensure that if the user does not press the STOP button after a reasonable amount of time (101 iterations), the loop stops anyway. If the user has not pressed the STOP button AND the loop has executed fewer than 101 times, the loop continues. If either of those conditions changes, the loop stops.

Why does it execute 101 times, instead of 100? Remember that the While Loop checks the conditional terminal at the end of each iteration. At the end of the 100th iteration, i = 99 (since it starts at zero), which is less than 100, and the loop continues. At the end of the 101st iteration, the count is no longer less than 100 and the loop stops (assuming the STOP button hasn't already stopped it).

4. Now that you understand how it works, run the VI. Remember to hit STOP to halt the While Loop, especially since the front panel indicator does not update until the entire array has been generated (remember, controls and indicators outside of loops are not read or updated while the loop is executing).

5. Close the VI and don't save any changes.

Functions for Manipulating Arrays

By now, you know a little bit about what arrays are, so let's talk about all the cool stuff you can do with them. LabVIEW has many functions to manipulate arrays in the **Array** subpalette of the **Functions** palette. To avoid a common pitfall, always keep in mind that arrays (and all other LabVIEW structures) are zero indexed—the first element has an index of zero, the second has an index of one, and so on. Some common functions are discussed here, but you might also want to browse through the **Array** subpalette just to see what else is built in for you.

◆ **Initialize Array** will create and initialize an n-dimensional array with the value of your choice. You can configure it for larger dimensions by "growing" it with the Positioning tool to provide more **dimension size** inputs. This function is useful for allocating memory for arrays of a certain size or for initializing shift registers with array-type data.

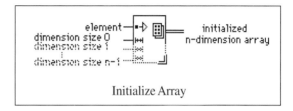

Initialize Array

In the following illustration, **Initialize Array** shows how to initialize a ten-element, one-dimensional array, with each element of the array containing a zero.

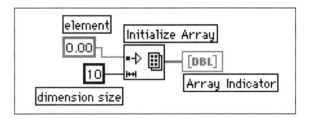

◆ **Array Size** returns the number of elements in the input array. If the input array is n-dimensional, **Array Size** returns an n-element, one-dimensional array, with each element containing the size of one of the array's dimensions.

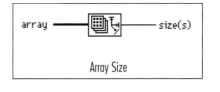

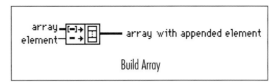

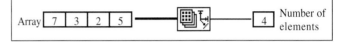

◆ Depending on how you configure it, **Build Array** concatenates, or combines, two arrays into one, or adds extra elements to an array. The function looks like the icon at left when first placed in the diagram window. You can resize or "grow" this function to increase the number of inputs. **Build Array** has two types of inputs, *array* and *element*, so that it can assemble an array from both array and single-valued inputs.

Build Array

For example, the **Build Array** function shown in the following illustration has been configured to concatenate two arrays and one element into a new array.

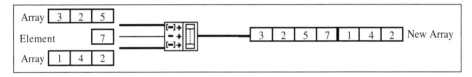

*Pay special attention to the inputs of a **Build Array** function. Array inputs have brackets, while element inputs do not. They are NOT interchangeable and can cause lots of confusing bad wires if you're not careful.*

To change the type of an input, pop up on it and select **Change to Array** or **Change to Element**.

As you get more advanced, you'll find that **Build Array** can also build or add elements to multidimensional arrays. To add an element to a multidimensional array, the element must be an array of one size smaller dimension (i.e., you can add a 1D element to a 2D array).

◆ **Array Subset** returns a portion of an array starting at **index** and containing **length** elements. Notice that the third element's index is two because the index starts at zero; that is, *the first element has an index of zero.*

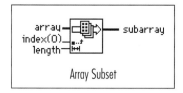

Array Subset

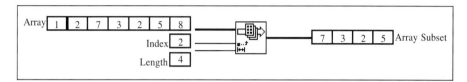

◆ **Index Array** accesses a particular element of an array. An example of an **Index Array** function accessing the third element of an array is shown in the following illustration.

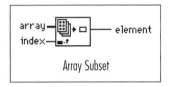

Array Subset

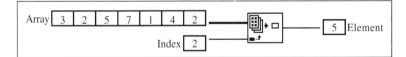

Here, the **Index Array** function is extracting a scalar element from an array. You also can use this function to *slice off* a row, column, or scalar element of a 2D array. To do this, stretch the **Index Array** function to include two index inputs. To extract a single scalar element, wire the desired element's row index to the top input and its column index to the bottom input. To slice of a row or a column from the 2D array, select the **Disable Indexing** command on the pop-up menu of the index terminal as shown.

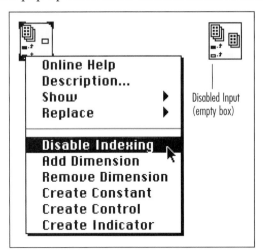

Notice that the index terminal symbol changes from a solid to an empty box when you disable indexing. You can extract subarrays along any combination of dimensions. The following illustration shows how to extract 1D row or column arrays from a 2D array. To extract a column, disable indexing on the row input (the top one) and wire the desired column index to the column input (the bottom one). To extract a row, disable indexing on the column input and wire the desired row index to the row input.

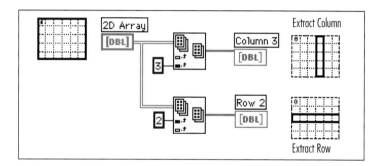

You can restore a disabled index with the **Enable Indexing** command from the same menu if you just want a single element from a 2D array.

Activity 7-2
Array
Acrobatics

Confused yet? You'll get a better feel for arrays as you work with them. In this exercise, you will finish building a VI that concatenates two arrays and then indexes out the element in the middle of the new concatenated array.

1. Open **Array Exercise.vi**, located in EVERYONE\CH7.LLB.

The front panel contains two input arrays (each showing three elements), two digital controls, and an output array (showing eight elements). The VI will concatenate the arrays and the control values in the following sequence to create the new array.

Initial Array + Element 1 + Element 2 + Terminal Array

The front panel is already built. You will finish building the diagram. Note that an array control or indicator will appear grayed out until you or the program assigns it some data.

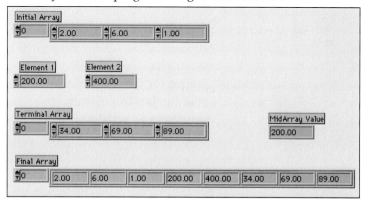

2. Build the block diagram shown. Use the Help window to find the correct terminals on the functions.

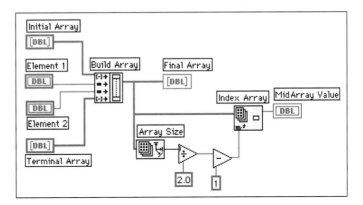

Build Array Function

Build Array

Array Input

■

Element Input

Array Size Function

Index Array Function

Build Array function (**Array** palette). In this exercise, concatenates the input data to create a new array in the following sequence: Initial Array + Element 1 + Element 2 + Terminal Array.

The function looks like the icon at the left when placed in the diagram window. Place the Positioning tool on the lower right corner and resize the function until it includes four inputs.

Change the top input of the **Build Array** function from an element input to an array input by popping up on it and choosing **Change to Array**. Repeat for the bottom input. The symbols at the left indicate whether an input is an array or an element.

Array Size function (**Array** palette). Returns the number of elements in the concatenated array.

Index Array function (**Array** palette). In this exercise, returns the element in the middle of the array.

LabVIEW builds the array using the **Build Array** function, then calculates the index for the middle element of the array by taking the length of the array, dividing it by two, and subtracting one (to account for the zero-based array index). Since the array has an even number of elements, the middle element will actually be one of the two middle elements.

3. Return to the front panel and run the VI. Try several different number combinations.

4. If you have the full version of LabVIEW, save the finished VI in your MYWORK directory and close the VI. If you have the sample software, don't bother to save it.

You might also want to look through the programs in EXAM-PLES/GENERAL/ARRAYS.LLB if you have the full version, to see some other things you can do with arrays.

Polymorphism

Another handy feature you'll find useful is the *polymorphism* of the LabVIEW arithmetic functions, **Add**, **Multiply**, **Divide**, and so on. *Polymorphism* is just a big word for a simple principle: the inputs to these functions can be of different size and representation. For example, you can add a scalar to an array or add two arrays together using the same function. The following illustration shows some of the polymorphic combinations of the **Add** function.

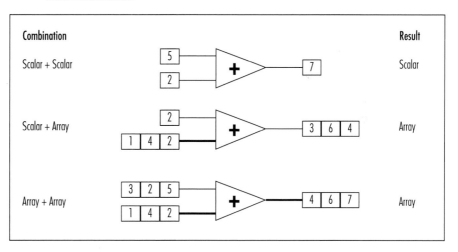

In the first combination, the result is a scalar number. In the second combination, the scalar is added to *each* element of the array. In the third combination, each element of one array is added to the corresponding element of the other array. In all instances, the same **Add** function is used, but it performs a different type of operation.

In the next illustration, each iteration of the For Loop generates one random number (valued between 0 and 1) that is stored in the array created at the border of the loop. After the loop finishes execution, the **Multiply** function multiplies each element in the array by the scaling factor you set. The front panel array indicator then displays the scaled array.

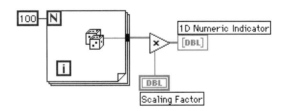

The next figure shows some of the possible polymorphic combinations of the **Add** function. You'll learn about the clusters it depicts in the next section.

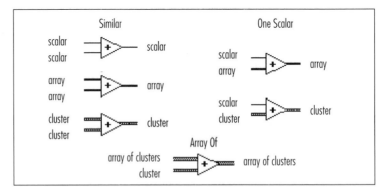

Note

If you are doing arithmetic on two arrays with a different number of elements, the resulting array will be the size of the smaller of the two. In other words, LabVIEW operates on corresponding elements in the two arrays until one of the arrays runs out of elements. The remaining elements in the longer array are ignored.

Activity 7-3
Polymorphism

You will build a VI that demonstrates polymorphism on arrays.

1. Open a new panel and recreate the one shown.

First, create the arrays. Remember, to create an array, you must first select the **Array** shell from the **Array & Cluster** subpalette of the **Controls** palette. Then put a numeric indicator into the shell's data object window. To see more than one element in the array, you must grab and drag the corner of the filled element display window with the grid cursor of the Positioning tool. You can drag out multiple elements in a 1D array either horizontally or vertically.

Grid Cursor
(Positioning Tool)

All four arrays in this exercise contain indicator elements. Make sure and give them unique labels so that you don't get them confused. If you ever do lose track of which front panel object corresponds to which block diagram terminal, simply pop up on one of them and choose **Find Terminal** or **Find Indicator**; LabVIEW will highlight the corresponding object.

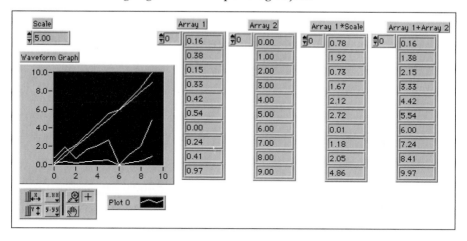

2. After you've created the arrays, select a **Waveform Graph** from the **Graph** subpalette of the **Controls** palette. Although you'll learn more about graphs in the next chapter, we wanted to give you a preview and spice up this exercise.

3. Don't forget to create the Scale control.

4. Build the diagram shown—be careful because the wiring can get a little tricky!

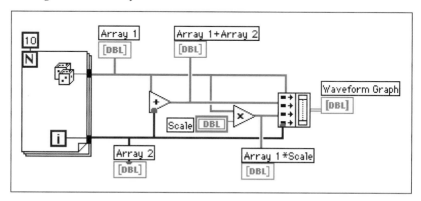

Auto-indexing is enabled by default on the For Loop, so your arrays will be generated automatically.

5. You can find **Add**, **Multiply**, and **Random Number (0-1)** in the **Numeric** palette.

6. Select **Build Array** from the **Array** palette. You will have to grow it using the Positioning tool so that it has four inputs. Leave all inputs as element inputs (the default), since you want the input arrays to become rows of an array of greater dimension size. As a result, the output from **Build Array** is a 2D array; each input array becomes a row, so the output 2D array is composed of four rows and ten columns.

7. Run the VI. Your graph will plot each array element against its index for all four arrays at once: <u>Array 1 data</u>, <u>Array 2</u> data, <u>Array 1* Scale</u>, and <u>Array 1 + Array 2</u>.

The results demonstrate several applications of polymorphism in LabVIEW. For example, <u>Array 1</u> and <u>Array 2</u> could be incoming waveforms that you wish to scale.

8. Save the VI as **Polymorphism Example.vi** and place it in your MYWORK directory or VI library. Close the VI.

■ Compound Arithmetic

While we're on the subject of arithmetic, we should mention the **Compound Arithmetic** function. While the previous polymorphism example showed you how to add and multiply different-sized data together, this function lets you operate on more than two numbers simultaneously. The **Compound Arithmetic** function eliminates the need for multiple **Add**, **Multiply**, **AND**, and **OR** terminals should you need to perform one of these functions on several numbers at once (**AND** and **OR** are Boolean arithmetic operations).

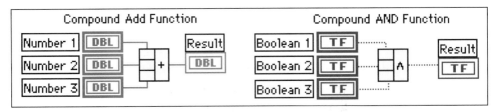

The **Compound Arithmetic** function can be found in both the **Numeric** and **Boolean** subpalettes of the **Functions** palette. Like many other functions, you can use the Positioning tool to grow it and provide more input terminals. The **Compound Arithmetic** function has only one form; you configure it to perform the arithmetic function of your choice. To change the function (choices are **Add, Multiply, AND,** or **OR**), pop up on the output terminal and select **Change Mode**▸. You can also click on the function with the Operating tool to change the mode.

Select the **Invert** pop-up option to invert the sign of numeric inputs and outputs or the value of Booleans (from FALSE to TRUE or vice versa). A small circle at the input or output symbolizes an inverted value.

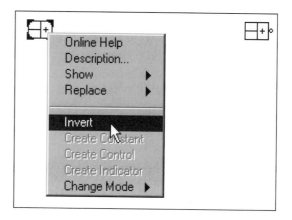

■ A Word About Boolean Arithmetic

LabVIEW's Boolean arithmetic functions, **And, Or, Not, Exclusive Or, Not Exclusive Or, Not And,** and **Not Or,** can be very powerful. You will see occasional Boolean arithmetic throughout this book. If you've never seen it before, we recommend reading about it in a good logic or digital design book. But just as a refresher, we'll mention a few basics. If you can't remember which function does what, use the Help window!

Not is probably the easiest to describe, since it simply inverts the input value. If the input value is TRUE, **Not** will output FALSE; if the input is FALSE, **Not** returns TRUE.

The **And** function outputs a TRUE only if all inputs are TRUE. The **Or** function outputs a TRUE if at least one input is TRUE. The **And** and **Or** functions have the following outputs, given the inputs shown:

FALSE **And** FALSE = FALSE   FALSE **Or** FALSE = FALSE
TRUE **And** FALSE = FALSE   TRUE **Or** FALSE = TRUE
FALSE **And** TRUE = FALSE   FALSE **Or** TRUE = TRUE
TRUE **And** TRUE = TRUE   TRUE **Or** TRUE = TRUE

All About Clusters

Now that you've got the concept of arrays under your belt, clusters should be easy. Like an array, a *cluster* is a data structure that groups data. However, unlike an array, a cluster can group data of different types (i.e., numeric, Boolean, etc.); it is analogous to a *record* in Pascal or a *struct* in C. A cluster may be thought of as a *bundle* of wires, much like a telephone cable. Each wire in the cable represents a different element of the cluster. Because a cluster has only one "wire" in the block diagram (even though it carries multiple values of different data types), clusters reduce wire clutter and the number of connector terminals that subVIs need. You will find that the cluster data type appears frequently when you plot your data on graphs and charts.

Bundling Data

You can access cluster elements by *unbundling* them all at once or by indexing one at a time, depending on the function you choose; each method has its place. Think of unbundling a cluster as unwrapping a telephone cable and having access to the different-colored wires. Unlike arrays, which can change size dynamically, clusters have a fixed size, or a fixed number of wires in them.

Unbundling Data

You can connect cluster terminals with a wire *only* if they have exactly the same type; in other words, both clusters must have the same number of elements, and corresponding elements must match in both data type and order. The principle of polymorphism applies to clusters as well as arrays, as long as the data types match.

You will often see clusters used in error handling. The following illustration shows the error clusters, **Error In.ctl** and **Error Out.ctl**, used by LabVIEW VIs to pass a record of errors between multiple VIs in a block diagram (for example, many of the data acquisition and file I/O VIs have error clusters built into them). These error clusters are so frequently used that they appear in the **Array & Cluster** subpalette of the **Controls** palette for easy access.

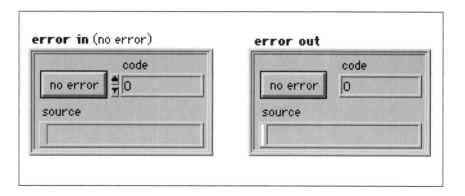

■ Creating Cluster Controls and Indicators

Create a cluster by placing a **Cluster** shell (**Array & Cluster** subpalette of the **Controls** palette) on the front panel. You can then place any front panel objects inside the cluster. Like arrays, you can deposit objects directly inside when you pull them off of the **Controls** palette, or you can drag an existing object into a cluster. *Objects inside a cluster must be all controls or all indicators.* You cannot combine both controls and indicators inside the same cluster, because the cluster itself must be one or the other. The cluster will be a control or indicator based on the status of the first object you place inside. Resize the cluster with the Positioning tool if necessary. The following illustration shows a cluster with four controls.

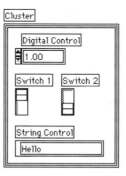

You can create block diagram cluster constants in a similar two-step manner.

If you want your cluster to conform exactly to the size of the objects inside it, pop up on the *border* (NOT inside the cluster) and choose **Autosizing**.

■ Cluster Order

Cluster elements have a logical order unrelated to their position within the shell. The first object placed in the cluster is element zero, the second is element one, and so on. If you delete an element, the order adjusts automatically. *You must keep track of your cluster order if you want to connect your cluster to another cluster—the order and data types must be identical.* Also, if you choose to unbundle the cluster all at once, you'll want to know which value corresponds to which output on the cluster function (more about that in a minute).

Change the order within the cluster by popping up on the cluster *border* and choosing **Cluster Order...** from the pop-up menu. A new set of buttons appears in the Toolbar, and the cluster appearance changes as shown in the next figure.

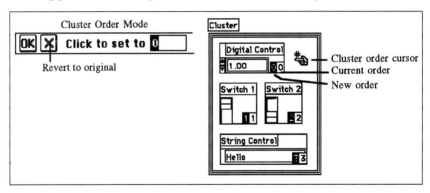

The white boxes on the elements show their current places in the cluster order. The black boxes show the new places. Clicking on an element with the cluster order cursor sets the element's place in the cluster order to the number displayed on the Toolbar. You can type a new number into that field before you click on an object.

Revert Button

OK Button

If you don't like the changes you've made, revert to the old order by clicking on the Revert button. When you have the order the way you want, you can set it and return to your regular front panel by clicking on the OK button and exiting the cluster order edit mode.

■ Using Clusters to Pass Data to and from SubVIs

The connector pane of a VI can have a maximum of 28 terminals. You probably don't want to pass information to all 28 terminals when calling a subVI anyway—the wiring can be very tedious and you can easily make a mistake. By bundling a number of controls or indicators into a cluster, you can use a single terminal and wire to pass several values into or out of the subVI. You can use clusters to work around the 28-terminal limit for the connector or just enjoy simplified wiring with fewer (and therefore larger and easier to see) terminals.

■ Bundling Your Data

Bundle Function

The **Bundle** function (**Cluster** palette) assembles individual components into a single new cluster or allows you to replace elements in an existing cluster. The function appears as the icon at the left when you place it in the diagram window. You can increase the number of inputs by dragging a corner of the function with the Positioning tool. When you wire to each input terminal, a symbol representing the data type of the wired element appears on the empty terminal. The order of the resultant cluster will be the order of inputs to the **Bundle**.

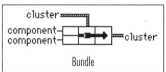

If you just want to create a new cluster, you do not need to wire an input to the center **cluster** input of the **Bundle** function. You do need to wire this input if you are replacing an element in the cluster.

Be careful with this function—if you add an element to your cluster without adding an element to the block diagram **Bundle** to match, your program will break!

■ Replacing a Cluster Element

If you want to *replace* an element in a cluster, first size the **Bundle** function to contain the same number of input terminals as there are elements in this cluster (it must be the same size or you will get a bad wire). Then wire the cluster to the middle terminal of the **Bundle** function (symbols for data types inside the cluster will appear in the **Bundle** inputs) and wire the new value(s) of the element(s) you want to replace to the corresponding inputs. You need to wire only to those terminals whose values you want to change.

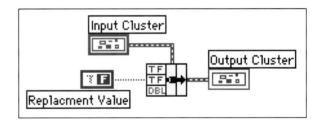

■ Unbundling Your Clusters

Unbundle Function

The **Unbundle** function (**Cluster** palette) splits a cluster into each of its individual components. *The output components are arranged from top to bottom in the same order they have in the cluster.* If they have the same data type, the elements' order in the cluster is the only way you can distinguish between them. The function appears as the icon at left when you place it in the diagram window. You can increase the number of outputs by dragging a corner of the function with the Positioning tool. The **Unbundle** function must be resized to contain the same number of outputs as there are elements in the input cluster, or it will produce bad wires. When you wire an input cluster to the correctly sized **Unbundle**, the previously blank output terminals will assume the symbols of the data types in the cluster.

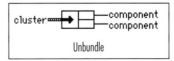

Unbundle

*Knowing the cluster order is essential when accessing cluster data with **Bundle** and **Unbundle**. For example, if you have two Booleans in the same cluster, it would be easy to mistakenly access <u>Switch 2</u> instead of <u>Switch 1</u>, since they are referenced in the **Unbundle** function by order, NOT by name. Your VI will wire with no errors, but your results will be incorrect.*

Be careful with this function—if you add an element to your cluster without resizing the block diagram **Unbundle** to match, your program will break!

LabVIEW does have a way to bundle and unbundle clusters using element names; we'll talk about it in a minute.

Activity 7-4
Cluster
Practice

In this activity, you will build a VI to teach you how to work with clusters. You will create a cluster, unbundle it, then re-bundle it and display the values in another cluster.

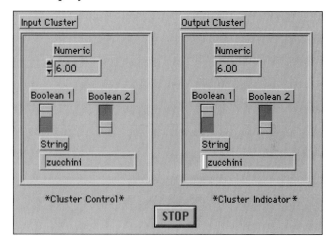

1. Open a new panel and place a **Cluster** shell (**Array & Cluster** palette) on it. Label it Input Cluster. Enlarge the shell (make sure to grab the cluster border or nothing will happen).

2. Place a digital control, two Boolean switches, and a string control inside the Input Cluster shell.

3. Now create Output Cluster by cloning Input Cluster. Then pop up on an object in the cluster (or on the cluster border) and select **Change to Indicator**. Also change the new cluster's label. (Remember, <control>-drag under Windows, <option>-drag on the Mac, <meta>-drag on Sun, and <alt>-drag on HP to clone an object. This technique ensures correct cluster order as well as being very efficient.)

You could also create Output Cluster the same way you created Input Cluster, using indicators instead of controls (make sure to put elements into the cluster in the same order).

4. Verify that Input Cluster and Output Cluster have the same cluster order by popping up on each cluster border and choosing **Cluster Order…**. If the cluster orders are different, change one of them so both are the same.

5. Finally, place a **Rectangular Stop Button** (**Boolean** palette) on the front panel. Note that this button is FALSE by default. Do not change its state.

6. Build the block diagram shown. Notice that even though each cluster contains four objects, you see only one terminal per cluster on the block diagram.

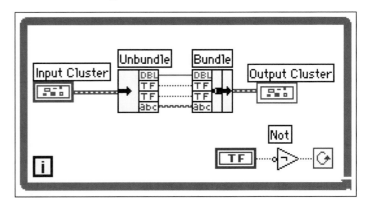

Unbundle Function

Unbundle function (**Cluster** palette). This function will break out the cluster so you can access individual elements. Resize it so that it has four outputs. The data-type labels will appear on the **Unbundle** function after you wire <u>Input Cluster</u> to it.

Bundle Function

Bundle function (**Cluster** palette). This function reassembles your cluster. Resize it so that it contains four inputs.

Not Function

Not function (**Boolean** palette). Remember that Booleans default to FALSE. The **Not** function inverts the <u>STOP</u> button's value so you don't have to click the button to TRUE before starting the program. In other words, if the button's value is FALSE, the **Not** function sends a TRUE value to the conditional terminal of the While Loop; when the button changes state and becomes TRUE, the conditional terminal will read a FALSE value from the **Not** function and stop the loop.

*You can also access **Bundle** and **Unbundle** by popping up the cluster terminal you plan to wire to and choosing the function you want from the **Cluster Tools▶** menu. The chosen function will appear containing the correct number of input or output terminals.*

7. Return to the front panel and run the VI. Enter different values for the control cluster and watch how the indicator cluster echoes the values. Press the <u>STOP</u> button to halt execution.

You might have noticed that you could really just wire from the <u>Input Cluster</u> terminal to the <u>Output Cluster</u> terminal and the VI would do the same thing, but we wanted you to practice with **Bundle** and **Unbundle**.

8. Close and save the VI as **Cluster Exercise.vi** in your MYWORK directory or VI library.

■ Bundling and Unbundling by Name

Sometimes you don't need to assemble or disassemble an entire cluster—you just need to operate on an element or two. You can use **Bundle By Name** and **Unbundle By Name** to do just that!

Bundle By Name, found in the **Cluster** palette, references elements by name instead of by position (as **Bundle** does). Unlike **Bundle**, you can access only the elements you need. However, **Bundle By Name** cannot create new clusters; it can only replace an element in an existing cluster. Unlike **Bundle,** you must always wire to **Bundle By Name**'s middle input terminal to tell the function in which cluster to replace the element.

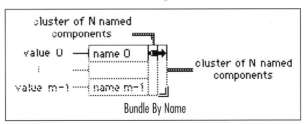

Bundle By Name

Unbundle By Name, also located in the **Cluster** palette, returns the cluster elements whose name(s) you specify. You don't have to worry about cluster order or correct **Unbundle** function size.

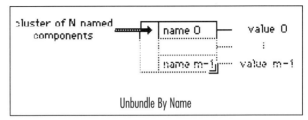

Unbundle By Name

*Make sure all cluster elements have owned labels when you are using the **By Name** functions. Obviously, you can't access by name if you don't have names—LabVIEW won't know what you want!*

For example, if you wanted to replace the value of <u>Boolean 2</u> in the last exercise, you could use the **Bundle By Name** function without having to worry about cluster order or size.

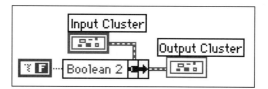

Similarly, if you only needed to access the value of <u>String</u>, you would want to use the **Unbundle By Name** function.

As soon as you wire the **cluster** input of **Bundle By Name** or **Unbundle by Name**, the name of the first element in the cluster appears in the **name** input or output. To access another element, click on the **name** input or output with the Operating or Labeling tool. You should see a list of the names of all labeled elements in the cluster. Select your element from this list, and the name will appear in the **name** terminal. You can also access this list by popping up on **name** and choosing **Select Item►**.

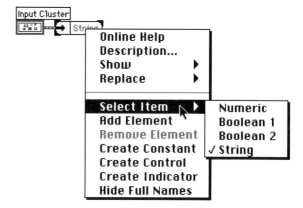

Both functions can be resized to accommodate as many elements as you need; select each **component** you want to access individually. As an added plus, you no longer have to worry about your program breaking if you resize your cluster. The **By Name** functions do not break unless you remove an element they reference.

Activity 7-5
More Fun
with Clusters

In this activity, you will build a VI that checks whether the value of the Numeric 1 digital control in the input cluster is greater than or equal to zero. If it is less than zero, the VI will take the absolute value of all controls. If Numeric 1 is greater than or equal to zero, the VI does not take the absolute value of any controls. Regardless of the value of Numeric 1, the VI multiplies all values by 0.5 and displays the results in Output Cluster, demonstrating how you can use polymorphism with clusters.

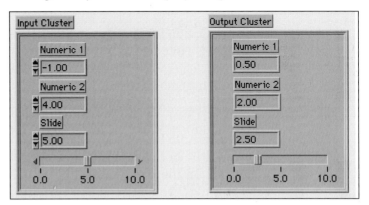

1. Open a new panel and place a **Cluster** shell (**Array & Cluster** palette) on it. Label it Input Cluster.

2. Create the Numeric 1, Numeric 2, and Slide controls from the **Numeric** palette. As you select them from the palette, click to place them inside the cluster shell. Make sure to create them in the order specified (since you will be wiring the Input Cluster to the Output Cluster) and give them labels.

3. Now create Output Cluster the same way using indicators (and make sure to put elements into the cluster in the same order). Or, you might create Output Cluster by cloning Input Cluster and then changing its label.

4. Build the block diagram shown. Make sure to build both the TRUE and FALSE cases of the Case Structure.

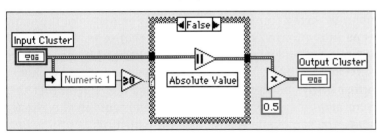
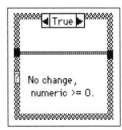

Unbundle By Name Function

Greater or Equal to 0? Function

Absolute Value

Unbundle By Name function (**Cluster** palette) extracts <u>Numeric 1</u> from <u>Input Cluster</u> so you can compare it to zero. If "Numeric 1" isn't showing in the output name area, click with the Operating tool to choose it from the list of cluster elements.

Greater Or Equal to 0? function (**Comparison** palette)

Absolute Value (**Numeric** palette) returns the input number if that number is greater than or equal to zero; returns the inverse of the input number if the input number is less than zero. In this activity, takes the absolute value of the entire cluster.

5. Run the VI. Try both positive and negative values of <u>Numeric 1</u>. Note the use of polymorphism to multiply all values in the cluster by 0.5 at the same time and to find the absolute value of the entire cluster.

6. Save the VI as **Cluster Comparison.vi** in your MYWORK directory or VI library.

■ Interchangeable Arrays and Clusters

Sometimes you will find it convenient to change arrays to clusters, and vice versa. This trick can be extremely useful, especially since LabVIEW includes many more functions that operate on arrays than clusters. For example, maybe you have a cluster of buttons on your front panel and you want to reverse the order of the buttons' values. Well, **Reverse 1D Array** would be perfect, but it only works on arrays. Have no fear—you can use the **Cluster to Array** function to change the cluster to an array, use **Reverse 1D Array** to switch the values around, then use **Array to Cluster** to change back to a cluster.

Array To Cluster

Cluster To Array

Cluster to Array converts a cluster of N elements of the same data type into an array of N elements of that type. Array index corresponds to cluster order (i.e., cluster element 0 becomes the value at array index 0). You cannot use this function on a cluster containing arrays as elements, because LabVIEW won't let you create an array of arrays. Note that all elements in the cluster must have the same data type to use this function.

Array To Cluster converts an N element, 1D array into a cluster of N elements of the same data type; you must pop up on the still **Array To Cluster** terminal and choose **Cluster Size...** to specify the size of the output cluster since clusters don't size automatically like arrays do. The cluster size defaults to 9; if your array has fewer than the number of elements specified in the cluster size, LabVIEW will automatically fill in the extra cluster values with the default value for the data type of the cluster. However, if the input array has a greater number of elements than the value specified in the cluster size window, the block diagram wire going to the output cluster will break until you adjust the size.

Both functions are very handy if you want to display elements in a front panel cluster control or indicator but need to manipulate the elements by index value on the block diagram. They can be found in both the **Array** and **Cluster** subpalettes of the **Functions** palette.

Wrap It Up!

An *array* is collection of ordered data elements of the same type. In LabVIEW, arrays can be of any data type, except chart, graph, or another array. You must create an array using a two-step process: first, place an array shell (**Array & Cluster** subpalette of the **Controls** palette) in the window, and then add the desired control or indicator to the shell.

LabVIEW offers many functions to help you manipulate arrays, such as **Build Array** and **Index Array**, in the **Array** subpalette of the **Functions** palette. Most often, you will use these array functions to work with only 1D arrays; however, these functions are smart and will work similarly with multidimensional arrays (although sometimes you need to resize them first).

Both the For Loop and the While Loop can accumulate arrays at their borders using *auto-indexing*, a useful feature for creating and processing arrays. Remember that by default, LabVIEW enables indexing in For Loops and disables indexing for While Loops.

Polymorphism is a fancy name for the ability of a function to adjust to inputs of different-sized data. We talked about polymorphic capabilities of arithmetic functions; however, many other functions are also polymorphic.

Clusters also group data, but unlike arrays, they will accept data of different types. You must create them on the front panel in a two-step process: first, place a cluster shell (**Array & Cluster** sub-palette of the **Controls** palette) on the front panel, and then add the desired controls or indicators to the shell. Keep in mind that objects inside a cluster must be all controls or all indicators. You cannot combine both controls and indicators within one cluster.

Clusters are useful for reducing the number of wires or terminals associated with a VI. For example, if a VI has many front panel controls and indicators that you need to associate with terminals, it is easier to group them as a cluster and have only one terminal.

The **Unbundle** function (**Cluster** palette) splits a cluster into each of its individual components. **Unbundle by Name** works similarly to **Unbundle** but accesses elements by their label. You can access as many or as few elements as you like using **Unbundle By Name**, while you have access to the whole cluster using **Unbundle** (and consequently have to worry about correctly sized terminals and cluster order).

The **Bundle** function (**Cluster** palette) assembles individual components into a single cluster or replaces an element in a cluster. **Bundle By Name** can't assemble clusters, but it can replace individual elements in a cluster without accessing the entire cluster. In addition, with **Bundle By Name**, you don't have to worry about cluster order or correct **Bundle** function size. Just make sure all cluster elements have names when using **Bundle By Name** and **Unbundle By Name**!

Additional Activities

■ Activity 7-6 **Reversing the Order Challenge**

Build a VI that reverses the order of an array containing 100 random numbers. For example, array[0] becomes array[99], array[1] becomes array[98], and so on. Name the VI **Reverse Random Array.vi**.

■ Activity 7-7 **Taking a Subset**

Build a VI that generates an array containing 100 random numbers and displays a portion of the array; for example, from index 10 to index 50. Name the VI **Subset Random Array.vi**.

*Use the **Array Subset** function [**Array** palette] to extract the portion of the array.*

■ Activity 7-8 **Dice! Challenge**

Build a VI that simulates the roll of a die (possible values 1–6) and keeps track of the number of times that the die rolls each value. Your input is the number of times to roll the die, and the outputs include (for each possible value) the number of times the die fell on that value. Name the VI **Die Roller.vi**.

You will need to use a shift register in order to keep track of values from one iteration of a loop to the next.

■ Activity 7-9 **Multiplying Array Elements**

Build a VI that takes an input 1D array, then multiplies pairs of elements together (starting with elements 0 and 1) and outputs the resulting array. For example, the input array with values 1, 23, 10, 5, 7, 11 will result in the output array 23, 50, 77. Name the VI **Array Pair Multiplier.vi**.

OVERVIEW

LabVIEW's charts and graphs let you display plots of data in a graphical form. Charts interactively plot data, appending new data to old so you can see the current value in the context of previous data, as the new data becomes available. Graphs plot pregenerated arrays of values in a more traditional fashion, without retaining previously-generated data. In this chapter, you will learn about charts and graphs, their appropriate data types, several ways to use them, and some of their special features. You will also learn about LabVIEW's special intensity charts and graphs, which have the unique ability to plot three dimensions of data on a two-dimensional (2D) plot, using color to represent the third dimension.

YOUR GOALS

- Understand the uses of charts and graphs
- Be able to recognize a chart's three modes: strip, scope, and sweep
- Understand mechanical action of Boolean switches
- Recognize the difference in functionalities of charts and graphs
- Know the data types accepted by charts and graphs for both single and multiple plots
- Customize the appearance of charts and graphs by changing the scales and using the palette, legend, and cursors
- Use the intensity chart or graph to plot three dimensions of data

KEY TERMS

- plot
- waveform chart
- legend
- strip mode
- scope mode
- sweep mode
- cursor
- palette
- mechanical action
- waveform graph
- XY graph
- intensity charts and graphs

LabVIEW's Exciting Visual Displays: Charts and Graphs

8

Waveform Charts

A *plot* is simply a graphical display of X versus Y values. Often, Y values in a plot represent the data value, while X values represent time. The *waveform chart*, located in the **Graph** subpalette of the **Controls** palette, is a special numeric indicator that can display one or more plots of data. Most often used inside loops, charts retain and display previously acquired data, appending new data as it becomes available in a continuously updating display. In a chart, the Y values represent the new data, and X values represent time (often, each Y value is generated in a loop iteration, so the X value represents the time for one loop). LabVIEW has only one kind of chart, but the chart has three different update modes for interactive data display. The following figure shows an example of a multiple-plot waveform chart.

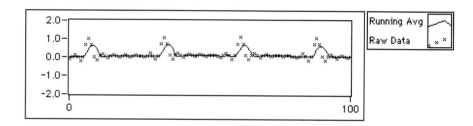

■ Chart Update Modes

The waveform chart has three update modes—*strip chart mode, scope chart mode,* and *sweep chart mode,* shown in the following illustration. The update mode can be changed by popping up on the waveform chart and choosing one of the options from the **Data Operations▶Update Mode** menu. If you want to change modes while the VI is running (and is subsequently in run mode, where the menus are slightly different), select **Update Mode** from the chart's runtime pop-up menu.

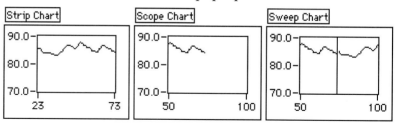

Waveform Chart Update Modes

The strip chart has a scrolling display similar to a paper strip chart. The scope chart and the sweep chart have retracing displays similar to an oscilloscope. On the scope chart, when the plot reaches the right border of the plotting area, the plot erases, and plotting begins again from the left border. The sweep chart acts much like the scope chart, but the display does not blank when the data reaches the right border. Instead, a moving vertical line marks the beginning of new data and moves across the display as new data is added. These distinctions are much easier to understand when you actually see the different modes in action, so don't worry if it sounds confusing now. You'll get to experiment with them in the next activity.

Because there is less overhead in retracing a plot, the scope chart and the sweep chart operate significantly faster than the strip chart.

■ Single-Plot Charts

The simplest way to use a chart is to wire a scalar value to the chart's block diagram terminal, as shown in the following illustration. Each loop iteration draws one more point on the displayed waveform.

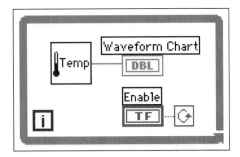

■ Wiring a Multiple-Plot Chart

Waveform charts can also accommodate more than one plot. However, since you can't wire from multiple block diagram sources to a single chart terminal, you must first bundle the data together using the **Bundle** function (**Cluster** palette). In the next figure, the **Bundle** function "bundles" or groups the outputs of the three different VIs that acquire temperature into a cluster so they can be plotted on the waveform chart. Notice the change in the waveform chart terminal's appearance when it's wired to the **Bundle** function. To add more plots, simply increase the number of **Bundle** input terminals by resizing using the Positioning tool.

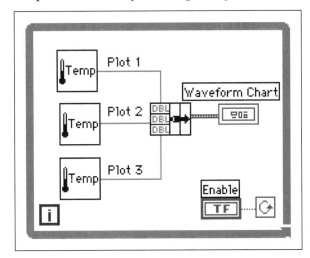

*For an online example of charts, their modes, and their expected data types, open and run **Charts.vi** in* CH8.LLB *in the* EVERYONE *directory.*

■ Show the Digital Display?

Like many other numeric indicators, charts have the option to show or hide the digital display (pop up on the chart to get the **Show▶** option). The digital display shows the most recent value displayed by the chart.

■ The Scrollbar

Charts also have scrollbars that you can show or hide. You can use the scrollbar to display older data that has scrolled off the chart.

■ Clearing the Chart

Sometimes you will find it useful to remove all previous data from the chart display. Select **Data Operations▶Clear Chart** from the chart's pop-up menu to clear a chart from edit mode (remember, you are usually in edit mode if your VI is not running. To switch between modes when the VI is not running, choose **Change to Run/Edit Mode** from the **Operate** menu). If you are in run mode, **Clear Chart** is a pop-up menu option instead of being hidden under **Data Operations**.

■ Stacked and Overlaid Plots

If you have a multiple-plot chart, you can choose whether to display all plots on the same Y-axis, called an *overlaid* plot; or you can give each plot its own Y scale, called a *stacked* plot. You can select **Stack Plots** or **Overlay Plots** from the chart's pop-up menu to toggle the type of display. The following figure illustrates the difference between stacked and overlaid plots.

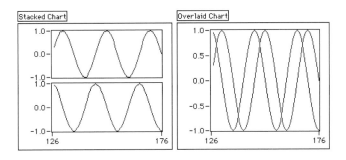

■ Chart History Length

By default, a chart can store up to 1,024 data points. If you want to store more or less data, select **Chart History Length…** from the pop-up menu and specify a new value of up to 100,000 points. Changing the buffer size does not affect how much data is shown on the screen—resize the chart to show more or less data at a time. However, increasing the buffer size does increase the amount of data you can scroll back through.

Activity 8-1
Temperature
Monitor

You will build a VI to measure temperature and display it on the waveform chart. This VI will measure the temperature using the **Thermometer** VI you built as a subVI in a previous lesson.

1. Open a new front panel. You will recreate the panel shown (but feel free not to type in the comments—they're for your benefit).

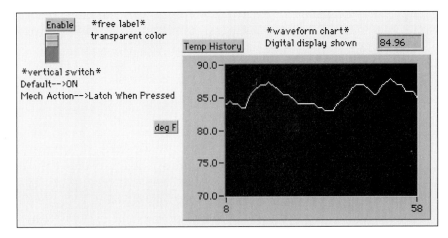

2. Place a vertical switch (**Boolean** palette) in the front panel window. Label the switch <u>Enable</u>. You will use the switch to stop the temperature acquisition.

3. Place a waveform chart (**Graph** palette) in the panel window. Label the waveform chart <u>Temp History</u>. The waveform chart will display the temperature in real time.

4. The waveform chart has a digital display that shows the latest value. Pop up on the waveform chart and choose **Show▶Digital Display** from the pop-up menu.

5. Because the temperature sensor measures room temperature, rescale the waveform chart so you can see the temperature (otherwise it will be "off the chart"). Using the Labeling tool, double-click on "10.0" in the waveform chart scale, type 90, and click outside the text area. The click enters the value. You also can press <enter> to input your change to the scale. Change "0.0" to 70 in the same way.

6. Open the block diagram window and build the diagram shown.

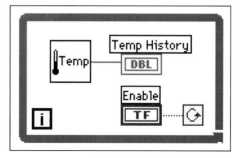

7. Place the **While Loop** (**Structures** palette) in the block diagram window and size it appropriately.

8. Place the two terminals inside the While Loop if they aren't already there.

9. Import the **Thermometer** subVI.

Thermometer.vi

Thermometer.vi. This VI returns one temperature measurement from a temperature sensor (or a simulation, depending on your setup). You should have written it in Chapter 4 and modified it in Chapter 5. Load it using the **Select A VI…** button on the **Functions** palette. It should probably be in your MYWORK directory. If you don't have it or you are using the sample version of LabVIEW, you can use **Thermometer.vi** in CH5.LLB or **Digital Thermometer.vi**, located in the **Tutorial** subpalette of the **Functions** palette.

10. Wire the block diagram as shown in the preceding illustration.

11. Return to the front panel and turn on the vertical switch by clicking on it with the Operating tool. Run the VI.

Remember, the While Loop is an indefinite looping structure. The subdiagram within its border will execute as long as the specified condition is TRUE. In this example, as long as the switch is *on* (TRUE), the **Thermometer** subVI will return a new measurement and display it on the waveform chart.

12. To stop the acquisition, click on the vertical switch. This action gives the loop conditional terminal a FALSE value and the loop ends.

13. The waveform chart has a display buffer that retains a number of points after they have scrolled off the display. You can show this scrollbar by popping up on the waveform chart and selecting **Show➤Scrollbar** from the pop-up menu. You can use the Positioning tool to adjust the scrollbar's size and position.

To scroll through the waveform chart, click on either arrow in the scrollbar.

To clear the display buffer and reset the waveform chart, pop up on the waveform chart and choose **Data Operations➤Clear Chart** from the pop-up menu. If you want to clear the chart while you're in run mode, select **Clear Chart** from the runtime pop-up menu.

14. Make sure the switch is TRUE and run the VI again. This time, try changing the update mode of the chart. Pop up and choose **Update Mode➤Scope Chart** from the chart's runtime menu. Notice the difference in chart display behavior. Now choose **Sweep Chart** and see what happens.

▥ Using Mechanical Action of Boolean Switches

Take a step out of this activity for a second. You've certainly noticed by now that each time you run this VI, you first must turn on the vertical <u>Enable</u> switch before clicking the run button, or the loop will only execute once. You can modify the *mechanical action* of a Boolean control to change its behavior and circumvent this inconvenience. LabVIEW offers six possible choices for the mechanical action of a Boolean control.

Switch When Pressed

Switch When Released

Switch Until Released

Latch When Pressed

Latch When Released

Latch Until Released

Switch When Pressed action changes the control's value each time you click on the control with the Operating tool. This action is the default for Booleans and is similar to that of a ceiling light switch. It is not affected by how often the VI reads the control.

Switch When Released action changes the control's value only when you release the mouse button during a mouse click within the graphical boundary of the control. The action is not affected by how often the VI reads the control. This mode is similar to what happens when you click on a check mark in a dialog box; it becomes highlighted but does not change until you release the mouse button.

Switch Until Released action changes the control's value when you click on the control. It retains the new value until you release the mouse button, at which time the control reverts to its original value. The action is similar to that of a door buzzer and is not affected by how often the VI reads the control.

Latch When Pressed action changes the control's value when you click on the control. It retains the new value until the VI reads it once, at which point the control reverts to its default value. This action happens whether or not you continue to press the mouse button. **Latch When Pressed** is similar in functionality to a circuit breaker and is useful when you want the VI do something only once for each time you set the control, such as to stop a While Loop when you press a STOP button.

Latch When Released action changes the control's value only after you release the mouse button. When your VI reads the value once, the control reverts to the old value. This action guarantees at least one new value. As with **Switch When Released**, this mode is similar to the behavior of buttons in a dialog box; the button becomes highlighted when you click on it, and latches a reading when you release the mouse button.

Latch Until Released changes the control's value when you click on the control. It retains the value until your VI reads the value once or until you release the mouse button, whichever occurs last.

For example, consider a vertical switch—its default value is *off* (FALSE).

15. Modify the vertical switch in your VI so that you do not need to turn the switch to TRUE each time you run the VI.

a. Turn the vertical switch to *on* (TRUE).

b. Pop up on the switch and choose **Data Operations▶Make Current Value Default** from the pop-up menu to make the *on* position the default value.

c. Pop up on the switch and choose **Mechanical Action▶Latch When Pressed** from the pop-up menu.

16. Run the VI. Click on the vertical switch to stop the acquisition. The switch will move to the *off* position briefly, and then automatically change back to *on* after the While Loop conditional terminal reads one FALSE value.

You cannot use mechanical action on a Boolean object if you will be modifying the object's value using local variables. We'll tell you why when we talk about locals in Chapter 12.

▦ Adding Timing

When you run the VI in this activity, the While Loop executes as quickly as possible. You may want to take data at certain intervals, however, such as once per second or once per minute. You can control loop timing using the **Wait Until Next ms Multiple** function (**Time & Dialog** menu).

17. Modify the VI to take a temperature measurement about once every half-second by placing the code segment shown here into the While Loop.

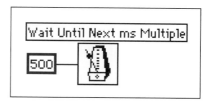

Wait Until Next ms
Multiple Function

Wait Until Next ms Multiple function (**Time & Dialog** menu) ensures that each iteration waits the specified amount of time (in this case, a half-second or 500 milliseconds) before continuing.

18. Run the VI. Run it several more times, trying different values for the number of milliseconds.

19. Save and close the VI. Name it **Temperature Monitor.vi** and place it in your MYWORK directory or VI library. Excellent job!

Graphs

Unlike charts, which plot data interactively, graphs plot pre-generated arrays of data all at once; they do not have the ability to append new values to previously generated data. LabVIEW provides three types of graph for greater flexibility: *waveform graphs*, *XY graphs*, and *intensity graphs*. We'll talk about waveform and XY graphs now and cover intensity graphs in the next section. Both waveform graphs and XY graphs look identical on the front panel of your VI but have very different functionality. An example of a graph with several graph options enabled is shown in the following picture.

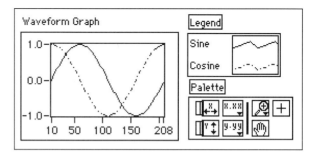

You can obtain both types of graph indicators from the **Graph** subpalette of the **Controls** palette. The *waveform graph* plots only single-valued functions (only one Y value for every X) with uniformly spaced points, such as acquired time-sampled, amplitude-varying waveforms. The waveform graph is ideal for plotting arrays of data in which the points are evenly distributed. The *XY graph* is a general-purpose, Cartesian graph, ideal for plotting data with varying timebases or data with several Y values for every X value, such as circular shapes. The two types of graph look the same but take different types of input, so you must be careful not to confuse them.

■ Single-Plot Waveform Graphs

For basic single-plot graphs, you can wire an array of Y values directly to a waveform graph terminal, as shown in the following illustration. This method assumes the initial X value is zero, and the delta X value (i.e., the increment between X values) is one. Notice that the graph terminal in the block diagram appears as an array indicator here.

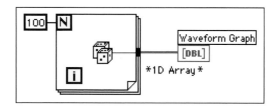

Sometimes you will want the flexibility to change the time-base for the graph. For example, you start sampling at a time other than "initial X=0" (or X_0 =0), or your samples are spaced more than one unit apart, or "delta X=1" (also written ΔX=1). To change the timebase, bundle the X_0 value, ΔX value, and the data array into a cluster; then wire the cluster to the graph. Notice in the following illustration that the graph terminal now appears as a cluster indicator.

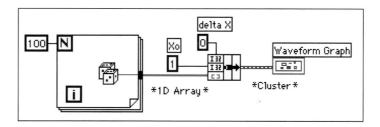

■ Multiple-Plot Waveform Graphs

You can show more than one plot on a waveform graph by creating an array (or a 2D array) of the data types used in the single-plot examples. Notice how graph terminals change appearance depending on the structure of data wired to them (array, cluster, array of clusters, etc.) and the data type (I16, DBL, etc.).

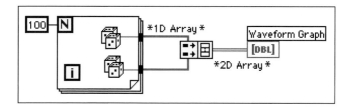

The preceding illustration assumes the initial X value is 0 and the delta X value is 1 for both arrays. The **Build Array** function creates a 2D array out of two 1D arrays. Notice that this 2D array has two rows with 100 columns per row—a 2 × 100 array. By default, graphs plot each *row* of a 2D array as a separate waveform. If your data is organized by column, you must make sure to transpose your array when you plot it! Transposing means simply switching row values with column values; for example, if you transpose an array with three rows and ten columns, you end up with an array with ten rows and three columns. LabVIEW makes it easy to do this—simply pop up on the graph and select **Transpose Array** (this menu option is grayed out if the graph does not have a 2D array wired to it). You can also use the **Transpose 2D Array** function found in the **Array** subpalette of the **Functions** menu.

In the following illustration, the X_0 value and a ΔX (or delta X) value for each array is specified. These X parameters do not need to be the same for both sets of data.

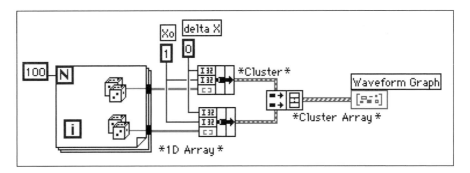

In this diagram, the **Build Array** function (**Array** palette) creates an array out of its two cluster inputs. Each input cluster consists of an array and two scalar numbers. The final result is an array of clusters, which the graph can accept and plot. In this case, the specified X_0 and ΔX are the same as the default values, but you can always use other values as well. You're now starting to learn about complex data structures—as long as you take them bit by bit, they're not too confusing!

Activity 8-2
Graphing a Sine on a Waveform Graph

You will build a VI that generates a sine wave array and plots it in a waveform graph. You will also modify the VI to graph multiple plots.

1. Open a new VI, and build the front panel shown.

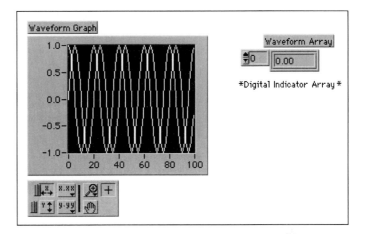

2. Place an array shell (**Array & Cluster** palette) in the front panel window. Label the array shell <u>Waveform Array</u>. Place a digital indicator (**Numeric** palette) inside the Data Object window of the array shell to display the array contents.

3. Place a waveform graph (**Graph** palette) in the front panel window. Label the graph <u>Waveform Graph</u> and enlarge it by dragging a corner with the Positioning tool.

Hide the legend by popping up on the graph and selecting **Show▶✓Legend.**

Disable autoscaling by popping up and choosing **Y Scale▶ ✓AutoScale Y**. Modify the Y axis limits by selecting the scale limits with the Labeling tool and entering new numbers; change the Y axis minimum to -1.0 and the maximum to 1.0. We'll talk more about autoscaling in the next section.

4. Build the block diagram shown.

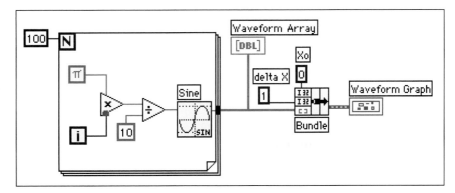

Sine Function

Pi Constant

Bundle Function

Sine function (**Numeric►Trigonometric** palette) computes `sin(x)` and returns one point of a sine wave. The VI requires a scalar index input that it expects in radians. In this exercise, the x input changes with each loop iteration and the plotted result is a sine wave.

Pi constant (**Numeric►Additional Numeric Constants** palette).

Bundle function (**Cluster** palette) assembles the plot components into a single cluster. The components include the initial X value (0), the delta X value (1), and the Y array (waveform data). Use the Positioning tool to resize the function by dragging one of the corners.

Each iteration of the For Loop generates one point in a waveform and stores it in the waveform array created at the loop border. After the loop finishes execution, the **Bundle** function bundles the initial value of X, the delta value for X, and the array for plotting on the graph.

5. Return to the front panel and run the VI. Right now the graph should show only one plot.

6. Now change the delta X value to 0.5 and the initial X value to 20 and run the VI again. Notice that the graph now displays the same 100 points of data with a starting value of 20 and a delta X of 0.5 for each point (shown on the X axis).

7. Place the Positioning tool on the lower right corner of the array until the tool becomes a grid, and drag. The indicator now displays several elements with indices ascending as you go from left to right (or top to bottom), beginning with the element that corresponds to the specified index, as illustrated here. Don't forget that you can view any element in the array simply by entering the index of that element in the index display. If you enter a number greater than the array size, the display will dim.

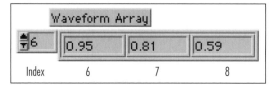

In the previous block diagram, you specified an initial X and a delta X value for the waveform. Frequently, the initial X value will be zero and the delta X value will be 1. In these instances, you can wire the waveform array directly to the waveform graph terminal, as shown in the following illustration, and take advantage of the default delta X and initial X values for a graph.

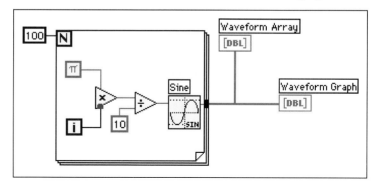

8. Return to the block diagram window. Delete the **Bundle** function and the **Numeric Constants** wired to it, then select **Remove Bad Wires** from the **Edit** menu or use the keyboard shortcut. Finish wiring the block diagram as shown in the preceding illustration.

9. Run the VI. Notice that the VI plots the waveform with an initial X value of 0 and a delta X value of 1, just as it did before, with a lot less effort on your part to build the VI.

■ Multiple-Plot Graphs

You can also create multiple-plot waveform graphs by wiring the data types normally passed to a single-plot graph first to a **Build Array** function, then to the graph. Although you can't build an array of arrays, you can create a 2D array with each input array as a row.

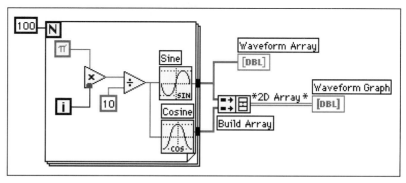

10. Create the block diagram shown.

Build Array Function

Build Array function (**Array** palette) creates the proper data structure to plot two arrays on a waveform graph. Enlarge the

Build Array function to include two inputs by dragging a corner with the Positioning tool. You will want to use *element inputs* to the **Build Array** (the default value) rather than *array inputs* so the output will be a 2D array. If you used array inputs, the function would create one long 1D array by appending the second array to the end of the first.

Cosine Function

Cosine function (**Numeric▶Trigonometric** palette) computes cos(x) and returns one point of a cosine wave. The VI requires a scalar index input that it expects in radians. In this exercise, the x input changes with each loop iteration and the plotted result is a cosine wave.

11. Switch to the front panel. Run the VI. Notice that the two waveforms both plot on the same waveform graph. The initial X value defaults to 0 and the delta X value defaults to 1 for both data sets.

12. Just to see what happens, pop up on the graph and choose **Transpose Array**. The plot changes drastically when you swap the rows with the columns, doesn't it? Choose **Transpose Array** again to return things to normal.

13. Save and close the VI. Name it **Graph Sine Array.vi** in your MYWORK directory or VI library.

■ XY Graphs

The waveform graphs you have been using are designed for plotting evenly sampled waveforms. However, if you sample at irregular intervals or are plotting a math function that has multiple Y values for every X value, you will need to specify points using their (X,Y) coordinates. *XY graphs* plot this different type of data; they require input of a different data type than waveform graphs. A single-plot XY graph and its corresponding block diagram are shown in the following illustration.

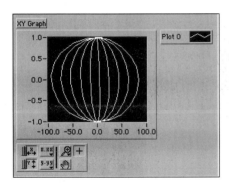

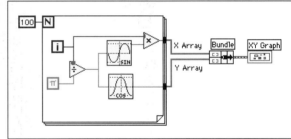

The XY graph expects an input of a bundled X array (the top input) and a Y array (the bottom input). The **Bundle** function (**Cluster** palette) combines the X and Y arrays into a cluster wired to the XY graph. The XY graph terminal now appears as a cluster indicator.

For a multiple-plot XY graph, simply build an array of the clusters of X and Y values used for single plots, as shown.

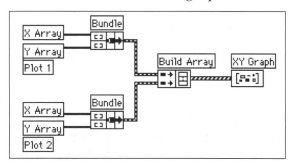

*It is very easy to confuse the **Bundle** and **Build Array** functions when assembling data for graphs! Pay close attention to which one you need to use!*

*For some online graph examples, look at **Waveform Graph.vi** and **X vs. Y Graph.vi** in* CH8.LLB *in the* EVERYONE *directory.*

Chart and Graph Components

Graphs and charts have many powerful features that you can use to customize your plots. This section covers how to configure these options.

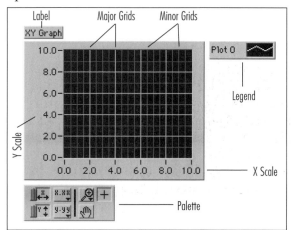

■ Playing with the Scales

Charts and graphs can automatically adjust their horizontal and vertical scales to reflect the points plotted on them; that is, the scales adjust themselves to show all points on the graph at the greatest resolution. You can turn this *autoscaling* feature on or off using the **AutoScale X** and **AutoScale Y** options from the **Data Operations** menu, the **X Scale** menu, or the **Y Scale** menu of the object's pop-up menu. You can also control these autoscaling features from the palette (which we'll get to in a minute). LabVIEW defaults to autoscaling on for graphs and off for charts. However, using autoscaling may cause the chart or graph to update more slowly, depending upon the computer and video system you use, because new scales must be calculated with each plot.

If you don't want to autoscale, you can change the horizontal or vertical scale directly using the Operating or Labeling tool to type in a new number, just as you can with any other LabVIEW control or indicator, and turn autoscaling off.

■ X and Y Scale Menus

The X and Y scales each have a submenu of options.

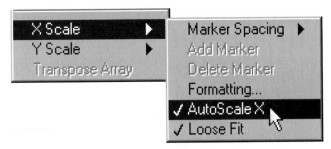

Use **AutoScale** to turn the autoscaling option on or off.

Normally, the scales are set to the exact range of the data when you perform an autoscale operation. You can use the **Loose Fit** option if you want LabVIEW to round the scale to "nicer" numbers. With a loose fit, the numbers are rounded to a multiple of the increment used for the scale. For example, if the markers increment by five, then the minimum and maximum values are set to a multiple of five instead of the exact range of the data.

The **Formatting...** option brings up a dialog box, shown in the following illustration, that allows you to configure the following things:

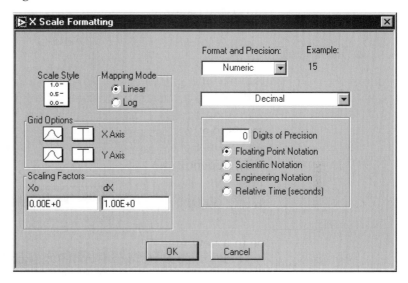

◆   The **Scale Style** menu lets you select major and minor tick marks for the scale, or none at all. Click on this icon to see your choices. A major tick mark corresponds to a scale label, while a minor tick mark denotes an interior point between labels. This menu also lets you select the markers for a given axis as either visible or invisible.

◆   The **Mapping Mode** lets you select either a linear or a logarithmic scale for the data display.

◆   The **Grid Options** lets you choose between no gridlines, gridlines only at major tick mark locations, or gridlines at both major and minor tick marks. You can also change the color of the gridlines here. Click on the grid buttons to see a pullout menu of your options.

You can set X_0, the X value you want to start at, and **dX**, the increment between X values (same as delta X), in the **Scaling Factors** section.

The **Format and Precision** section lets you choose **Numeric** format or **Time & Date** format. If you choose **Numeric**, you can then pick the number of digits of precision as well as the notation (**Floating Point**, **Scientific, Engineering,** or **Relative Time**) of the scale display. You can also specify your numbers in **Decimal, Unsigned Decimal, Hexadecimal, Octal,** or **Binary** format. If you choose **Time & Date** format, you can choose how you would like the time and date displayed.

■ The Legend

Charts and graphs use a default style for each plot unless you have created a custom plot style for it. The *legend* lets you label, color, select line style, and choose point style for each plot. Show or hide the legend using the **Show** submenu of the chart or graph pop-up menu. You can also specify a name for each plot in the legend. An example of a legend is shown in the following illustration.

When you select **Legend**, only one plot shows in the box that appears. You can show more plots by dragging down a corner of the legend with the Positioning tool. After you set plot characteristics in the **Legend**, the plot retains those settings, regardless of whether the legend is visible. If the chart or graph receives more plots than are defined in the legend, LabVIEW draws the extra plots in default style.

When you move the chart or graph body, the legend moves with it. You can change the position of the legend relative to the graph by dragging only the legend to a new location. *Resize the legend on the left to give labels more room in the window or on the right to give plot samples more room.*

By default, each plot is labeled with a number, beginning with zero. You can modify this label the same way you modify other LabVIEW labels—just start typing with the Labeling tool. Each plot sample has its own menu to change the plot, line, color, and point styles of the plot. You can also access this menu by clicking on the legend with the Operating tool.

◆ The **Common Plots** option lets you easily configure a plot to use any of six popular plot styles, including a scatter plot, a bar plot, and a fill to zero plot. Options in this subpalette configure the point, line, and fill styles of the graph in one step (instead of setting these options individually, as listed next). Look at Activity 8-6 for an example of a bar graph.

◆ The **Point Style**, **Line Style**, and **Line Width** options display different plot styles you can choose from. The **Line Width** sub-palette contains a hairline option, which prints each line that is one pixel wide on the screen as a very thin line on a printer that supports hairline printing.

◆ The **Bar Plots** option lets you create bar plots of 100%, 75%, or 1% width, either horizontally or vertically. The **Fill Baseline** option controls the baseline of the bars; plots can have no fill or they can fill to zero, negative infinity, or infinity.

◆ The **Interpolation** option determines how LabVIEW draws lines between data points. The first option does not draw any lines, making it suitable for a scatter plot (in other words, you get points only). The option at the bottom left draws a straight line between points. The two stepped options link points with a right-angled elbow, useful for histograms. The step option at the top right plots the Y axis first, while the step option at the bottom right plots the X axis first.

• The **Color** option displays the Color palette so that you can select the plot color. You can also color the plots on the legend with the Color tool. You can change the plot colors while running the VI.

Activity 8-3
Using an XY Graph to Plot a Circle

You will build a VI that plots a circle on an XY graph using in-dependent X and Y arrays.

1. Open a new front panel. You will recreate the panel shown.

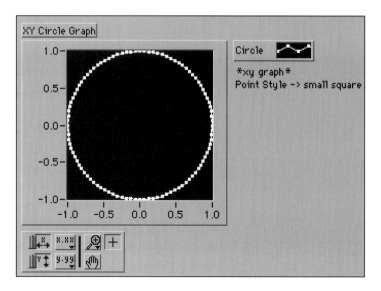

2. Place an **XY Graph** (**Graph** palette) in the panel window. Label the graph XY Circle Graph.

3. Enlarge the graph by dragging a corner with the Positioning tool. Try to make the plot region approximately square.

You can maintain the aspect ratio while dragging by holding down the <shift> key and dragging diagonally.

4. Pop up on the graph and select **Show▶Legend**. Resize the legend from the left side and enter the Circle label using the Labeling tool. Pop up on the line in the legend and select the small square from the **Point Style** palette. Then select a new plot color from the **Color** palette.

5. Build the block diagram shown.

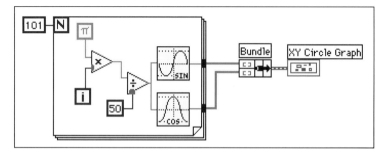

Sine Function

Sine and **Cosine** functions (**Numeric▶Trigonometric** palette) calculate the sine and cosine, respectively, of the input number. In this exercise, you use the function in a For Loop to build an array

Cosine Function

Bundle Function

Pi Constant

of points that represents one cycle of a sine wave and one cycle of a cosine wave.

Bundle function (**Cluster** palette) assembles the sine array (X values) and the cosine array (Y values) to plot the sine array against the cosine array.

Pi constant (**Numeric▶Additional Numeric Constants** palette) is used to provide radian input to **Sine** and **Cosine** functions.

Using a **Bundle** function, you can graph the one-cycle sine array versus the one-cycle cosine array on an XY graph, which produces a circle.

6. Return to the front panel and run the VI. Save it as **Graph Circle.vi** in your MYWORK directory or VI library. Congratulations! You're getting the hang of this stuff!

■ Using the Palette

The *palette* is the little box that shows up under the graph or chart when you first drop it on the front panel. With the palette, you can change several plot characteristics while the VI executes. You can clear the chart or graph, scale the X or Y axis, and change the display format of the scale at any time (remember, you can also scale axes and change display format using the **X Scale** or **Y Scale** pop-up menu). You can also pan (i.e., scroll the display area), and focus in on a specific area using palette buttons (called zooming). The palette, which you access from the **Show** menu of the chart or graph pop-up menu, is shown in the following illustration.

Autoscale X Data

Autoscale Y Data

Lock Switch

If you press the autoscale x data button, LabVIEW will autoscale the X data of the graph. If you press the autoscale Y data button, LabVIEW will autoscale the Y data of the graph. If you want the graph to autoscale either of the scales continuously, click on the corresponding lock switch to lock autoscaling on (the *on* position of the switch is to the right, the *off* position is to the left).

Format and Precision
Controls

Standard Operate Mode

Pan Button

Pan Cursor

Zoom Button

The two buttons at the left give you runtime control over the format and precision of the X and Y scale markers, respectively.

The remaining three buttons let you control the operation mode for the graph. Normally, you are in standard mode, meaning that you can click on the graph cursors to move them around. If you press the pan button, then you switch to a mode where you can scroll the visible data by dragging sections of the graph with the pan cursor. If you click on the zoom button, you get a pop-up menu that lets you choose from several methods of zooming (focusing on specific sections of the graph by magnifying a particular area), shown in the following illustration.

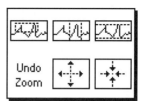

Here's how these zoom options work:

Zoom by Rectangle. Drag the cursor to draw a rectangle around the area you want to zoom in on. When you release the mouse button, your axes will rescale to show only the selected area.

Zoom by Rectangle

Zoom by Rectangle
(restricted to X data)

Zoom by Rectangle
(restricted to Y data)

**Undo
Zoom**
Undo Last Zoom

Zoom In About a Point

Zoom Out About a Point

Zoom by Rectangle in X, with zooming restricted to X data (the Y scale remains unchanged).

Zoom by Rectangle in Y, with zooming restricted to Y data (the X scale remains unchanged).

Undo Last Zoom is pretty obvious. It also resets the scales to their previous setting.

Zoom In about a Point. If you hold down the mouse on a specific point, the graph will continuously zoom in until you release the mouse button.

Zoom Out about a Point. If you hold down the mouse on a specific point, the graph will continuously zoom out until you release the mouse button.

*For the last two modes, **Zoom In** and **Zoom Out about a Point**, shift-clicking will zoom in the other direction.*

■ Graph Cursors

LabVIEW graphs have *cursors* to mark data points on your plot to further animate your work. The following illustration shows a picture of a graph with cursors and the **Cursor** palette visible. You won't find cursors on charts, so don't bother to look.

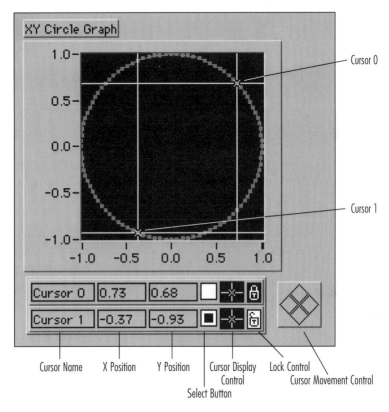

You can view the **Cursor** palette by selecting **Show►Cursor Display** from the graph's pop up menu. When the palette first appears, it is grayed out. Click on the Select button to make a cursor active (or type text in one of the Cursor Name fields with the Labeling tool). You can move cursors around manually, or programmatically using the attribute node, which you'll learn about in Chapter 12.

To manually move a cursor, drag it around the graph with the Operating tool. If you drag the intersection point, you can move in all directions. If you drag the horizontal or vertical lines of the cursor, you can drag only horizontally or vertically, respectively.

You can also use the Cursor Movement Control to move a cursor up, down, left, and right.

A graph can have as many cursors as you want. The **Cursor** palette helps you keep track of them, and you can stretch it to display multiple cursors.

What does this **Cursor** palette do, you ask? You can label the cursor in the first box at the left of the **Cursor** palette. The next box shows X position, and the next after that shows Y position. Click on the little white Select button to make the cursor on that line active and select it for movement (a black box inside denotes a selected cursor). When you click a button on the Cursor Movement Control, all active cursors will move.

Click with the Operating tool (do not pop up!) on the Cursor Display Control (which looks like a crosshair) to specify things like cursor style, point style, and color.

The little lock at the far right will lock the cursor to a plot. Locking restricts cursor movement so it can reside only on plot points, but nowhere else in the graph. Click on the lock with the Operating tool to access its menu (don't pop up or you'll get the wrong one!).

To delete a cursor, execute the **Start Selection** and **End Selection** commands on the **Data Operations** pop-up menu, then choose **Cut** from the same menu.

Activity 8-4
Temperature
Analysis

You will build a VI that measures temperature approximately every 0.25 seconds for 10 seconds. During the acquisition, the VI displays the measurements in real time on a waveform chart. After the acquisition is complete, the VI plots the data on a graph and calculates the minimum, maximum, and average temperatures.

1. Open a new front panel and build the VI shown.

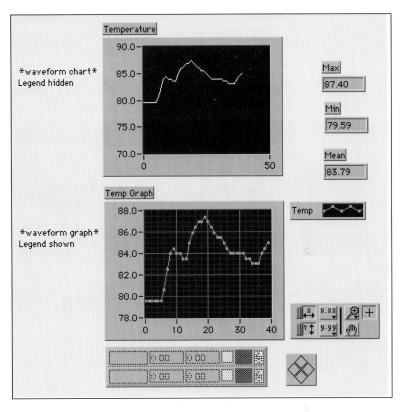

2. Rescale the chart so that it ranges from 70.0 to 90.0. Also, make sure autoscaling is on for both axes of the graph (it should be, because that is the default).

3. Add grid lines to the graph by popping up on the X or Y scale and selecting **Formatting...**. Choose major and minor grid lines for both X and Y axes, then color the grid lines by clicking on the boxes next to the grid buttons.

4. Show the legend on the graph. Resize it from the left side to make the text area larger. Using the Labeling tool, type in Temp as shown. Now pop up (or click with the Operating tool) on the Temp plot representation in the legend and change the **Point Style** to small squares. Feel free to color your plots as well.

The Temperature chart displays the temperature as it is acquired. After acquisition is finished, the VI plots the data in Temp Graph. The Mean, Max, and Min digital indicators will display the average, maximum, and minimum temperatures, respectively.

5. Build the block diagram shown. Make sure to use the Help window to show you the inputs and outputs of these functions, and pay attention to the wire whiskers and tip strips, or you will almost certainly wire to the wrong terminal!

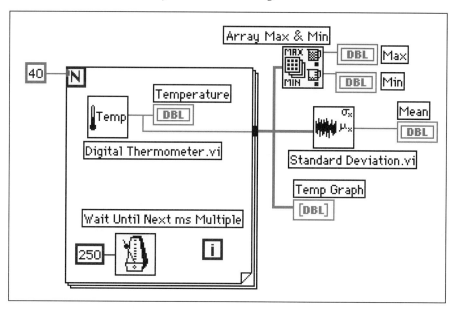

Thermometer.vi

Wait Until Next ms
Multiple Function

Array Max & Min
Function

Standard Deviation.vi

Thermometer.vi. Use the **Select A VI...** palette to access the one you built. It should probably be in your MYWORK directory. If you don't have it or you are using the sample version of LabVIEW, you can use **Thermometer.vi** in CH5.LLB or **Digital Thermometer.vi**, located in the **Tutorial** subpalette of the **Functions** palette. **Thermometer** returns one temperature measurement each time it is called.

Wait Until Next ms Multiple function (**Time & Dialog** palette) causes the For Loop to execute every 0.25 seconds (250 ms).

Array Max & Min function (**Array** palette) returns the maximum and minimum values in the array, in this case the maximum and minimum temperatures measured during the acquisition.

Standard Deviation.vi (**Analysis▶Probability and Statistics** palette) returns the average of the temperature measurements. Make sure you wire to the right output terminal to get the mean.

6. The For Loop executes 40 times. The **Wait Until Next ms Multiple** function causes each iteration to take place approximately every 250 ms. The VI stores the temperature measurements in an

array created at the For Loop border using auto-indexing. After the For Loop completes, the array passes to various nodes. The **Array Max & Min** function returns the maximum and minimum temperature. **Standard Deviation.vi** returns the average of the temperature measurements.

7. Return to the front panel and run the VI.

Zoom Button

8. Using the palette, change the precision so that the graph shows three decimal places on the Y scale. Now click on the Zoom button, select a zooming mode, and zoom in on the graph.

9. Pop up on the graph and **Show▶Cursor Display**. It will first appear grayed out, so click on the Select button on the top line to make the first cursor active.

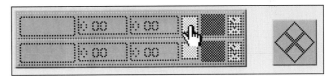

Use the Operating tool to drag the cursor around on the graph; pay attention to how the X and Y values in the cursor display change. These values can help you determine the value of a particular point on the graph. Now use the Cursor Movement Control buttons to move the cursor around. Make the second cursor active and use the cursor movement control to move both cursors at the same time. Click on the Cursor Display Control (the crosshairs) with the Operating tool (if you pop up, you will get a different menu) and change the color of one of your cursors.

Are you starting to see what cursors can do? If you want, select the Cursor 0 text and replace it with a cursor name of your own. Finally click on the lock button with the Operating tool and deselect **Allow Drag** from the menu that appears. Then choose **Lock to Plot** from the same menu. You can no longer drag the cursor around on the graph with the mouse. However, try using the cursor movement control and you'll find that the cursor will track the plot it's assigned to.

10. Close and save the VI. Name it **Temperature Analysis.vi** and place it in your MYWORK directory or VI library.

Intensity Charts and Graphs— Plotting in 'Three Dimensions'

So what do you do if you want to plot three variables against each other, instead of just two? Simple—make an intensity plot! *Intensity charts and graphs* display three dimensions of data on a 2D plot by using color to display values of the third dimension of data (the "Z" values). Like the waveform chart, the intensity chart features a scrolling display, while the intensity graph display is fixed. Intensity plots are extremely useful for displaying patterned data such as terrain, where color represents altitude over a two-dimensional area, or temperature patterns, where color represents temperature distribution.

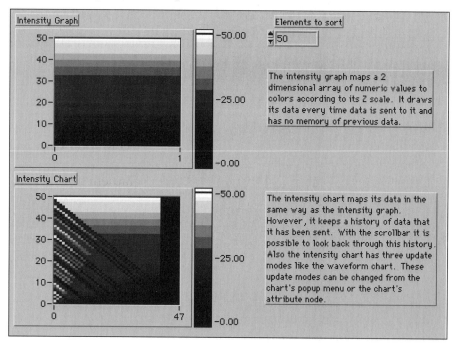

Intensity plots function much like two-dimensional charts and graphs in most ways, with the addition of color to represent the third variable. A Color Scale is available so you can set and display the color mapping scheme. Intensity graph cursor displays also include a Z value.

Intensity charts and graphs accept 2D arrays of numbers, where each number in the array is a color value. The indices of each array element represent the plot location for that color. Not only can you define the mapping of numbers to colors using the color scale (which works like a color ramp, if you've played with that), but you can also do it programmatically using attribute nodes (Chapter 12 will tell you all about them). The simple example below shows a 3×4 array plotted on an intensity graph. The colors mapped are red (1.0), blue (2.0), and green (3.0).

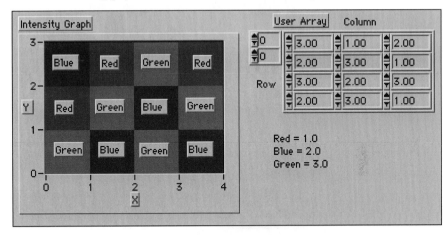

To assign a color to a value on the Color Scale, pop up on the corresponding marker and select **Marker Color▶**. The Z scale of intensity charts and graphs has Arbitrary Marker Spacing by default, so you can alter the "gradient" of colors by dragging a marker with the Operating tool. Create new markers by selecting **Add Marker** from the Color Scale pop-up menu, then drag them to a desired location and assign a new color. If you want to learn more about intensity charts and graphs, we suggest playing around with them. You can also look in LabVIEW's copious documentation for more exact instructions.

Activity 8-5
The Intensity Graph

In this activity, you will look at a VI that displays wave interference patterns. This VI will also show you how the intensity graph maps its input 2D array to the display.

1. To get an idea of how intensity graphs and charts work, open and run the **Intensity Graph Example** VI, located in EVERY-ONE\CH8.LLB. Run the VI. You will see an intricate interference waveform plotted on the graph. The color range is defined in the block diagram using the intensity graph attribute node. Modify the color range by clicking on the color boxes in the first frame of the Sequence Structure with the Operating tool, then choosing a new color from the palette that appears. Run the VI again.

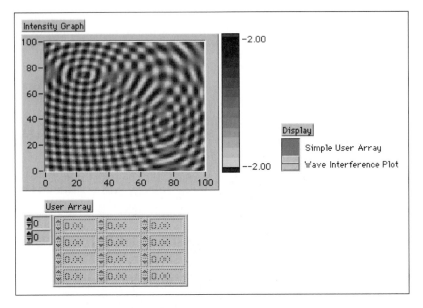

2. Switch the <u>Display</u> switch to *Simple User Array* and enter values between 0.0 and 10.0 in the <u>User Array</u> control (the color range for the graph has been defined in the block diagram using the intensity graph attribute node—blue (0.0) to red (10.0). After you enter all of the values, run the VI. Notice how the magnitude of each element is mapped to the graph. Now change your values and run it again.

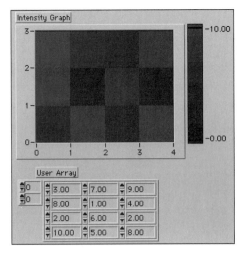

3. Take a look at the block diagram to see how the VI works.

4. Close the VI and don't save any changes.

For more examples of intensity plots, open some of the VIs that ship with LabVIEW (if you have the full version) in exam- ples\general\graphs\intgraph.llb. Also check out **Simulation of Tomography** and **Heat Equation Example** in ex- amples\analysis\mathxmpl.llb.

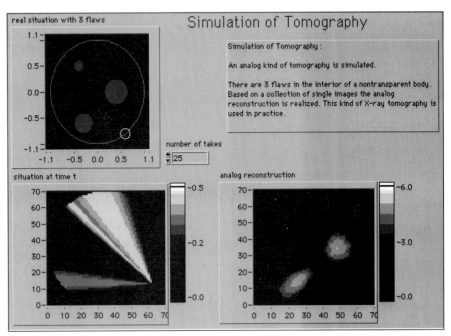

Wrap It Up!

You can create exciting visual displays of data using Lab- VIEW's charts and graphs. *Charts* append new data to old data, interactively plotting one point (or one set of points) at a time, so you can see a current value in context with previous values. *Graphs*, on the other hand, display a full block of data after it has been generated.

LabVIEW provides three kinds of graphs: *waveform* graphs, XY graphs, and *intensity* graphs.

The *waveform graph* plots only single-valued functions with points that are evenly distributed with respect to the X axis, such as time-varying waveforms. In other words, the waveform graph plots a Y array against a set timebase.

The *XY graph* is a general-purpose, Cartesian graph that lets you plot multivalued functions such as circular shapes. It plots a Y array against an X array.

Intensity plots are excellent for displaying patterned data since they can plot three variables of data against each other on a 2D display. Intensity charts and graphs use color to represent the third variable. They accept a 2D array of numbers, where each number is mapped to a color and the number's indices in the array specify location for the color on the graph or chart. In most other ways, intensity plots function like standard two-variable charts and graphs.

You can configure the appearance of charts and graphs using the *legend* and the *palette*. You can also change the scales to suit your data and bring up cursors to mark your plots.

Both charts and graphs can draw multiple plots at a time. Data types can get tricky, so you may want to refer to the examples in this chapter or those that ship in the `examples` directory (if you have the full version of LabVIEW) as a template while writing your own graphing VIs.

Mechanical action of Boolean switches allows you to control how they behave when you click on them. You can set a switch to return to its default value after its new value has been read once—that way it's all ready to be used again. This type of action is called *latch* action. You can also specify if you want the mouse click to register when you press the mouse button or release it.

Additional Activities

■ Activity 8-6 Bar Graphs

Open **Array to Bar Graph Demo**, found in EVERYONE\ CH8.LLB. Now run it to see what it does. Mess with the parameters and run it again, then take a look at the block diagram to see how it works.

■ Activity 8-7 Temperature Limit

Build a VI that continuously measures the temperature once per second and displays the temperature on a chart in scope mode. If the temperature goes above or below the preset limits, the VI turns on a front panel LED. The chart should plot the temperature as well as the upper and lower temperature limits. You should be able to set the limits from the front panel. Take a look at the front panel shown for a start. Name the VI **Temperature Limit.vi**.

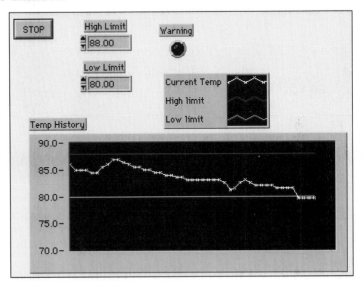

■ Activity 8-8 **Max/Min Temperature Limit**

Modify the VI you created in Activity 8-7 to display the maximum and minimum values of the temperature trace. Name the VI **Temp Limit (max/min).vi**.

*You must use shift registers and the **Array Max & Min** function (**Array** palette).*

■ Activity 8-9 **Plotting Random Arrays**

Build a VI that generates a 2D array (three rows by ten columns) containing random numbers. After generating the array, index off each row and plot each row on its own graph. Your front panel should contain three graphs. Name the VI **Extract 2D Array.vi**.

OVERVIEW

This chapter introduces some of the powerful things you can do with strings. LabVIEW has many built-in string functions, similar to its array functions, that let you manipulate string data for screen display, instrument control, or any number of reasons. You will also learn how to save data to and retrieve data from a disk file.

YOUR GOALS

- Learn more about options for string controls and indicators
- Understand how to use LabVIEW's string functions
- Convert numeric data to string data, and vice versa
- Use the file input and output (I/O) VIs to save data to a disk file and then read it back into LabVIEW

KEY TERMS

- scrollbar
- table
- spreadsheet file

Exploring Strings and File I/O

9

More about Strings

We introduced strings in Chapter 4—a string is simply a collection of ASCII characters. Often, you may use strings for more than simple text messages. For example, in instrument control, you pass numeric data as character strings. You then convert these strings to numbers to process the data. Storing numeric data to disk can also use strings; in many of the file I/O VIs, LabVIEW first converts numeric values to string data before it saves them to a file.

■ Choose Your Own Display Type

String controls and indicators have several options you might find useful. For example, they can display and accept characters that are normally nondisplayable, such as backspaces, carriage

251

returns, and tabs. If you choose '\' **Codes Display** (instead of **Normal Display**) from a string's pop-up menu, nondisplayable characters appear as a backslash (\) followed by the appropriate code. The following table shows what these codes mean.

LabVIEW '\' Codes

Code	LabVIEW Interpretation
\00–\FF	Hexadecimal value of an 8-bit character; alphabetical characters must be uppercase
\b	Backspace (ASCII BS, equivalent to \08)
\f	Formfeed (ASCII FF, equivalent to \OC)
\n	New Line (ASCII LF, equivalent to \OA)
\r	Return (ASCII CR, equivalent to \0D)
\t	Tab (ASCII HT, equivalent to \09)
\s	Space (equivalent to \20)
\\	Backslash (ASCII \, equivalent to \5C)

You must use uppercase letters for hexadecimal characters and lowercase letters for the special characters, such as formfeed and backspace. LabVIEW interprets the sequence \BFare as hex BF followed by the word "are," whereas LabVIEW interprets \bFare and \bfare as a backspace followed by the words "Fare" and "fare." In the sequence \Bfare, \B is not the backspace code, and \Bf is not a valid hex code. In a case like this, when a backslash is followed by only part of a valid hex character, LabVIEW assumes a zero follows the backslash, so LabVIEW interprets \B as hex 0B. Any time a backslash is not followed by a valid character code, LabVIEW ignores the backslash character.

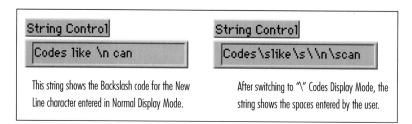

This string shows the Backslash code for the New Line character entered in Normal Display Mode.

After switching to "\" Codes Display Mode, the string shows the spaces entered by the user.

Don't worry, the data in the string does not change when the display mode is toggled; only the display of certain characters changes. '\' **Codes Display** mode is very useful for debugging programs and for specifying nondisplayable characters required by instruments, the serial port, and other devices.

Strings also have a **Password Display** option, which sets the string control or indicator to display a "*" for every character entered into it, so that no one can see what you type. While the front panel shows only a stream of "****," the block diagram reads the actual data in the string. Obviously, this display can be useful if you need to programmatically password-protect all or part of your VIs.

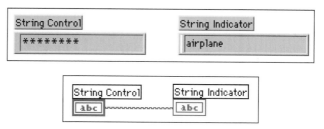

If you want to see your string as hexadecimal characters instead of alphanumeric characters, use the **Hex Display** option.

■ Single Line Strings

If you choose **Limit to Single Line** from a string's pop-up menu, your string cannot exceed one line of text; that is, no carriage returns are allowed in the string. If you hit <enter> or <return>, text entry will be automatically terminated. If strings are not limited to a single line, hitting <return> causes the cursor to jump to a new line to let you type more.

■ The Scrollbar

If you choose the **Show►Scrollbar** option from the string pop-up **Show** submenu, a vertical scrollbar appears on the string control or indicator. You can use this option to minimize the space taken up on the front panel by string controls that contain a large amount of text. Note that this option will be grayed out unless you've increased the size of your string enough for a scrollbar to fit.

■ Tables

A table is a flat piece of furniture...no—wait—wrong context! In LabVIEW, a table is a special structure that displays a two-dimensional (2D) array of strings. You can find it in the **String &**

Table subpalette of the **Controls** palette. A table in all its glory is shown in the following illustration.

Index Display	Table	columns header		
rows	0			
columns	0	x	x^2	sqrt(x)
		0	0.0000	0.0000	0.0000

Tables have row and column headings that you can show or hide; the headings are separated from the data space by a thin open border space. You can enter text headings using the Labeling tool or Operating tool (like everything else). You can also update or read headings using attribute nodes, which you've heard so much about and will learn how to use soon enough.

Like an array index display, a table index display indicates which cell is visible at the upper left corner of the table.

For a good example of a table and how to use it, open and run **Building Tables.vi**, located in EVERYONE\CH9.LLB.

Using String Functions

Like arrays, strings can be much more useful when you take advantage of the many built-in functions provided by LabVIEW. This section examines a few of the functions from the **String** subpalette of the **Functions** palette. You might also browse through the rest of this palette to see what other functions are built in.

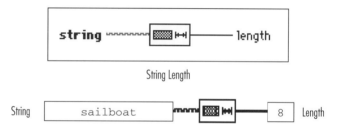

String Length

String Length

String Length returns the number of characters in a given string.

Concatenate Strings concatenates all input strings into a single output string.

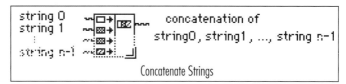

Concatenate Strings

Concatenate Strings
Function

The function appears as the icon at the left when you place it on the block diagram. You can resize the function with the Positioning tool to increase the number of inputs.

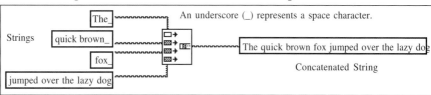

In addition to simple strings, you can also wire a one-dimensional (1D) array of strings as input; the output will be a single string containing a concatenation of strings in the array.

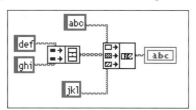

In many instances, you must convert strings to numbers or numbers to strings. The **Format Into String** and **Scan From String** functions have these capabilities (as do various other functions, but we'll concentrate on these). We'll talk about **Format Into String** now and **Scan From String** in a little while.

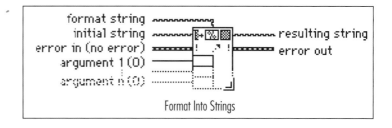

Format Into Strings

Simply put, **Format Into String** converts numeric data into string data.

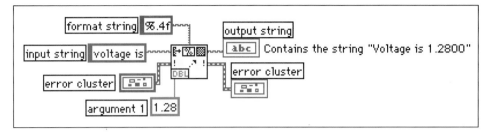

In this example, the function converts the floating-point number 1.28 to the 6-byte string "1.2800."

Format Into String formats the input **argument** (which is in numeric format) as a string, according to the format specifications in **format string**. These specifications are listed in detail in LabVIEW manuals and Online Reference (search on "String Function Overview"). The function appends the newly converted string to the input wired to **initial string**, if there is one, and outputs the results in **resulting string**. The following table gives some examples of **Format Into String's** behavior.

Initial String	Format String	Number	Resulting String
(empty)	score=%2d%%	87	score=87%
score=_	%2d%%	87	score=87%
(empty)	level=%7.2eV	0.03642	level=3.64E-2V
(empty)	%5.3f	5.67 N	5.670 N

In these examples, the underline character (_) represents a space character at the end of a string. The "%" character begins the formatting specification. Given "%number1.number2," *number 1* specifies field width of the resulting string and *number 2* specifies the precision (i.e., number of digits after the decimal point). An "f" formats the input number as a floating-point number with fractional format, "d" formats it as a decimal integer, and "e" formats it as a floating-point number with scientific notation.

Format Into String can be resized to convert multiple values to a single string simultaneously.

Get Date/Time String (found in the **Time & Dialog** palette) outputs **date string**, which contains the current date, and **time string**, which contains the current time. This function is useful for time stamping your data. Note that you don't have to wire any inputs to **Get Date/Time String**; it can use the default values.

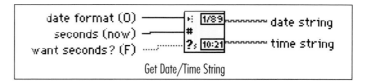

Get Date/Time String

Activity 9-1
String
Construction

It's time to practice with strings now. You will build a VI that converts a number to a string and concatenates that string with other strings to form a single output string. The VI also determines the length of the output string.

1. Build the front panel shown.

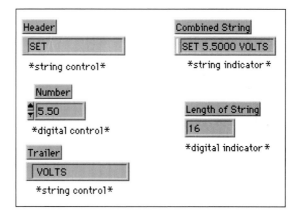

The VI will concatenate the input from the two string controls and the digital control into a single output string, which is displayed in the string indicator. The digital indicator will display the string's length.

2. Build the block diagram pictured.

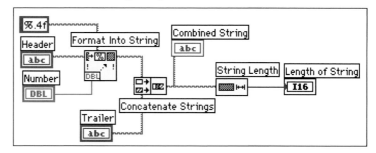

Format Into String
Function

Concatenate String
Function

String Length
Function

Format Into String (**String** palette) converts the number you specify in the <u>Number</u> digital control to a string with fractional format and four digits of precision.

Concatenate Strings function (**String** palette) combines all input strings into a single output string. To increase the number of inputs, stretch the icon using the Positioning tool.

String Length function (**String** palette) returns the number of characters in the concatenated string.

3. Return to the front panel and type text inside the two string controls and a number inside the digital control. Make sure to add spaces at the end of the header and the beginning of the trailer strings, or your output string will run together. Run the VI.

4. Save and close the VI. Name it **Build String.vi** and place it in your MYWORK directory or VI library. Do you feel like a LabVIEW expert yet? You're getting there!

■ Parsing Functions

Sometimes you will find it useful to take strings apart or convert them into numbers, and these parsing functions can help you accomplish these tasks.

String Subset accesses a particular section of a string. It returns the substring beginning at **offset** and containing **length** number of characters. Remember, the first character's offset is zero.

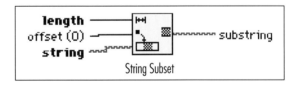

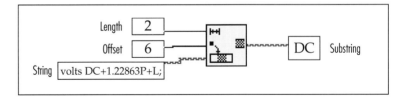

Match Pattern is used to look for a given pattern of characters in a string. It searches for and returns a **matched substring**. **Match Pattern** looks for the **regular expression** in **string**, beginning at **offset**; if it finds a match, it splits the string into three substrings. If no match is found, the match substring is empty and **offset past match** is set to –1.

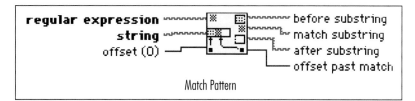

Match Pattern

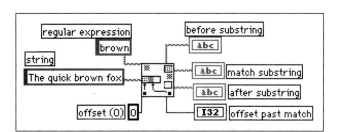

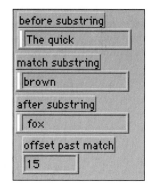

Scan From String, the "opposite" of **Format Into String**, converts a string containing valid numeric characters (0 to 9, +, -, e, E, and period) to numeric data. This function starts scanning the **input string** at **initial search location** and converts the data according to the specifications in **format string** (to learn more about the specifications, see the LabVIEW manuals or "String Function Overview" in Online Reference). **Scan From String** can be resized to convert multiple values simultaneously.

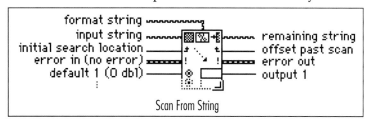

Scan From String

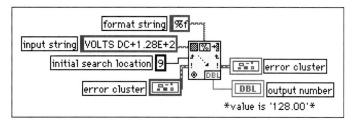

In this example, **Scan From String** converts the string "VOLTS DC+1.28E+2" to the number 128.00. It starts scanning at the eighth character of the string (which is the + in this case—remember that the first character offset is zero).

Both **Format Into String** and **Scan From String** have an **Edit Scan String** interface that you can use to create the **format string.** In this dialog box, you can specify format, precision, data type, and width of the converted value. Double click on the function or pop up on it and select **Edit Format String** to access the **Edit Scan String** or **Edit Format String** dialog box.

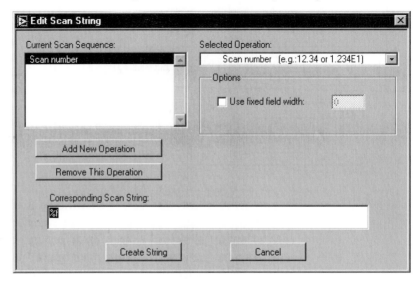

After you create the format string and click the Create String button, the dialog box creates the string constant and wires it to the **format string** input for you.

Activity 9-2
More String
Parsing

You will create a VI that parses information out of a longer string by taking a subset of a string and converting the numeric characters in that subset into a numeric value.

1. Build the front panel shown.

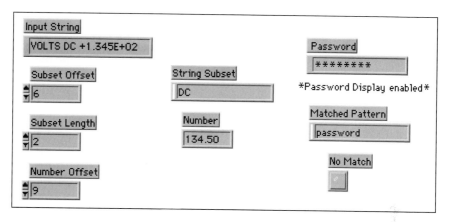

2. Set the <u>Password</u> string to display only asterisks by selecting **Password Display** from its pop-up menu.

3. Create the block diagram pictured.

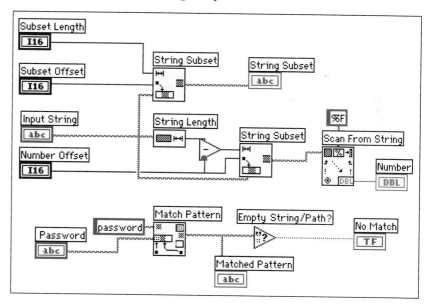

String Subset
Function

The **String Subset** function, found in the **String** palette, returns a subset of given length from the input string, according to the offset specified.

Scan From String
Function

The **Scan From String** function, located in the **String** palette, converts a string containing valid numeric characters (0 to 9, +, –, e, E, and period) to a number.

Match Pattern
Function

Match Pattern, also found in the **String** palette, compares the user's input password string to a given password string. If there is a match, it is displayed; if not, the string indicator shows an empty string.

Empty String/Path?
Function

String Length
Function

Empty String/Path?, from the **Comparison** palette, returns a Boolean TRUE if it detects an empty string from the **match substring** output of **Match Pattern**.

String Length function (**String** palette) returns the number of characters in the string.

4. Run the VI with the inputs shown. Notice that the string subset of "DC" is picked out of the input string. Also notice that the numeric part of the string was parsed out and converted to a number. You can try different control values if you want, just remember that strings, like arrays, are indexed starting at zero.

Also note how the Password string shows only "**.**." **Match Pattern** checks the input password against a password string (which in this case contains the characters, "password"), then returns a match if it finds one. If it finds no match, it returns an empty string.

5. Close the VI by selecting **Close** from the **File** menu. Save the VI in your MYWORK directory or VI library as **Parse String.vi**.

File Input/ Output

File input and output (I/O) operations retrieve information from and store information in a disk file. LabVIEW has a number of very versatile file I/O functions, as well as some simple functions that take care of almost all aspects of file I/O in one shot. We'll talk about the simple file functions in this chapter. All are located in the **File I/O** subpalette of the **Functions** palette.

■ How They Work

The File functions expect a file path input, which looks kind of like a string. A path is a specific data type that provides a platform-specific way to enter a path to a file. We talked about them briefly in Chapter 4, and they'll come up again in Chapter 14. If you don't wire a file path, the File functions will pop up a dialog box asking you to select or enter a filename. When called, the File functions open or create a file, read or write the data, and then close the file. The files created with the VIs we'll talk about now are just ordinary text files. Once you have written data to a file, you can open the file using any word processing program to see your data.

One very common application for saving data to file is to format the text file so that you can open it in a spreadsheet program. In most spreadsheets, tabs separate columns and EOL (End of Line) characters separate rows. **Write To Spreadsheet File** and **Read From Spreadsheet File** deal with files in spreadsheet format.

Write Characters To File writes a character string to a new file or appends the string to an existing file.

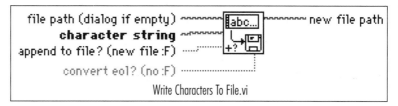
Write Characters To File.vi

Read Characters From File reads a specified number of characters from a file beginning at a specified character offset.

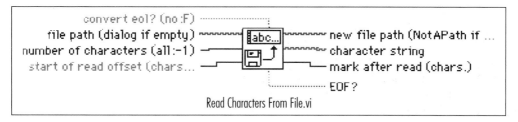
Read Characters From File.vi

Read Lines From File reads a specified number of lines from a file beginning at a specified character offset.

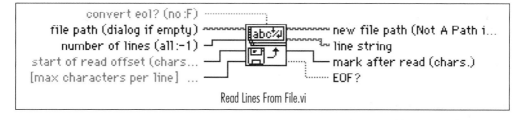
Read Lines From File.vi

Write To Spreadsheet File converts a 2D or 1D array of single-precision numbers to a text string, then writes the string to a new file or appends the string to an existing file. You can optionally transpose the data. Do not wire inputs for both 1D and 2D data (or one will be ignored). The text files created by this VI are readable by most spreadsheet applications.

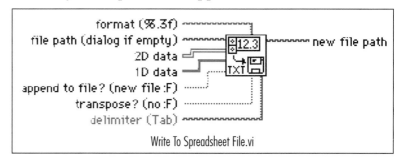

Write To Spreadsheet File.vi

Read From Spreadsheet File reads a specified number of lines or rows from a numeric text file, beginning at a specified character offset, and converts the data to a 2D single-precision array of numbers. You can optionally transpose the array. This VI will read spreadsheet files saved in text format.

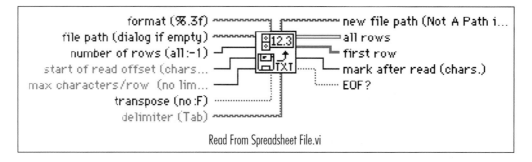

Read From Spreadsheet File.vi

These file functions are very high-level and easy to use. All are found in the **File I/O** palette. LabVIEW contains other file functions that are much more versatile but more complicated, and we'll tell you about them in Chapter 14, *Communications and Advanced File I/O.*

Read From Spreadsheet File.vi is not available in the sample software.

▪ HiQ Compatibility

Write To HiQ Text File.vi and **Read From HiQ Text File.vi**, located in the **Communications➤HiQ➤HiQ File Transfer** sub-palette, function similarly to the spreadsheet file VIs. They are

useful if you need to import data from or export data to HiQ®
analysis software.

Activity 9-3
Writing to a
Spreadsheet
File

You will modify an existing VI to save data to a new file in
ASCII format. Later you can access this file from a spreadsheet
application.

1. Open **Graph Sine Array.vi**, which you built in Chapter 8. If
you have the sample version or didn't finish building that VI,
you can find a finished version in EVERYONE\CH8.LLB. As you
recall, this VI generates two data arrays and plots them on a
graph. You will modify this VI to write the two arrays to a file in
which each column contains a data array.

2. Open the diagram of **Graph Sine Array.vi** and modify the VI
by adding the diagram code shown inside the oval.

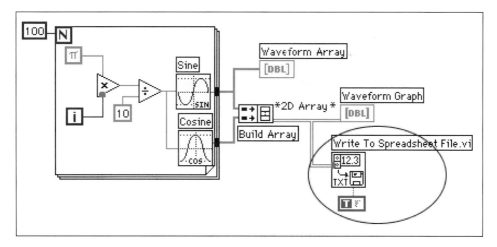

Write to Spreadsheet
File VI

Boolean Constant

The **Write To Spreadsheet File** VI (**File I/O** palette) converts the
2D array to a spreadsheet string and writes it to a file. If no path
name is specified (as in this activity), then a file dialog box will
pop up and prompt you for a file name.

The **Boolean Constant** (**Boolean** palette) controls whether or not
the 2D array is transposed before it is written to file. To change it
to TRUE, click on the constant with the Operating tool. In this
case, you do want the data transposed because the data arrays
are row specific (each row of the 2D array is a data array). Since
you want each column of the spreadsheet file to contain data for
one waveform, the 2D array must first be transposed.

3. Return to the front panel and run the VI. After the data arrays have been generated, a file dialog box will prompt you for the file name of the new file you are creating. Type in a file name and click the OK button. Remember the name and location of the file, as you will read in the data in the next exercise.

Do not attempt to write data files in VI libraries with the File I/O VIs. Doing so may overwrite your library and destroy your previous work.

4. Save the VI in your MYWORK directory or VI library, name it **Graph Sine Array to File.vi**, and close the VI.

5. Use spreadsheet software if you have it, or a simple text editor, to open and view the file you just created. You should see two columns of 100 elements each.

Activity 9-4
Reading from a Spreadsheet File

You will write a VI to read in the data from the file written in the last exercise and plot it on a graph. If you are using the sample software, you will not be able to do this activity.

1. Open a new VI and place a waveform graph on its front panel. Make sure autoscaling is on.

2. Create the little block diagram shown in the following illustration. Use the **Read From Spreadsheet File** function to bring in data and display it on the graph.

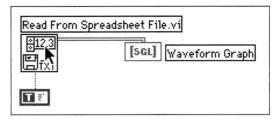

3. Using the TRUE **Boolean Constant**, you must transpose the array when you read it in, because graphs plot data by row and it has been stored in the file by column. Note that if you hadn't transposed the data in the last exercise to store it in columns in the file, you wouldn't have to transpose it back now.

4. Run the VI. Since you are not providing a file path, a dialog box will prompt you to enter a filename. Select the file you created in Activity 9-3.

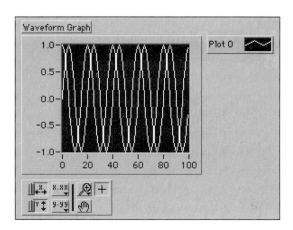

The VI will read the data from the file and plot both waveforms on the graph.

5. Save the VI in your MYWORK directory or VI library as **Read File.vi**.

Wrap it Up!

LabVIEW contains many functions for manipulating strings. These functions can be found in the **String** subpalette of the **Functions** palette. With them, you can determine string length, combine two strings, peel off a string subset, convert a string to a number (or vice versa), and many other useful things.

Using the functions in the **File I/O** subpalette of the **Functions** palette, you can write data to or read data from a disk file. **Write Characters To File** will save a text string to a file. **Read Characters From File** and **Read Lines From File** can then read that file back into LabVIEW. If you want to save an array of numbers, you must use the **Write To Spreadsheet File** function. You can read that data back in and convert it to numeric format using the **Read From Spreadsheet File** function.

Congratulations! You've mastered the fundamentals of Lab-VIEW! You have a strong foundation now and should have the background to investigate almost any LabVIEW topic that interests you. The Advanced section of this book, coming up next, will teach you about many of the very cool, more complex features LabVIEW contains that make your programming job easier, so stay tuned for more exciting LabVIEW adventures!

Additional Activities

■ Activity 9-5 **Temperatures and Time Stamps**

Build a VI that takes 50 temperature readings inside a loop, once every 0.25 second, and plots each on a chart. It also converts each reading to a string, then concatenates that string with a Tab character, a time stamp, and an End of Line character. The VI writes all of this data to a file. Save the VI as **Temperature Log.vi**.

- *Use the **Tab** and **End of Line** constants in the **String** palette*
- *Use **Concatenate Strings** to put all of the strings together.*
- *Use **Write Characters To File** to save the data.*
- *You can write data to file one line at a time, but it is much faster and more efficient to collect all of the data in one big string using shift registers and **Concatenate Strings**, and then write it all to file at one time.*

You can look at your file using any word processing program, but it should look something like this:

78.9	11:34:48
79.0	11:34:49
79.0	11:34:50

■ Activity 9-6 **Spreadsheet Exercise**

Build a VI that generates a 2D array (three rows x 100 columns) of random numbers and writes the transposed data to a spreadsheet file. The file should contain a header for each column as shown. Use the VIs from the **String** and **File I/O** palettes for this activity. Save the VI as **Spreadsheet Exercise.vi**.

Use the **Write Characters To File** VI to write the header and then the **Write to Spreadsheet File** VI to write the numerical data to the same file.

	A	B	C	
1	Waveform 1	Waveform 2	Waveform 3	— Header
2	0.281	0.078	0.874	
3	0.402	0.647	0.597	
4	0.011	0.62	0.731	
5	0.605	0.435	0.889	
6	0.049	0.259	0.78	
.				
.				
99	0.89	0.933	0.54	
100	0.864	0.312	0.343	
101	0.541	0.134	0.487	

Preconditioning Automobile Evaporative Canisters

By David Garter, Environmental Protection Agency National Vehicle and Fuels Emission Laboratory (NVFEL)

The Challenge

Developing and building an automated, flexible, state-of-the-art Canister Preconditioning System.

The Solution

Using DAQ boards and RS-232 under control of a LabVIEW program for data acquisition and control.

From left to right: Tom Rhodes, David Garter, and Manish Patel, all members of the Canister Preconditioning System (CPS) team, stand before CPS control cabinets, which contain their computer-based DAQ system.

Introduction

NVFEL needed an automated system to precondition the evaporative canisters of automobiles. Typically canisters are 0.5 to 2.0 liter containers filled with charcoal granules. The purpose of the canister in an automobile is to reduce evaporative emissions by storing vapors generated by gasoline in the fuel tank until the vapors are burned in the engine.

System Requirements

A Canister Preconditioning System (CPS) is necessary to perform the Federal Test Procedure (FTP) emission test sequences for 1996 and later vehicles, including automobiles and light trucks. Vehicles undergo test sequences at NVFEL that determine conformity with the FTP Emission Standards as required by the Code of Federal Regulations (CFR). In the CFR section regarding Vehicle Preconditioning, two different canister preconditioning test procedure sequences are specified—the three-day sequence, which tests vehicles for all sources of evaporative emissions; and the two-day sequence, which is designed to verify that vehicles sufficiently purge their evaporative canisters during the exhaust emission test.

◆   The three-day preconditioning sequence requires that the vehicle canister be purged with ambient air of controlled humidity and then loaded to 1.5 times its capacity with a 50/50 mixture of butane and nitrogen gas, by volume.

◆   The two-day preconditioning sequence requires that the vehicle canister be loaded with a 50/50 mixture (by volume) of butane and nitrogen gas until breakthrough, which is determined by measuring the weight gain of an auxiliary canister connected to the outlet of the vehicle canister.

System Design

Each CPS unit controls purge air, butane gas, and nitrogen gas by means of a mass flow controller (MFC) and solenoid valve pairs. Digital relays control the solenoid valves. Each CPS is equipped with an explosion-proof, digital balance for measuring breakthrough. All CPS units receive ambient temperature and dew-point data from a central dew-point hygrometer. Electrical power for the MFCs, solenoid valves, and digital relay backplane is provided by three separate DC power supplies.

We selected LabVIEW software running on a Power Macintosh computer as the data acquisition and control system for the CPS. A National Instruments NB-MIO-16 multifunction board reads analog flow signals from the mass flow controllers and switches digital relays that control the solenoid valves. An NB-AO-6 analog output board establishes analog set point inputs for the mass flow controllers. The balance is monitored and controlled by RS-232 through the modem port. The central dew-point hygrometer sends dew-point and ambient temperatures via an RS-232 serial broadcaster to the serial port on the computer. We use rail-mounted, clamping-type, screw terminal blocks to connect the components to each other. The terminal blocks for the analog input signals have integral fuses and surge-suppressing diodes to protect the boards and the computer from overvoltage and current surge.

> **V**isitors from the automotive industry have asked where we bought this system. They are surprised and impressed when we reply, "We built it in-house!"

System Performance

Using the LabVIEW graphical user interface (GUI), our programmer, Tom Rhodes, developed a custom virtual instrument (VI) that incorporated several displays to monitor the CPS process. The ambient conditions are displayed continuously; the displays that monitor the purging and loading processes are displayed only when the respective process is operating. Displays are updated once per second. At the LabVIEW diagram level, test action is best described as a "three-ring circus." In the main ring, data acquisition and process control is taking place. In the second ring, weight measurements are requested and received from the balance. In the third ring, ambient temperature and humidity are extracted from the free running dew-point hygrometer.

Prior to initiating the preconditioning sequence, the CPS operator enters vehicle information and selects the type of preconditioning. Typically, it takes from three to 13 hours to purge and

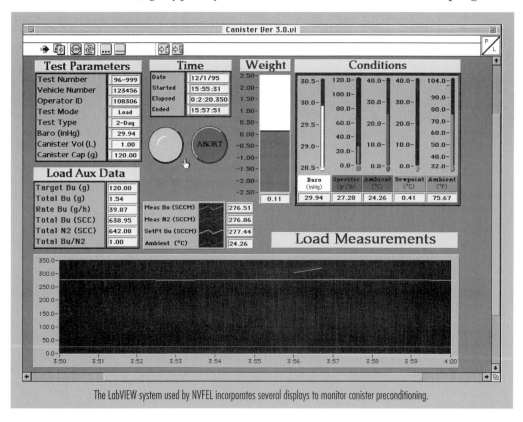

The LabVIEW system used by NVFEL incorporates several displays to monitor canister preconditioning.

load a canister. Upon initiation of the preconditioning sequence, the CPS VI creates an Excel spreadsheet file and writes the data to disk once per minute. When the sequence is completed, the CPS VI creates and automatically prints a summary report of the data. Quality of the data is reported by a subVI that monitors for faults and out-of-tolerance events. Any faults and out-of-tolerance events found are displayed on the front panel of the CPS VI and are listed in the spreadsheet file and on the summary report.

Results

The CPS installation at NVFEL will have four stations when it is completed. NVFEL staff, including technicians, engineers, and management, are very impressed with the CPS. Visitors from the automotive industry have asked where we bought this system. They are surprised and impressed when we reply, "We built it in-house!"

The real benefits of this system are its expandability and flexibility, so that we can quickly accommodate the inevitable changes needed for new test programs and regulations. The National Instruments DAQ boards in the system have more channel capacity than the basic system required, and the Power Macintosh has room for one more board if necessary. This hardware flexibility complements the inherent flexibility of LabVIEW; we can make additions and changes to the system with relative ease.

At NVFEL, we are now developing a modified CPS to precondition canisters for special tests on in-use vehicles. This system requires heated purge air, additional temperature measurement, and gasoline vapor loading to replace butane loading. We can add these changes to the basic CPS with minimal hardware and software modifications. FTP canister preconditioning regulations are likely to change in the near future. With our LabVIEW-based CPS, we are confident that we can easily accommodate future changes.

A Bit About the Advanced Section

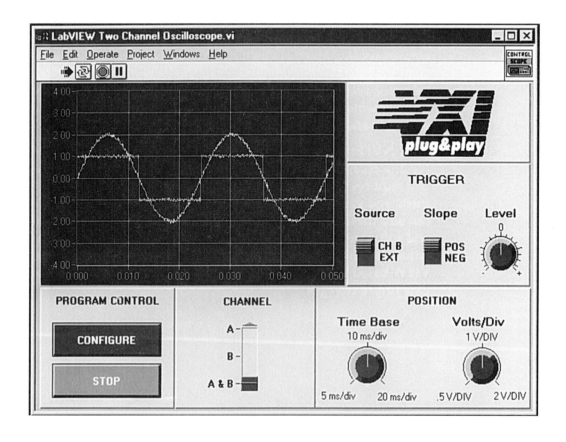

If you've just completed the Fundamentals section of this book, congratulations! You've learned the basics of creating a virtual instrument in LabVIEW—choosing front panel controls and indicators, wiring the block diagram, using structures such as the While Loop or Case Structure, and making simple, usable VIs. Are we having fun yet? Now that you're getting a grasp on graphical dataflow programming, it's time to delve further into some of the more powerful functions and features LabVIEW offers.

The advanced section of this book, Chapters 10–15, expands on most of the remaining palette functions, and focuses on techniques and tools for writing better programs. It also teaches you the basics of data acquisition from LabVIEW using plug-in data acquisition boards. You won't find as many walk-through exercises in this section—we've oriented the material more towards informing ("Wow, I didn't know you could do that in LabVIEW!") and writing applications ("How do I acquire data into my PC with LabVIEW?"). Because LabVIEW's functions and features are so extensive (the boxed manual set is close to $^3/_4$-foot long), you should find this section to be a valuable overview of what you can do with LabVIEW—leaving many of the gory details to the manuals.

Two reminders before we start:

◆ *Don't forget that all the examples and activities in this section require the full version of LabVIEW; they won't work with the sample software that comes with this book.*

◆ Feel free to skip sections that you won't be needing. Unlike the first section of the book, this material doesn't build on itself in most areas.

OVERVIEW

This chapter will give you a deeper look at what we touched on in Chapter 2. Data acquisition and instrument control are some of the main reasons why people use LabVIEW—turning their computers into virtual instruments by gathering data from the real world. We'll take a look at the various options you have for taking or making data, including the use of existing instruments: serial communications, GPIB interfaces, and plug-in DAQ cards. You'll also learn some signal theory and about the kind of hardware used for these systems.

GOALS

- Finally find the meaning of all those acronyms that everyone thinks you know
- Become familiar with the hardware options you have for acquiring or sending data
- Learn some signal theory, including the classification of signals, measurement types, signal conditioning, and sampling
- Get some hints on picking and installing a DAQ board that suits your needs
- Discover the GPIB interface
- Investigate serial communications

KEY TERMS

- DAQ
- GPIB
- serial
- signals
- analog
- digital

- frequency
- grounded signal
- floating signal
- ground reference
- sampling rate
- Nyquist frequency

- signal conditioning
- NI-DAQ
- differential measurement
- single-ended measurement
- instrument driver

Getting Data into and out of Your Computer: Data Acquisition and Instrument Control

10

Acronyms Unlimited

"Let's go ahead and apply VHDL tools to designing the PCMCIA interface using that new MXR standard."

Admit it, how many times has someone mentioned an acronym in a technical discussion and everyone pretends to understand because nobody wants to ask what it stands for and look ignorant? Well, here's your chance to see what all the acronyms in this chapter stand for and where they came from. To our knowledge, the following list is the only one of its kind in print, so use it to put your colleagues to the test!

AC: <u>A</u>lternating <u>C</u>urrent. This acronym originally referred to how a device was powered, AC being the plug in the wall and DC (direct current) being batteries. Now it's used more generally to refer to any kind of signal (not just current) that varies "rapidly" (whatever you want that to mean) with time.

ADC: <u>A</u>nalog-to-<u>D</u>igital <u>C</u>onversion. This conversion takes a real-world analog signal, and converts it to a digital form (as a series of bits) that the computer can understand. Very often just abbreviated as **A/D**. Many times the chip used to perform this operation is called "the ADC."

DAQ: <u>D</u>ata Ac<u>Q</u>uisition. This little word just refers to collecting data in general, usually by performing an A/D conversion. Its meaning is sometimes expanded to include, as in this book, data generation. Don't confuse DAQ with **DAC**, which sound the same when pronounced in English. (**DAC**, or **D/A**, stands for <u>D</u>igital-to-<u>A</u>nalog <u>C</u>onversion, usually referring to the chip that does this).

DC: <u>D</u>irect <u>C</u>urrent. The opposite of AC. No longer refers to current specifically. Sometimes people use DC to mean a constant signal of zero frequency. In other cases, such as in DAQ terminology, DC also refers to a very low-frequency signal, such as something that varies less than once a second. Obviously the border between an AC and DC signal is subjective.

DMA: <u>D</u>irect <u>M</u>emory <u>A</u>ccess. You can use plug-in DAQ boards that have built-in DMA, or buy a separate DMA board. DMA lets you throw the data you're acquiring directly into the computer's RAM (there we go, another acronym), thus increasing data transfer speed. Without DMA you still acquire data into memory, but it takes more steps and more time because the software has to direct it there.

GPIB: <u>G</u>eneral <u>P</u>urpose <u>I</u>nterface <u>B</u>us. Also known as HP-IB (Hewlett-Packard Interface Bus) and IEEE 488.2 bus (Institute of Electrical and Electronic Engineers standard 488.2), it has become the world standard for almost any instrument to communicate with a computer. Originally developed by Hewlett-Packard in the 1960s to allow their instruments to be programmed in BASIC with a PC, now IEEE has helped define this bus with strict hardware protocols that ensure uniformity across instruments.

MXI: <u>M</u>ultisystem e<u>X</u>tension <u>I</u>nterface is a standard for connecting the VXI mainframe chassis and conventional computers in a similar fashion to GPIB.

RS-232: <u>R</u>ecommended <u>S</u>tandard #232. A standard proposed by the Instrument Society of America for serial communications.

It's used interchangeably with the term "serial communication," although serial communication more generally refers to communicating one bit at a time. A few other standards you might see are RS-485, RS-422, and RS-423.

SCXI: Signal Conditioning eXtensions for Instrumentation. A high-performance signal conditioning system devised by National Instruments, using an external *chassis* that contains I/O modules for signal conditioning, multiplexing, etc. The chassis is wired into a DAQ board in the PC.

SISTA: Sometimes I'm Sick of These Acronyms. Just kidding, sort of.

VISA: Virtual Instrument Standard Architecture, a recent driver software architecture developed by National Instruments. Its purpose is to try to unify instrumentation software standards, whether the instrument uses GPIB, DAQ, VXI, or RS-232.

VXI: Talk about acronym abuse; this is an acronym for an acronym: VME eXtensions for Instrumentation. VME stands for Versa-Modular Eurocard. VXI is a very high-performance system for instrumentation. You can buy VXI instruments that are small modules (instead of the regular big instrument with a front panel) that plug into a VXI chassis. The VXI chassis often has an embedded computer motherboard, so you don't have to use an external PC. VXI is an open industry standard, which means that companies besides National Instruments support it.

How to Connect Your Computer to the Real World

You've got a good PC, you're excited about using LabVIEW, and you're ready to build an application that will do something *outside* your computer. Maybe you need to monitor the electroencephalogram (EEG, brain waves) of some research subjects, or plot force-versus-displacement curves to test the strength of some new plastic. Or perhaps you need something more elaborate, like a whole process control system for a semiconductor manufacturing facility, and you need to provide control signals to the plant.

Whatever your application, you need a way to gather data into your computer. Several solutions are usually possible, but the best solution will decidedly depend on what trade-offs you can afford. Before you run out and buy hardware, you need to analyze and understand what kind of signals you're trying to measure (in some cases such as serial communication, you may not even need any additional hardware). The following diagram shows you the most common ways you can get data into your computer.

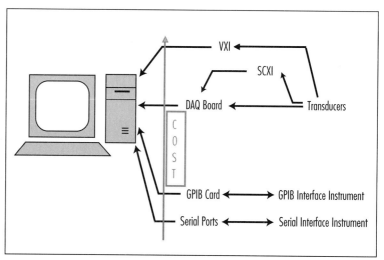

One of the first things you should decide when designing your data acquisition system is whether you're going to use traditional external instruments, such as a multimeter. Do you want a "physical" instrument to do some of the data acquisition and processing, or do you want to write a LabVIEW virtual instrument that will do everything via a plug-in DAQ board?

Issues such as cost, scheduling, and flexibility will play a part in this decision. For example, if you wanted to read some low-voltage data, you might use a simple plug-in DAQ card. You could write a VI that is very specific to your application and creates the exact virtual instrument you need. And generally speaking, it's much cheaper to buy a plug-in DAQ card than a stand-alone instrument. On the other hand, if you already have an existing instrument you want to use (a multimeter for example), it might be cheaper to use your existing meter to acquire the voltages and send the data via the GPIB bus to the computer, if your meter has a GPIB interface.

Sometimes there's no question about using an external instrument. If you need to perform mass spectroscopy, for example, we don't know of any mass spectrometer available as a card that plugs into your PC!

One more factor that might influence your decision to use an external instrument is the availability of an *instrument driver*. Contrary to a popular perception, an instrument driver is not a "virtual" copy of the instrument's front panel in LabVIEW; in fact, most instrument drivers look pretty boring. An instrument driver usually consists of a collection of subVIs, each of which sends a specific command or group of commands to the instrument. These subVIs have all the low-level code written in them so you can quickly put together an application just by using these subVIs. National Instruments has a database of instrument drivers for hundreds of popular instruments that use the GPIB, serial, or VXI interface. You can get these for free from NI! Other instrument drivers are often available from third-party vendors (such as Alliance members) or the instrument manufacturer itself.

You may keep seeing references to the term "Alliance member." Several companies offer software and hardware products, as well as services (such as managing LabVIEW projects), that closely relate to or use LabVIEW. National Instruments has developed an Alliance Program for these companies that work with National Instruments' products. You can request a catalog of the Alliance member companies in your area from National Instruments.

Finally, if you are planning to buy or already have bought a plug-in DAQ card, make good use of it! Most people don't fully realize the potential their computer has when a DAQ card is running and humming inside it. Need to view an AC signal? Wait, don't go borrow that oscilloscope; just look at your signal right on your screen with one of the DAQ example VIs that comes with LabVIEW! With one DAQ card, you can create as many virtual instruments as you need. And, when it's time to upgrade your plug-in card or move to another platform, you may not even need to change a thing in your block diagram. That's right, LabVIEW's DAQ VIs work (with a few exceptions) *independently of whatever board you have in your computer.*

The rest of this chapter is essentially divided into two parts: DAQ and instrument control. The DAQ sections will cover hardware considerations such as signal theory, types of hardware, and configuration instructions. The instrument control section will give you some more details about serial and GPIB protocols.

Signals 101

Before we delve completely into data acquisition, we want to talk a little about *what* you'll be acquiring. A signal is simply any physical quantity whose magnitude and variation with time (or occasionally some other variable) contain information.

■ Timing is Everything

Although it may not be obvious at first, *time* is usually the most critical aspect of almost any measurement. Whether we want to observe how an engine's temperature changes over time, see what a filtered audio signal looks like, or close some valves when a gas mixture reaches its optimum ratio, time is the deciding factor in data acquisition and control. We want to know not just *what* happens, but *when*. Even so-called "DC" signals are not really steady-state; if they never changed over time, we would know their value, so why would we want to measure them?

Timing is important in designing your data acquisition software for a couple of reasons. First, you need to figure out how to set the *sampling rate*, or how often your computer takes a measurement. Secondly, you need to be able to allocate processor time to other tasks such as file I/O.

If you only need to read data a few times a second or less and you don't need much timing accuracy between the sample points, you can probably use LabVIEW's timing functions to control the sampling rate "directly" from your program, by putting a **Wait** function in a VI that acquires one point, for example, in a loop. For more precise applications or AC signals, you'll let the hardware and low-level software set the sampling rate by configuring your measurement on the DAQ board accordingly. We'll discuss these details along with some examples in Chapter 11.

■ Signal Classification

Let's say you want to take a measurement. For signal conditioning hardware to condition a signal, or for the DAQ board to measure it directly, you must first convert it to an electrical signal such as voltage or current. A *transducer* performs this conversion. For example, if you wish to measure temperature, you must somehow represent temperature as a voltage that the DAQ board can read. A variety of temperature transducers exist that use some physical properties of heat and materials to convert the temperature to an electrical signal.

Once the physical quantity is in an electrical signal form, you can then measure the signals to extract some type of useful information conveyed through one or more of the following parameters: state, rate, level, shape, and frequency content.

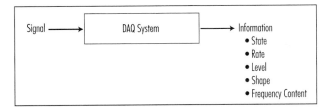

Strictly speaking, all signals are analog time-varying signals. However, to discuss signal measurement methods, you should classify a given signal as one of five signal types. Classify the signal by the way it conveys the needed information. First, you can classify any signal as *analog* or *digital*. A digital, or binary, signal has only two possible discrete levels—a high (on) level or low (off) level. An analog signal, on the other hand, contains information in the continuous variation of the signal with respect to time.

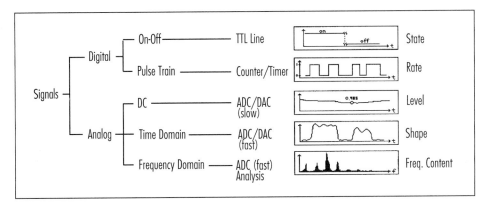

Engineers often classify digital signals into two more types and analog signals into three more types. The two digital signal types are the *on-off* signal and the *pulse train* signal. The three analog signal types are the *DC* signal, the *time domain* (or *AC*) signal, and the *frequency domain* signal. The two digital and three analog signal types are unique in the information each conveys. You will see that the five signal types closely parallel the five basic types of signal information: state, rate, level, shape, and frequency content.

Digital Signals

The first type of digital signal is the on-off signal. An on-off signal conveys information concerning the digital state of the signal. Therefore, the instrument needed to measure this signal type is a simple digital state detector. The output of a transistor-transistor logic (TTL) switch is an example of a digital on-off signal.

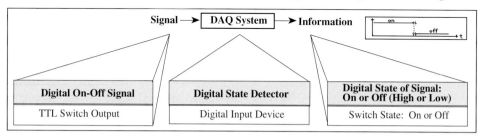

Signal → DAQ System → Information		
Digital On-Off Signal	**Digital State Detector**	**Digital State of Signal: On or Off (High or Low)**
TTL Switch Output	Digital Input Device	Switch State: On or Off

The second type of digital signal is the pulse train signal. This signal consists of a series of state transitions. Information is contained in the number of state transitions occurring, the rate at which the transitions occur, or the time between one or more state transitions. The output signal of an optical encoder mounted on the shaft of a motor is an example of a digital pulse train signal. In some instances, devices require a digital input for operation. For example, a stepper motor requires a series of digital pulses as an input to control the motor position and speed.

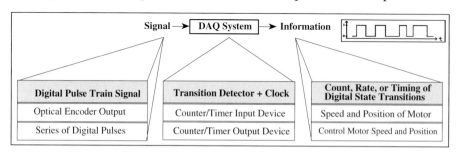

Signal → DAQ System → Information		
Digital Pulse Train Signal	**Transition Detector + Clock**	**Count, Rate, or Timing of Digital State Transitions**
Optical Encoder Output	Counter/Timer Input Device	Speed and Position of Motor
Series of Digital Pulses	Counter/Timer Output Device	Control Motor Speed and Position

Analog DC Signals

Analog DC signals are static or slowly varying analog signals. The most important characteristic of the DC signal is that the level, or amplitude, of the signal at a given instant conveys information of interest. Because the analog DC signal varies slowly, the accuracy of the measured level is of more concern than the time or rate at which you take the measurement. The instrument or plug-in DAQ board that measures DC signals operates as an analog-to-digital converter (ADC), which converts the analog electrical signal into a digital value for the computer to interpret.

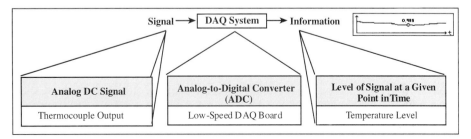

As shown in the following figure, common examples of DC signals include temperature, battery voltage, flow rate, pressure, strain gauge output, and fluid level. In each case, the DAQ system monitors the signal and returns a single value indicating the magnitude of the signal at that time. Therefore, these signals are often displayed through LabVIEW indicators such as meters, gauges, strip charts, and numerical readouts.

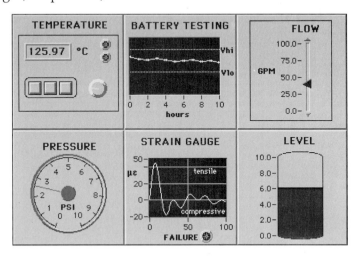

Your DAQ system should meet the following specifications when acquiring analog DC signals:

- *Accuracy/resolution—accurately measure the signal level*

- *Low bandwidth—sample the signal at low rates (software timing should be sufficient)*

Analog Time Domain Signals

Analog time domain signals differ from other signals in that they convey useful information not only in the signal level, but also in how this level varies with time. When measuring a time domain signal, often referred to as a waveform, you are interested in some characteristics of the waveform shape, such as slope, locations and shapes of peaks, and so on.

To measure the shape of a time domain signal, you must take a precisely timed sequence of individual amplitude measurements, or points. These measurements must be taken at a rate that will adequately reproduce the shape of the waveform. Also, the series of measurements should start at the proper time, to guarantee that the useful part of the signal is acquired. Therefore, the instrument or plug-in DAQ board that measures time domain signals consists of an ADC, a sample clock, and a trigger. A sample clock accurately times the occurrence of each analog-to-digital (A/D) conversion. To ensure that the desired portion of the signal is acquired, the trigger starts the measurement at the proper time according to some external condition.

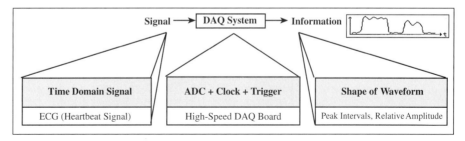

Your DAQ system should meet the following specifications when acquiring analog time domain signals:

◆ *Higher bandwidths—sample the signal at high rates*

◆ *Accurate sample clock—sample the signal at precise intervals (hardware timing needed)*

◆ *Triggering—start taking the measurements at a precise time*

There are an unlimited number of different time domain signals, a few of which are shown in the following figure. What they all have in common is that the shape of the waveform (level versus time) is the main feature of interest.

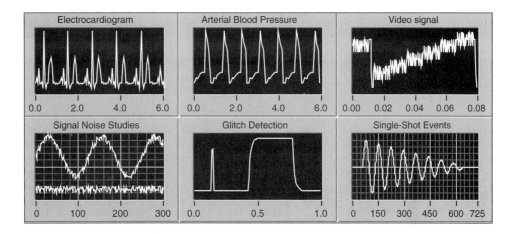

Analog Frequency Domain Signals

Analog frequency domain signals are similar to time domain signals because they also convey information on how the signals vary with time. However, the information extracted from a frequency domain signal is based on the signal frequency content, as opposed to the shape or time-varying characteristics of the waveform.

Like the time domain signal, the instrument used to measure a frequency domain signal must include an ADC, a sample clock, and a trigger to accurately capture the waveform. Additionally, the instrument must include the necessary analysis capability to extract frequency information from the signal. You can perform this type of digital signal processing (DSP) using application software or special DSP hardware designed to analyze the signal quickly and efficiently.

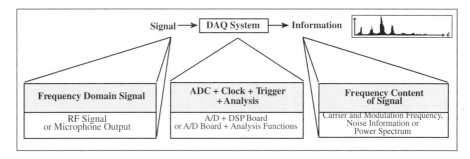

Your DAQ system should meet the following specifications when acquiring analog frequency domain signals:

◆ *Higher bandwidths—sample the signal at high rates*

◆ *Accurate sample clock—sample the signal at precise intervals (hardware timing needed)*

◆ *Triggering—start taking the measurements at a precise time*

◆ *Analysis functions—convert time information to frequency information*

The next figure shows a few examples of frequency domain signals. Each example includes a graph of the originally measured signal as it varies with respect to time, as well as a graph of the signal frequency spectrum. While you can analyze any signal in the frequency domain, certain signals and application areas lend themselves especially to this type of analysis. Among these areas are speech and acoustics, geophysical signals, vibration, and system transfer functions.

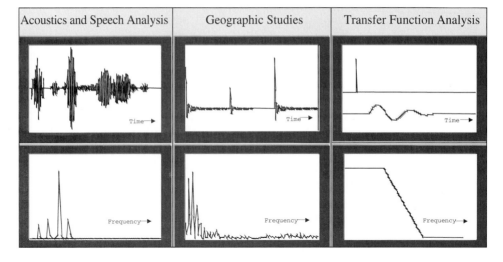

One Signal—Five Measurement Perspectives

The five classifications of signals presented in this section are not mutually exclusive. A particular signal may convey more than one type of information. Therefore, a signal can be classified as more than one type of signal, and thus you can measure it in more than one way. You can use simpler measurement techniques with the digital on-off, pulse train, and DC signals because they are just simpler cases of the analog time domain signals.

You can measure the same signal with different types of systems, ranging from a simple digital input board to a sophisticated frequency analysis system. The measurement technique you choose depends on the information you want to extract from the signal. Look at the following figure. It demonstrates how one signal—a series of voltage pulses—can provide information for all five signal classes.

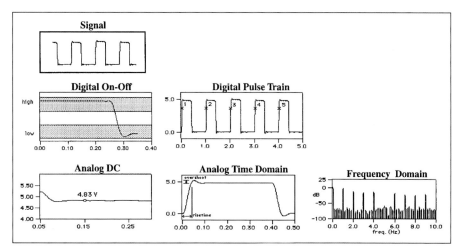

Activity 10-1

Classify the following signals into one of the five signal types described earlier—from the perspective of data acquisition. In some cases, a signal can have more than one classification. Choose the type that you think best matches the signal. Circle the number on the left margin as follows:

1 *Analog DC*

2 *Analog AC*

3 *Digital on/off*

4 *Digital pulse/counter*

5 *Frequency*

1 2 3 4 5   Voltage level of a battery

1 2 3 4 5   State of a solid-state relay

1 2 3 4 5   Data at your PC's parallel port during printing

1 2 3 4 5   Glitch or spike noise in a power source

1 2 3 4 5   Transfer function of a digital filter

1   2   3   4   5   EEG, brain waves

1   2   3   4   5   Speech through a microphone

1   2   3   4   5   Number of machine cycles in an hour

1   2   3   4   5   Absolute pressure in an engine cylinder

1   2   3   4   5   Pressure in the atmosphere

You'll find the answers at the end of the chapter.

◆   ◆   ◆

▇ Transducers

When you're setting up your DAQ system, remember that ultimately everything you're going to measure will have to become an electrical voltage or current. The way you convert measurable phenomena such as temperature, force, sound, light, stupidity, etc., to an electrical signal is by using a *transducer*. The following table lists some common transducers used to convert physical phenomena into a measurable quantity.

Phenomena	Transducer
Temperature	Thermocouple
	Resistance temperature detector (RTD)
	Thermistor
	Integrated circuit sensor
Light	Photomultiplier tube
	Photoconductive cell
Sound	Microphone
Force and pressure	Strain gauge
	Piezoelectric transducer
	Load cell
Position (displacement)	Potentiometer
	Linear voltage differential transformer (LVDT)
	Optical encoder
Fluid flow	Differential pressure flowmeter
	Rotational flowmeter
	Ultrasonic flowmeter
pH	pH electrode

▇ Signal Conditioning

All right, so now that you've figured out what kind of signals you need to acquire, you can just plug the output of your transducers directly into the DAQ board, right? In perhaps 50% of the cases or more, NO! We may not always be aware of it, but we live in a very electrically noisy world. By the time your signal makes it to the DAQ board, it may have picked up so much noise or have so many other problems that it renders your measurement useless.

You usually need to perform some type of signal conditioning on analog signals that represent physical phenomena. What is signal conditioning, anyway? Simply put, it is a manipulation of your signal to prepare it for digitizing at the DAQ board. Your signal has to arrive as clean as possible, within the voltage (usually ±5 or 0 to 10 V) and current (usually 20 mA) limits of your DAQ board, with enough precision for your application. It's hard to be more specific unless you can specify what kind of transducers you're going to use. For example, signal conditioning for audio data from a microphone may involve nothing more than grounding the system properly and perhaps using a lowpass filter. On the other hand, if you want to measure ionization levels in a plasma chamber sitting at 800 V, and you don't want to fry your computer, you'd need to provide some more complex circuitry that includes isolation amplifiers with a step-down gain.

For signals that need special conditioning, or for systems that have very many signals, National Instruments devised *SCXI* (Signal Conditioning eXtensions for Instrumentation). An SCXI system provides a chassis where modular units can be inserted to build a custom system. These modular units include analog input multiplexers, analog output boards, "blank" breadboards, signal conditioning modules for thermocouples, etc. For more information on SCXI, see the National Instruments catalog. Remember that you don't necessarily need SCXI to do signal conditioning.

The chassis to the right of the computer in the picture is an SCXI system. Notice the individual modules that plug in to the chassis. LabVIEW, of course, is running on the computer. The application shown is for an automotive testing application that acquires data through the SCXI system.

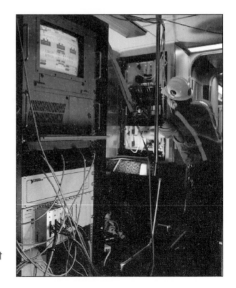

SCXI systems can also be used in industrial rack-mount systems, as shown here.

Some common types of signal conditioning are:

◆ *Amplification*

◆ *Transducer excitation*

◆ *Linearization*

◆ *Isolation*

◆ *Filtering*

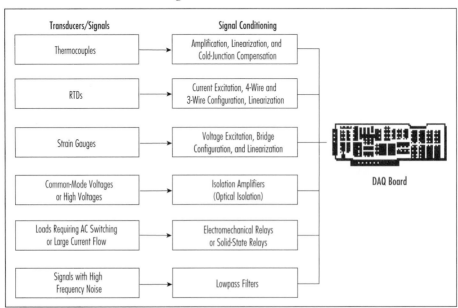

■ Finding A Common Ground

Some physical properties are absolute: luminosity or mass, for example (I know, I know, the physicists won't agree—things act weird when you approach the speed of light, and then almost nothing is absolute). But voltage is decidedly *not* absolute; it always requires a reference to be meaningful. Voltage is always the measure of a potential *difference* between two bodies. One of these bodies is usually picked to be the reference and is assigned "0 V." So to talk about a 3.47 V signal really means nothing unless we know with respect to what reference. If you've noticed, though, often a reference isn't specified. That's because the 0 V reference is usually the famous *ground*. Herein lies the source of much confusion, because "ground" is used in different contexts to refer to different reference potentials.

Earth ground refers to the potential of the earth below your feet. Most electrical outlets have a prong that connects to the earth ground, which is also usually wired into the building electrical system for safety. Many instruments also are "grounded" to this earth ground, so often you'll hear the term *system ground*. The main reason for this type of grounding is safety, and not because it is used as a reference potential. In fact, you can bet that no two sources that are connected to the earth ground are at the same reference level; the difference between them can easily be several hundred millivolts. Conclusion: we're usually not talking about earth, or safety ground, when we need to specify a reference voltage.

Reference ground, sometimes called a return path or signal common, is usually the reference potential of interest. The common ground may or may not be wired to earth ground. The point is that many instruments, devices, and signal sources provide a reference (the negative terminal, common terminal, etc.), which gives meaning to the voltages we are measuring.

The following ground symbols are used in this book when you see wiring diagrams. Be aware, however, that you will find these same symbols used inconsistently among engineers.

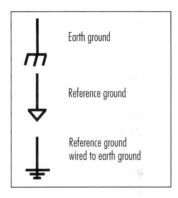

The DAQ boards in your computer are also expecting to measure voltage with respect to some reference. What reference should the DAQ board use? You have your choice, which will depend on the kind of signal source you're connecting. Signals can be classified into two broad categories:

◆ *Grounded*

◆ *Floating*

Let's examine these categories a bit further:

■ Grounded Signal Source

A grounded source is one in which the voltage signals are referenced to a system ground, such as earth or building ground. Because they use the system ground, they share a common ground with the DAQ board. The most common examples of grounded sources are devices that plug into the building ground through wall outlets, such as signal generators and power supplies.

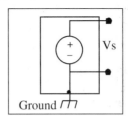

■ Floating Signal Source

A floating source is a source in which the voltage signal is not referenced to any common ground, such as earth or building ground. Some common examples of floating signal sources are batteries, thermocouples, transformers, and isolation amplifiers. Notice, as shown in the following figure, that neither terminal of the source is connected to the electrical outlet ground. Thus, each terminal is independent of the system ground.

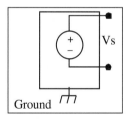

■ Measuring Differences

To measure your signal, you can almost always configure your DAQ board to make measurements that fall into one of these three categories:

◆ *Differential*

◆ *Referenced single-ended (RSE)*

◆ *Nonreferenced single-ended (NRSE)*

■ Differential Measurement System

In a differential measurement system, neither input is connected to a fixed reference such as earth or building ground. Most DAQ boards with instrumentation amplifiers[1] can be configured as differential measurement systems. The figure at right depicts the eight-channel differential

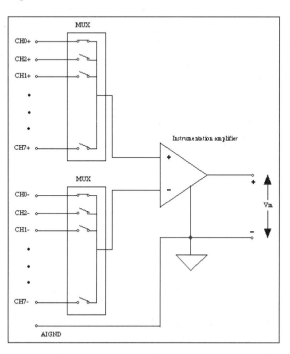

---

1. An *instrumentation amplifier* is a special kind of circuit (usually embedded in a chip) whose output voltage with respect to ground is proportional to the difference between the voltages at its two inputs.

measurement system used in the MIO-16 series boards. Analog multiplexers increase the number of measurement channels while still using a single instrumentation amplifier. For this board, the pin labeled AIGND (the analog input ground) is the measurement system ground.

Before we discuss single-ended systems, it is worth noting that SCXI systems always use the popular differential measurement system, while *most* plug-in DAQ boards give you a choice.

■ For Nerds Only

An ideal differential measurement system reads only the potential *difference* between its two terminals—the (+) and (-) inputs. Any voltage present at the instrumentation amplifier inputs with respect to the amplifier ground is referred to as a common-mode voltage. An ideal differential measurement system completely rejects (does not measure) common-mode voltage. Practical devices, however, limit this ability to reject the common-mode voltage. The common-mode voltage range limits the allowable voltage swing on each input with respect to the measurement system ground. Violating this constraint results not only in measurement error but also in possible damage to components on the board. The common-mode voltage range specification quantifies the ability of a DAQ board, operating in differential mode, to reject the common-mode voltage signal.

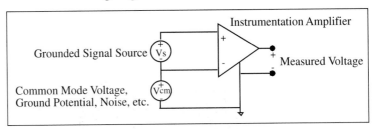

You measure the common-mode voltage, V_{cm}, with respect to the DAQ board ground, and calculate it using the following formula:

$$V_{cm} = \frac{V^+ + V^-}{2}$$

Where

V+ = Voltage at the *noninverting* terminal of the measurement system with respect to the instrumentation amplifier ground.

V- = Voltage at the *inverting* terminal of the measurement system with respect to the instrumentation amplifier ground.

You cannot have an arbitrarily high common-mode voltage when measuring with a DAQ board. All plug-in DAQ boards specify a maximum working voltage (MWV) that is the maximum common-mode voltage the board can tolerate and still make accurate measurements.

Referenced Single-Ended Measurement System

A referenced single-ended (RSE) measurement system, also called a grounded measurement system, is similar to a grounded signal source, in that the measurement is made with respect to earth ground. The following figure depicts a 16-channel RSE measurement system.

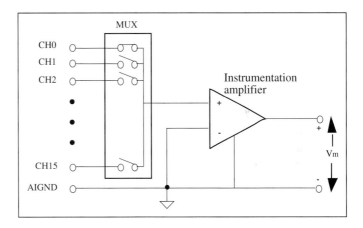

Be careful when connecting a voltage or current source to your DAQ board. Make sure the signal source will not exceed the maximum voltage or current that the DAQ board can handle. Otherwise, you could damage both the board and your computer.

NRSE Measurement System

DAQ boards often use a variant of the RSE measurement technique, known as the *nonreferenced single-ended* (NRSE) measurement system. In an NRSE measurement system, all measurements are made with respect to a common reference ground, but the voltage at this reference can vary with respect to the measurement system ground. The following figure depicts an NRSE measurement system where AISENSE is the common reference for taking measurements and AIGND is the system ground.

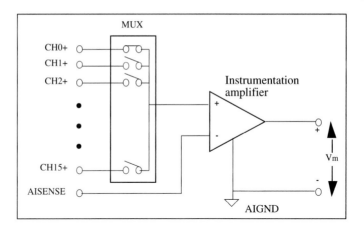

Incidentally, your measurement system is determined by how you configure your DAQ board. Most boards from National Instruments can be configured for differential, RSE, or NRSE from a software utility called NI-DAQ. Some of their older boards also have to be configured at the board by placing jumpers in certain position. When you configure a particular DAQ board for a particular measurement system type, all your input channels will follow that measurement type. You should note that you can't change this from LabVIEW—you have to decide ahead of time what kind of measurement you're making.

The general guideline for deciding which measurement system to pick is to measure grounded signal sources with a differential or NRSE system, and floating sources with an RSE system. The hazards of using an RSE system with a grounded signal source is the introduction of *ground loops*, a possible source of measurement error. Similarly, using a differential or NRSE system to measure a floating source will very likely be plagued by *bias currents*, which cause the input voltage to drift out of the range of the DAQ board (although you can correct this problem by placing bias resistors from the inputs to ground).

The following table summarizes the measurement configurations for each signal type.

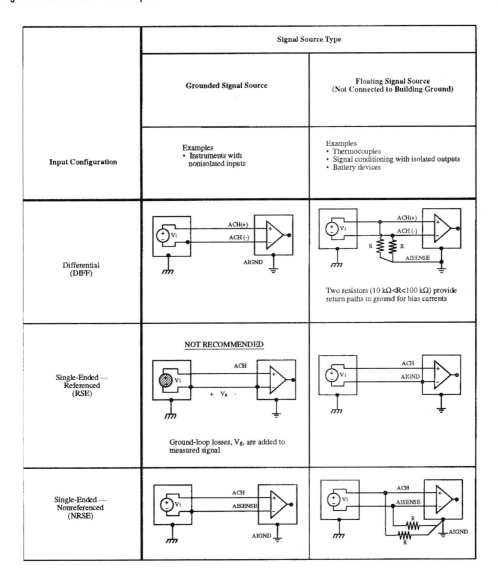

	Signal Source Type	
Input Configuration	**Grounded Signal Source**	**Floating Signal Source** **(Not Connected to Building Ground)**
	Examples • Instruments with nonisolated inputs	Examples • Thermocouples • Signal conditioning with isolated outputs • Battery devices
Differential (DIFF)	ACH(+) ACH (-) AIGND	ACH(+) ACH (-) R R AISENSE Two resistors (10 kΩ<R<100 kΩ) provide return paths to ground for bias currents
Single-Ended — Referenced (RSE)	NOT RECOMMENDED ACH + V₉ - Ground-loop losses, V₉, are added to measured signal	ACH AIGND
Single-Ended — Nonreferenced (NRSE)	ACH AISENSE AIGND	ACH AISENSE R R AIGND

Activity 10-2   Given the following signal sources, decide on the best measurement system. Circle the appropriate number:

1 *Differential*

2 *Referenced single-ended*

3 *Nonreferenced single-ended*

1   2   3   You have an instrument, such as a function synthesizer, which plugs into a standard wall outlet. The outputs of the instrument are referenced to the same ground as the instrument's power source.

1   2   3   To measure the vibration levels of a machine, you use a couple of accelerometers—in this case, transducers without any power source.

1   2   3   You have 14 thermocouples you need to connect to a board configurable to a maximum of 16 analog inputs.

You'll find the answers at the end of the chapter.

◆   ◆   ◆

■ Sampling, Mr. Nyquist, and All That

The last, and perhaps most important, part of signal theory we're going to cover here is sampling.

Zeno's arrow aside, real-world signals are continuous things. To represent these signals in your computer, the DAQ board has to check the level of the signal every so often and assign that level a discrete number that the computer will accept; this is called an analog-to-digital conversion. The computer then sort of "connects the dots" and, hopefully, gives you something that looks similar to the real-world signal (that's why we say it *represents* the signal).

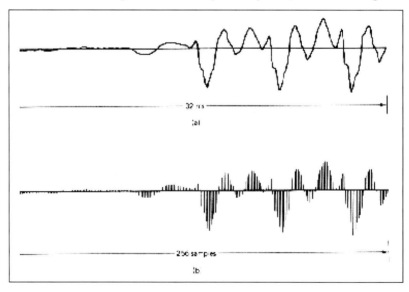

Example of a speech signal in analog form (top), and in digitized form (bottom).

The *sampling rate* of a system simply reflects how often an analog-to-digital conversion (ADC) takes place. Each vertical line in the previous picture represents one ADC. If the DAQ system is making one ADC every half a second, we say the sampling rate is 2 samples/second, or 2 Hz. Alternatively, we can specify the sampling period, which is the inverse of the sampling rate (in this example, 0.5 sec). It turns out that the sampling rate has a terribly important effect on whether your digitized signal looks anything like the real-world signal.

When the sampling rate isn't high enough, a scary thing called *aliasing* happens. Aliasing, while not intuitive, is easy to observe[2]:

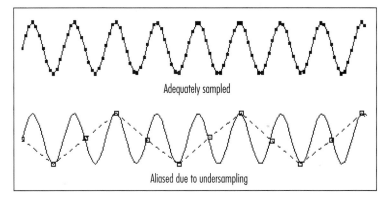

Adequately sampled

Aliased due to undersampling

Aliasing has the effect of introducing high-frequency components into your data that didn't exist in the real-world signal. Once you have aliased data, you can never go back: there is no way to remove the "aliases." That's why it's so important to sample at a high enough rate.

How do you determine what your sampling rate should be? A guy named Nyquist figured it out, and his principle, called **Nyquist's theorem**, is simple to state:

> *To avoid aliasing, the sampling rate must be greater than twice the maximum frequency component in the signal to be acquired.*

So, for example, if you know that the signal you are measuring is capable of varying as much as 1000 times per second

---

2. Another interesting example of a phenomenon akin to aliasing is observing fast circular motions with your eye. You may have noticed this phenomenon when looking at an airplane propeller, for example, in which after the propeller reaches a certain speed, it seems to actually be turning the *opposite* direction very slowly. Although not technically the same as aliasing in an A/D system, you can think of this curiosity as being related to the fact that the human eye can't "sample" fast enough to correctly observe the high circular speed.

(1000 Hz), you'd need to choose a sampling rate higher than 2 kHz. Notice that the whole Nyquist sampling theorem implies that you know what the highest frequency component will be. It is imperative that you find out if you don't know already; if you can't know ahead of time what the highest frequency component will be, you'll need to filter the signal to remove potential high-frequency components, as we describe next.

Another reason for knowing the frequency range of your signal is *anti-aliasing filters* (lowpass filters). In many real-world applications, signals pick up a great deal of high-frequency noise, glitches, or spikes that will greatly exceed the theoretical frequency limit of the frequency measurement you are making. For example, a common biomedical signal is the electrocardiogram (ECG or EKG), a voltage that is related to heart activity. Although these signals rarely have components beyond 250 Hz, the electrode leads easily pick up RF (radio-frequency) noise in the 100 kHz and MHz range! Rather than sample at extremely high frequencies, these DAQ systems implement some low-pass filters that cut out waveforms above the 250 Hz. The DAQ board can then breathe easier and sample only at, say, 600 Hz.

The only case where sampling rate is not important is in the so-called DC signals, such as temperature or pressure. The physical nature of these signals is such that they cannot vary by much more than perhaps once or twice a second. In these cases, a low sampling rate like 10 Hz should do.

■ In conclusion . . .

We've covered a lot of issues involving the path from the physical phenomena to the DAQ board. If you didn't grasp many or even most of these concepts at first, don't worry. DAQ theory is a complex subject, and unless you've had some experience in this area of electrical engineering, it can take some practice before you understand it all.

You've seen a summary of how signals are classified, what kind of transducers are often used, the importance of signal conditioning, the different measurement configurations for digitizing grounded or floating signal sources, and the necessity of using Nyquist's sampling theorem. You'd need to take a couple of electrical engineering courses to thoroughly cover the whole topic of data acquisition and instrumentation; we've just skimmed the surface in this section. Nonetheless, it should be enough to get you started with your measurements.

Selecting and Configuring a DAQ Board

■ Choosing Your Hardware

The PCI-MIO series boards are multifunction boards designed for Power Macintosh systems that use the PCI bus.

The NB-MIO series boards are multifunction boards designed for MacOS systems that use the NuBus slots.

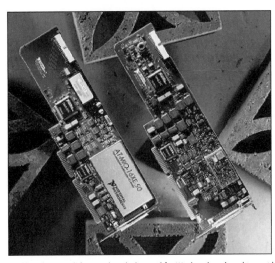

The AT-MIO series boards are multifunction boards designed for Windows-based machines with the AT bus.

Once you know what kind of signals you want to measure and/or generate, it's time to choose a plug-in DAQ board (assuming a DAQ board will meet your requirements). Generally speaking, we recommend always using a National Instruments board if you're going to use LabVIEW. Although it's theoretically possible to use boards from other vendors with LabVIEW, you're in for a long, low-level, device-driver programming marathon. National Instruments offers a huge selection of all types of boards with a good selection of platforms, performance, functionality, and price range. You can request one of their free catalogs on paper, CD-ROM, or go check it out online (http://www.natinst.com).

Another tool you might use to choose a board is NI's program called DAQ Designer. DAQ Designer (which is available upon request from NI for free) is an interactive program that will prompt you for information about your system, and then suggest the appropriate hardware. Only a Windows version of this program is available.

To pick the best hardware for your system, you need to understand well what your system requirements are, most noticeably the "I/O count" (how many inputs and how many outputs). The following checklist should be useful in determining if you have all the information you need to select a board:

- *What type of system am I using (MacOS, Windows, Sun, HP-UX)?*

- *What type of bus or connector is available (AT or EISA for x86 machines, PCI or NuBus for MacOS machines, PCMCIA for laptops, etc.)?*

- *How many analog inputs will I need (multiply by 2 if you need differential inputs)?*

- *How many analog outputs will I need?*

- *Are the analog inputs voltage, current, or both? What are the ranges?*

◆ *How many digital input and output lines will I need?*

◆ *Do I need any counting or timing signals? How many?*

◆ *Do any of the analog I/O signals require special signal conditioning (e.g., temperature, pressure, strain)?*

◆ *Will any of the analog I/O signals exceed ±10 V or 20 mA?*

◆ *What is the minimum sampling rate required on any one channel?*

◆ *What is the minimum scan rate for all the channels?*

◆ *What precision, or resolution (12-bit or 16-bit) will I need?*

◆ *Is portability, ruggedness, isolation, or cost an issue? If so, what are the trade-offs?*

◆ *Have I accounted for my needs in the future (expansion, new DAQ systems, etc.)?*

Now you're ready to pick your DAQ board(s). The most popular type of National Instruments boards are the so-called *multifunction (MIO) boards*, such as the AT-MIO-16E-10 board, which provides you with 16 analog inputs, 2 analog outputs channels, 8 digital I/O channels, and 2 counters. As the name implies, this board typically contains various combinations of ADCs, DACs, digital I/O lines, and counter/timer circuitry, making it useful for a wide range of applications. Interboard connections, such as the Real-Time System Integration (RTSI) bus, transfer timing and trigger signals between boards, permitting synchronization of the operations on multiple boards. This board fits well with most applications, which usually require several analog inputs and occasionally an output or digital signal. If you need more analog outputs, you can get a board such as the AT-AO-10, which provides 10 analog outputs. You can also obtain timing boards, high-count digital I/O boards, PCMCIA cards for laptops, parallel port DAQ adapters, and just about any combination or specialization of features you can imagine.

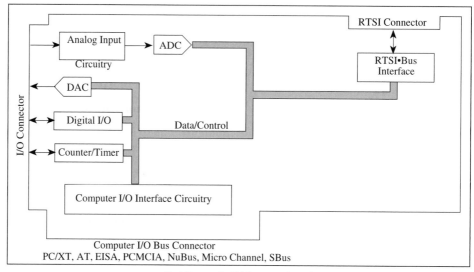

Block Diagram of a Multifunction DAQ Board

The analog input section of the MIO board consists of the ADC and the *analog input circuitry*. The analog input circuitry contains the analog multiplexer, the instrumentation amplifier, and the sample-and-hold circuitry. For more detailed information on the ADC, refer to the manual that accompanies your DAQ board.

In addition to the circuitry shown in the previous figure, older DAQ boards can also contain several components, such as jumpers and dip switches, that configure the DAQ hardware. These components, which configure parameters such as base address, direct memory access (DMA) channels, and interrupt levels, are discussed in the next section. The *Plug&Play* standard is making this obsolete.

What if you're concerned about cost? Most plug-in boards set you back about $700 to $2000. In general, two factors are directly proportional to and most influential on a board's price: the sampling rate and the number of analog I/O channels. So if you are making DC measurements, there's no need to shell out extra dough for a board with a 1-MHz sampling rate. Also, digital I/O is usually cheap. If you want something relatively inexpensive, for a student lab, for example, National Instruments offers low-cost boards such as the Lab-PC+. The board prices shouldn't really seem expensive once you pause to consider what you're getting, however. How much would an oscilloscope, spectrum analyzer, strain gauge meter, and hundreds of other instruments all together cost you if you had to buy them as "nonvirtual" instruments?

If, after perusing through the NI catalog, you're still confused as to which board to buy, here are a few common scenarios and recommendations that might help you.

1. I don't know exactly what my system will require yet, and I want a middle-of-the-road, flexible but powerful board to get me started. Choose one of the multifunction boards, such as the AT-MIO-16E-10 (x86 machines), NB-MIO-16 (Macintosh NuBus), or PCI-MIO-16XE-50 (PCI Power Macs). These give you mostly analog inputs, with a little of everything else (digital, counters, analog out) as well.

2. I have a lot (16-256) of analog channels to sample. You can purchase a standard MIO board, and add an external analog multiplexer, the AMUX-64T, which will multiplex 64 channels into one input on the plug-in board. Note that the maximum sampling rate for these channels will be reduced, however.

3. I really have A LOT (256-4096) of analog channels to sample. With this many channels, you probably need to move to an SCXI system. The SCXI chassis houses DAQ modules, signal conditioning modules, etc., and connects to your computer through a plug-in DAQ board. SCXI is ideal for applications where there is a large channel count for signals that require special conditioning, such as thermocouples or strain gauges.

4. I need a portable DAQ system to go with my laptop. Choose one of the DAQCard series PCMCIA cards, such as the DAQCard-1200, which provides you with an assortment of functions.

5. For a particular reason, I can't use any plug-in cards. Maybe you need to buy a new computer. But NI does offer a couple of DAQ systems that feed data through the parallel port of your PC, such as the DAQPad-1200.

6. I only need a couple of I/O channels for a simple system. Look over some of the low-cost boards, such as the PC-AO-2DC (2 analog outputs), or Lab-NB (multifunction I/O board for the Macintosh NuBus).

7. I need a lot of digital signals for a control system. You can get a board such as the PC-DIO-96 or NB-DIO-96, which give you 96 digital lines.

Please realize that the above are only very general guidelines, and that everybody's particular needs are different. Before making any purchases, we recommend that you speak with a National Instruments representative or with a third-party systems integrator (Alliance member) to advise you on what hardware system would best suit your system.

Finally, be aware that boards not made by National Instruments will only work with LabVIEW if the manufacturer of that board provides a LabVIEW driver (or if you write one yourself).

Activity 10-3

Following are a couple of more challenging signal measurement problems. Your objective is to specify the needed information.

Answer the following for each scenario:

1. *What kind of signals need to be measured?*

2. *What signal measurement type do you recommend?*

3. *What should the sampling rate be for the signals? What is the Nyquist frequency?*

4. *Is any signal conditioning needed? If so, what?*

5. *Choose the board(s) from the National Instruments catalog that should do the job (pick them for your favorite OS).*

A. Professor Harry Phace, of the biomedical engineering lab, wants to acquire heart signals from human subjects—hopefully without electrocuting them. He wants to measure electrocardiograms from two subjects at a time. Each subject has four electrodes connected to his body at different places. The objective is to measure in real-time the potential between each of three electrodes, with the fourth electrode designated as a reference. The leads from the electrodes to the DAQ system have no isolation or ground shields. The maximum amount of detail expected in the waveforms are segments 2 ms wide. The signals are within a 0.02–1 mV range.

B. Ms. I. M. Aynurd needs to measure how the resistance of a flexible material changes under stress and high temperatures. To do so, she has a special chamber with a machine that twists and stretches the material over and over. The chamber also functions as an oven with a variable temperature control. She wants to observe in real time how the resistance of each of 48 strands of this material changes inside the chamber. The resistance is measured by applying a known voltage to each strand and measuring the current. The temperature of the chamber is monitored through a thermocouple. The stress machine is turned on and off through a solid-state relay. Finally, the number of cycles from the stress machine needs to be measured.

Answers are at the end of the chapter.

◆　　　　◆　　　　◆

■ Installing the Boards

All plug-in boards use *drivers*, nasty and painful pieces of low-level code that convince your computer that the boards really are inside it and can be used. Fortunately, National Instruments boards come with the drivers, which can be used with LabVIEW (and other applications). NI's plug-in DAQ boards use a program called NI-DAQ as a driver and configuration utility. GPIB boards use a utility called NI-488. Both NI-DAQ and NI-488, once installed, can be configured from utilities on your PC.

If you bought any boards, you should have received the latest version of NI-DAQ with them. Older boards on Windows machines that aren't *Plug&Play* may require you to set certain jumpers on the boards to set the correct base address, IRQ, etc. All the E-series boards running under Windows 95 and all boards for MacOS systems can automatically configure themselves. In any case, you should still follow the instructions in the manual that came with the board and run NI-DAQ to make sure the board is recognized and working properly. With NI-DAQ, you assign each of your boards a *device number*, which is later used by LabVIEW to uniquely identify each board in the DAQ VIs.

As computer architectures and operating systems continue to rapidly evolve, so does the software that configures and manages National Instruments' plug-in boards. Much of the information presented in this section was still in a state of evolution at press time. We recommend that you compare the NI-DAQ manual and instructions with the information here, as it is likely that some features will have changed, especially for Windows systems.

This section discusses the various board parameters that you must configure in hardware if your board is not *Plug&Play*.

■ Computer Resource Parameters

There are three parameters that must be set on a DAQ board to determine how the DAQ board can communicate with the computer. These parameters are the base I/O address, interrupt level, and DMA channels.

◼ Base I/O Address

The DAQ board communicates with the computer primarily through its registers. The driver software writes to configuration registers on the board to configure the board, and the software reads data registers on the board to obtain the board's status or a signal measurement. The base I/O address setting determines where in the computer's I/O space the board's registers reside.

◼ Interrupt Level

Another way the DAQ board communicates with the computer is through processor interrupts. Interrupts are very important to the operation of any computer. They give the processor the ability to respond quickly to its peripherals. You can think of a processor interrupt as a doorbell. If your door did not have a doorbell, you would have to periodically go to the door to see if anyone happened to be there at that time. Of course, that would be very inefficient. With a doorbell, you only need to go to the door when the doorbell rings, and you are then confident that someone is there waiting. Likewise, in the case of a DAQ board, it is not always efficient for the processor to continually check whether data is ready to be read from the board. A DAQ board can use an interrupt as a "doorbell" to the processor to signal that it has data waiting to be read. Every device that uses processor interrupts must be assigned a different "interrupt level," or else the devices will conflict with each other.

◼ Direct Memory Access (DMA)

The third way the DAQ board can communicate with the computer is through direct memory access (DMA). DMA is a data transfer method in which data is transferred directly from the peripheral to computer memory, bypassing the processor. DMA is usually required to achieve maximum data transfer speed, and thus is useful for high-speed data acquisition devices. Every device that uses DMA must be assigned a separate DMA channel, or the devices will conflict with each other. Some devices, including MIO-series DAQ boards, can be assigned more than one DMA channel to achieve higher throughput rates.

▓ DAQ Board Configuration

Depending on the computer platform you are on, the above three parameters are set in different ways:

PC/XT/AT Bus
On non-Plug&Play DAQ boards, jumpers and dip switches determine the base I/O address, interrupt level, and DMA channel(s). You should verify that your system does not have other hardware at these settings. If you change the jumper settings, you must make a corresponding change in the software configuration. If your board is a Plug&Play device such as an E-Series MIO board, then the settings can only be changed in software.

NuBus/PCI
The board automatically configures the base I/O address, DMA channels(s), and interrupt level(s). You need to know only the number of the slot in which you installed the board. You can use the NI-DAQ Control Panel item to determine the slot number.

▓ Analog I/O Settings

A DAQ board has several analog input/output (I/O) parameters that control the operation of the ADC and DAC. Some boards use only software configuration utilities to set the I/O parameters, while others use jumpers. On a software configurable board, you configure the board settings using only the software configuration utility: the NI-DAQ Configuration Utility (*Windows 95*), WDAQCONF (*Windows 3.x*), or NI-DAQ (*MacOS*). On a jumper-configurable board, you must both physically position the jumpers on the board for the desired setting, and match those settings in the software configuration utility.

As an example, consider an AT-MIO-16 or an NB-MIO-16. Both boards use jumpers and software to configure the analog I/O settings listed here.

- ADC Input Range   Unipolar 0 V to +10 V
    Bipolar ±5 V
    Bipolar ±10 V (factory setting)
- ADC Input Mode   Ground-referenced single-ended
    Nonreferenced single-ended
    Differential (factory setting)

- DAC Reference   Internal (factory setting)
  External

- DAC Polarity   Unipolar—Straight binary mode
  Bipolar—Two's complement mode
  (factory setting)

In contrast, on a *Plug&Play* board such as an AT-MIO-16E-2, the above settings are configured only by using the software configuration utility.

Signal Connections

You can use a connector such as a CB-50 or an SC-207x to physically wire your signals to the DAQ board. If you use SCXI, you can use the appropriate SCXI terminal block. A typical wiring configuration and the pinout for the AT-MIO-16 is shown in the following figure.

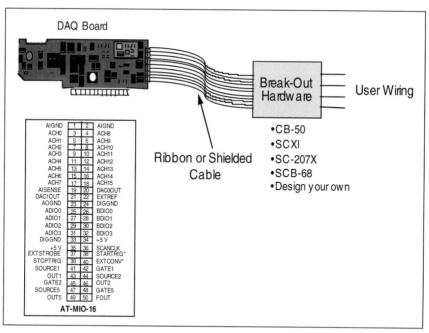

Typical Wiring and Connector Pinout for the AT-MIO-16

Software Configuration: Windows 95

LabVIEW for Windows 95 DAQ VIs use an intermediate driver, DAQDRV, to access the National Instruments standard

NI-DAQ for Windows 95 32-bit dynamic link library (DLL), as shown in the following diagram. The LabVIEW setup program installs NI-DAQ32 DLL in the WINDOWS\SYSTEM directory. NI-DAQ for Windows 95 supports all National Instruments AT/XT/EISA boards and SCXI.

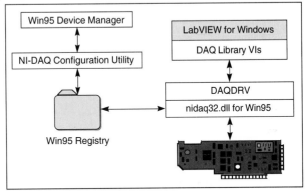

The Windows 95 Configuration Manager keeps track of all the hardware installed in your system, including National Instruments DAQ boards. If you have a *Plug&Play* board, such as an E-Series MIO board, the Windows 95 Configuration Manager will automatically detect and configure the board. If you have a non-*Plug&Play* board (known as a *Legacy* device) you have to configure the board manually using Windows 95's **Add New Hardware** option under the Control Panel.

The NI-DAQ Configuration Utility is a utility that configures the parameters for your DAQ boards. After installing a DAQ board in your computer and configuring the board with the Device Manager as described above, you must run this configuration utility. The utility reads the information recorded in the Windows 95 Registry by the Device Manager and assigns a logical device number to each DAQ board. You use the device number to refer to the board in LabVIEW.

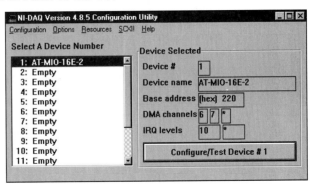

The board parameters that you can set using the configuration utility depend on the board. Some boards are fully software configurable, while others require you to set jumpers. The utility saves the logical device number and the configuration parameters in the Windows 95 registry.

You start the program by selecting NI-DAQ Configuration Utility from the LabVIEW program group under the **Start** menu. The previous figure displays the primary configuration utility window. This utility also configures your SCXI system.

After you configure your system, you do not need to run the configuration utility again unless you change the system parameters.

■ Software Configuration: Windows 3.1

LabVIEW for Windows DAQ VIs use an intermediate driver, DAQDRV, to access the National Instruments standard NI-DAQ for Windows dynamic link library (DLL), as shown. The LabVIEW setup program installs NI-DAQ DLL in the WINDOWS\ SYSTEM directory. NI-DAQ supports all National Instruments AT/XT/EISA boards and SCXI.

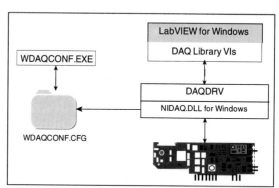

WDAQCONF is a utility that assigns logical device numbers and configures the parameters for your DAQ boards. The parameters that you can set using WDAQCONF depend on the board. After you assign a logical device number to a board, you use that number to refer to the board in LabVIEW. WDAQCONF saves the configuration parameters and logical device numbers in a file called WDAQCONF.CFG. You start WDAQCONF by double-clicking on its icon in Windows. The following figure displays the primary WDAQCONF window. WDAQCONF also configures your SCXI system.

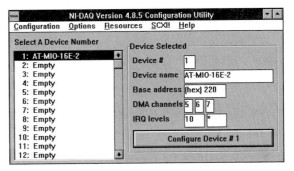

After you configure your system, you do not need to run WDAQCONF *again unless you change the system parameters.*

NI-PNP is a utility run by WDAQCONF to automatically detect and configure *Plug&Play* DAQ boards, such as the AT-MIO-16E-2. When you install an E-series board in your computer and run WDAQCONF for the first time, WDAQCONF detects that the card has been installed and prompts for the logical device number. The figure below shows this prompt.

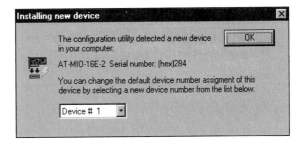

■ Software Configuration: MacOS

LabVIEW for Macintosh DAQ VIs use the National Instruments standard NI-DAQ, as shown in the next diagram. The Lab-VIEW setup program installs the NI-DAQ item in the Control Panels folder, as well as an NI-DMA/DSP extension in the Extensions folder. NI-DAQ for Macintosh works with all National Instruments NB, PCI, Lab Series boards and SCXI. If NI-DAQ is correctly installed, you will see a DAQ Board icon appear along the bottom of your Macintosh screen when your computer boots.

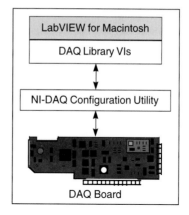

The NI-DAQ Control Panel item is also a utility that configures your DAQ boards. This utility lists the boards installed in your Macintosh using the **Devices** option, as shown in the following figure. National Instruments NB, PCI, and Lab Series boards are listed in boldface; all other boards are dimmed. The **Device Configuration** option allows you to configure the hardware settings for each board. The parameters that you can set using NI-DAQ depend on the board. Some boards are fully software configurable, while you must set jumpers on others. NI-DAQ also configures your SCXI hardware and displays a reference table of all DAQ and LabVIEW errors.

After you configure your system, you do not need to run NI-DAQ again unless you change the system parameters.

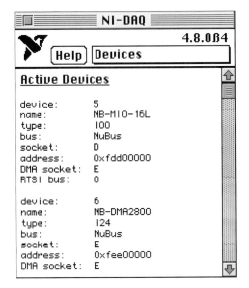

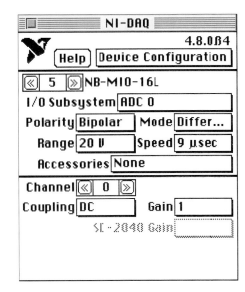

■ Summary: Installing a DAQ Board

1. If you are using a board that requires jumper settings, consult the documentation that came with the board to set the correct base I/O address, DMA, IRQ, and signal type settings.

2. Disconnect the power from your computer and place the board in its slot. The board might need a firm push to wedge it in; however, don't *force* it in!

ALWAYS disconnect the power from your computer before plugging in a board. Failure to do so could fry your board, your computer, and result in a warrant for your arrest.

3. Power the computer back on. On Windows 95 systems with *Plug&Play* boards, the computer should notify you that it has detected a new piece of hardware.

4. Run the corresponding NI-DAQ utility for your operating system (under Windows 3.1, you must "add" the board with this utility). Verify that the correct board type is installed.

5. Make a note of what *device number* has been assigned to the board. You will use this device number in the LabVIEW DAQ VIs to access the board. On the MacOS, the device number is the same as the *slot number*.

6. Configure the measurement type and gain (be sure these match your jumper settings if you are using a board with jumpers).

7. Run the configuration test (under the **Test** menu) to verify the correct configuration (you don't need to do this—and can't—on the MacOS).

You can also use NI-DAQ to test analog inputs, outputs, digital I/O, etc. using the other options under the **Test** menu (not available under the MacOS). For example, you can generate a +5-V signal at a channel you specify, and use a multimeter to see if the board is really providing a 5-V output.

Using a GPIB Board

A PC communicates with a Tektronix oscilloscope via GPIB.

A GPIB board is used to control and communicate with one or more external instruments that have a GPIB interface. The General-Purpose Interface Bus, invented by Hewlett-Packard (HP) in the 1960s, has become the most popular instrument communications standard. All HP instruments support it, as well as thousands of others. GPIB was also updated and standardized by the IEEE, and was duly named **IEEE 488.2**. Some nice features about GPIB are:

◆ *It transfers data in parallel, one byte (eight bits) at a time*

◆ *The hardware takes care of handshaking, timing, etc.*

◆ *Several instruments (up to 15) can be strung together on one bus*

◆ *Data transfer is fast: 800 Kbytes/sec or more*

You can obtain a plug-in GPIB board for almost every platform and bus, as well as external interfaces that will convert your serial port, parallel port, Ethernet connection, or SCSI port to GPIB.

Another feature that makes GPIB popular is that the computer and instrument talk to each other using simple, intuitive ASCII commands. For example, a PC connected to an HP 3458A Digital Multimeter might say something like this:

PC: `IDN?;` [Identity?—i.e., who are you]

HP: `HP3458A`

PC: `RMEM 1;` [Recall Memory register 1]

HP: `+4.23789`

A multitude of GPIB instrument drivers, available free of charge, make communication with an external instrument as simple as using a few subVIs that do all the ASCII commands for you.

Installing a GPIB board can be literally a snap (if you're using a MacOS system) or sometimes a major ordeal (if you're using Windows 3.1). The best advice is to just follow the installation procedures described in the manual that comes with the GPIB board.

An Instron materials strength tester communicates with LabVIEW over the GPIB interface.

Getting Ready for Serial Communications

Serial communication has the advantage that it's cheap and simple. You rarely need any additional hardware, since almost all computers have at least two serial ports available. It doesn't mean it will be easy to get the hardware hooked up right, however. Although popular standards such as RS-232 and RS-485 specify the cabling, connectors, timing, etc., manufacturers of serial devices more often than not abuse and ignore these standards. If you just plug a cable from your serial port on your PC to the instrument, roll dice to find out if it will work the first time.

It's important to become familiar with your serial instrument, its pinouts, and the parameters it uses, such as baud speed, parity, stop bits, etc. You should also know what the different lines of a serial port are for.

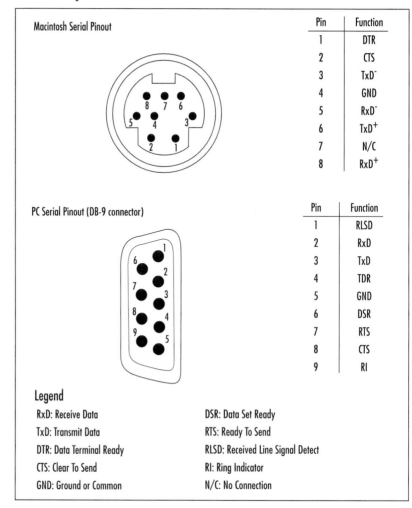

Macintosh Serial Pinout

Pin	Function
1	DTR
2	CTS
3	TxD⁻
4	GND
5	RxD⁻
6	TxD⁺
7	N/C
8	RxD⁺

PC Serial Pinout (DB-9 connector)

Pin	Function
1	RLSD
2	RxD
3	TxD
4	TDR
5	GND
6	DSR
7	RTS
8	CTS
9	RI

Legend

RxD: Receive Data

TxD: Transmit Data

DTR: Data Terminal Ready

CTS: Clear To Send

GND: Ground or Common

DSR: Data Set Ready

RTS: Ready To Send

RLSD: Received Line Signal Detect

RI: Ring Indicator

N/C: No Connection

In many cases, only a few of the lines are used. Chances are you'll end up using the following or fewer:

Transmit (TxD): Sends data *from* the PC *to* the instrument.

Receive (RxD): Sends data *to* the PC *from* the instrument.

Ground (GND): Ground reference. Never forget to hook this one up.

Clear-to-Send (CTS): The PC uses this line to tell the instrument when it's ready to receive data.

Ready-to-Send (RTS): PC uses this line to advise the instrument it's ready to send data.

If you experience problems getting your serial device to communicate, try a few of these things:

◆   Swap the transmit and receive lines. You can use a special cable, called a "null modem" cable to do this.

◆   Check the baud speed, parity, stop bits, handshaking (if any) and any other serial parameters on your PC. Then check what these parameters are on the instrument. If any of these differ, the two parties won't talk to each other.

◆   The serial instrument will always require a power supply. Make sure it's on.

◆   Make sure the CTS and RTS lines are connected properly. Some instruments require their use; others don't.

◆   Be sure the serial port is not being used by any other application.

◆   Check to see if you are using the serial port you think you're using (remember, there are usually at least two).

◆   Make sure you are sending the proper termination characters (EOL).

A handy way to know if your PC is set up correctly for serial communication is to get a second PC and connect their serial ports together. Then, using a dumb terminal program on each computer, you can verify that the data you type on one screen appears on the other and vice versa.

Wrap It Up!

Whew! This chapter has been a heavy one—if you read it end-to-end, you deserve to take a break.

We've covered basic signal and data acquisition theory. Different types of signal can be classified for measurement purposes into analog AC, analog DC, digital on/off, digital counter/pulse, and frequency. Signal sources can be *grounded* or *floating*. *Grounded signals* usually come from devices that are plugged in or somehow connected to the building ground. *Floating source* examples include many types of sensors, such as thermocouples or accelerometers. Depending on the signal source type, signal characteristics, and number of signals, three measurement systems can be used: *differential, referenced single-ended (RSE),* or *nonreferenced single-ended (NRSE).* The big no-no is a grounded source using a referenced single-ended system. The sampling rate of a data acquisition system (for AC signals) is very important. According to the *Nyquist theorem,* the sampling rate must be more than twice the maximum frequency component of the signal being measured.

Selecting a DAQ board or system that will do the right the job is the next step. Many kinds of DAQ boards exist for different platforms, applications, and budgets. National Instruments' SCXI systems provide ways to work with a very high number of channels, as well as perform elaborate *signal conditioning.* Installing a DAQ board is getting easier than before, but still requires a little knowledge of how to set parameters in the configuration utility NI-DAQ.

Finally, we looked briefly at some hardware aspects of communicating to external instruments through *GPIB* and *serial* interfaces. The GPIB interface is a widely accepted standard for many instruments, and you can often obtain an instrument driver in LabVIEW for your particular instrument. Serial communication is cheap and conceptually simple, but in practice requires much troubleshooting and tweaking.

Don't feel bad if much of the material in this chapter eluded you. DAQ is a complex subject and often requires the expertise of an instrumentation engineer. However, for relatively simple DAQ systems, if you have a willingness to experiment and a sense of adventure, you can assemble your own DAQ system at your PC. Remember to consult the manuals that come with your board and hardware for more details.

Solutions to Activities

10-1 1,3,4,2,5,2,2,5,2,1

10-2 3 or 1, 1, 2

10-3

A. 1. Analog AC signals, small amplitude, floating

2. Differential (because of small amplitudes relative to ground)

3. Nyquist frequency= $f_n = \dfrac{1}{2\,ms} = \dfrac{1}{2 \cdot 10^{-3}\,ms} = 500$ Hz

   Sample at more than 1000 Samples/sec.

4. Yes. Amplification (small signal amplitude), isolation (safety), and perhaps anti-aliasing filters (noise) are required.

5. Any MIO-series board should do.

B. 1. Analog input DC (resistance, thermocouple readings), analog output DC (voltage excitation), digital output on/off (relay to turn on machine), digital input counter (count machine cycles).

2. Referenced-single ended (RSE), instead of differential because of the high number of channels.

3. All DC measurements. 10 Hz should be a good sampling frequency.

4. Yes. Thermocouple conversions.

5. SCXI system: because of the high channel count and special signal conditioning needed.

OVERVIEW

"In this chapter, we'll get to the heart of what LabVIEW is supposed to do: data acquisition and instrument control. You'll become familiar with some of the VIs on the **DAQ** *palette and the* **Instrument I/O** *palette. Analog input and output, digital input and output, GPIB instrument control, and serial communications are covered. We'll go through the basic steps necessary to get you started with some data acquisition in LabVIEW, and point you in the right direction to do your own more advanced DAQ and instrument control programs.*

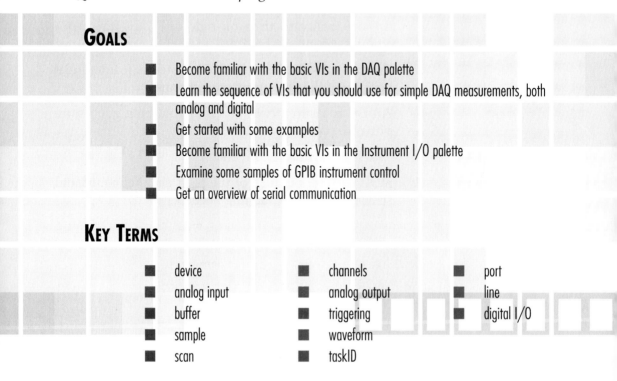

GOALS

- Become familiar with the basic VIs in the DAQ palette
- Learn the sequence of VIs that you should use for simple DAQ measurements, both analog and digital
- Get started with some examples
- Become familiar with the basic VIs in the Instrument I/O palette
- Examine some samples of GPIB instrument control
- Get an overview of serial communication

KEY TERMS

- device
- analog input
- buffer
- sample
- scan

- channels
- analog output
- triggering
- waveform
- taskID

- port
- line
- digital I/O

DAQ and Instrument Control in LabVIEW

11

Definitions, Drivers, and Devices

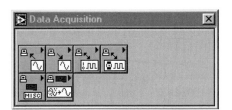

Take a good look at the above palettes. The VIs in these palettes go a level beyond what any other function in LabVIEW does—they let you communicate with, read from, write to, measure, control, turn on and off, or blow up stuff in the external world via DAQ and GPIB boards. We'll be taking a whirlwind tour of the analog, digital, GPIB, and serial functions available in LabVIEW. To do this effectively, you first need to understand a little bit about the interface between LabVIEW and the boards.

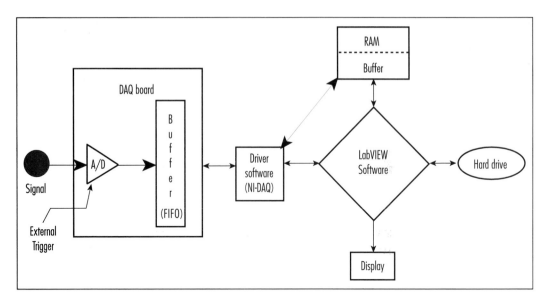

As you can see in the previous figure, initiating a DAQ operation involves LabVIEW calling NI-DAQ (which contains the
driver for the board in use), which in turn signals the hardware
to initiate the I/O operation. DAQ boards use on-board buffers
(called FIFOs, for "First-In, First-Out") and RAM buffers as an
intermediate place to store the data they acquire. Also note that
the software isn't the only place where an initiation of an I/O
operation takes place—an external piece of hardware can also
trigger the operation.

Two important characteristics help classify the type of DAQ
operation you are performing:

◆ *Whether you use a buffer*

◆ *Whether you use an external trigger to start, stop, or synchronize
an operation*

■ Buffers

A *buffer*, as used in this context, is an area of memory in the
PC (not the on-board FIFO) reserved for data to reside temporarily before it goes somewhere else. For example, you may want to
acquire a few thousand data samples in one second. It could be
difficult to display or graph all that data in one second. But by
telling your board to acquire the data into a buffer, you can

quickly store the data there first, and then later retrieve it for display or analysis. Remember, buffers are related to the speed and volume of the DAQ operation (generally analog I/O). If your board has DMA capability, analog input operations have a fast hardware path to the RAM in your computer, meaning the data can be acquired directly into the computer's memory.

Not using a buffer means you must do something with data (graph, save to disk, analyze, whatever) *one point at a time* as each is acquired, since there is no place to "keep" several points of data before you can handle them.

Use *buffered I/O* when:

◆ You need to acquire or generate many samples at a rate faster than is practical to display, store on a hard drive, or analyze in real time

◆ You need to acquire or generate AC data (>10 samples/sec) continuously and to analyze or display some of the data on the fly

◆ The sampling period must be precise and uniform throughout the data samples.

Use *nonbuffered I/O* when:

◆ The data set is small and short (for example, acquiring one data point from each of 2 channels every second)

◆ You need to reduce memory overhead (the buffer takes up memory).

We'll talk more about buffering soon.

■ Triggering

Triggering refers to any method by which you initiate, terminate, or synchronize a DAQ event. A trigger is usually a digital or analog signal whose condition is analyzed to determine the course of action. Software triggering is the easiest and most intuitive way to do it—you control the trigger directly from the software, such as using a Boolean front panel control to start/stop data acquisition. Hardware triggering lets the circuitry in the board take care of the triggers, adding much more precision and control over the timing of your DAQ events. Hardware triggering can be further subdivided into *external* and *internal* triggering. An example of an internal trigger is to program the board to output a digital pulse when an analog-in channel reaches a certain voltage level. An example of an external trigger is to have the board wait for a digital pulse from an external instrument before initiating an acquisition. All National Instruments DAQ boards have an external trigger pin, which is a digital input used for triggering. Many instruments provide a digital output (often called "trigger out") used specifically to trigger some other device or instrument, in this case, the DAQ board.

Use *software triggering* when:

◆   The user needs to have explicit control over all DAQ operations, *and*

◆   The timing of the event (such as when an analog input operation begins) needn't be very precise.

Use *hardware triggering* when:

◆   Timing a DAQ event needs to be very precise

◆   You want to reduce software overhead (for example, a While Loop that watches for a certain occurrence can be eliminated)

◆   The DAQ events must be synchronized with an external device.

   In the following sections, you will see how you can use the DAQ VIs to configure an I/O operation as buffered or non-buffered, and configure the type of triggering.

Analog I/O

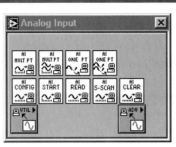 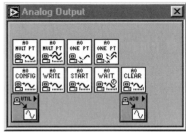

   The examples in this section, and selected examples in subsequent sections, assume you've been able to install and configure a DAQ board. If you need some help doing this, consult Chapter 10 and the documentation that came with your board. The examples and exercises also assume you have some signal source to measure and are using the LabVIEW Full Development System.

A few definitions are in order before we delve into the DAQ VIs. The following terms appear consistently in the inputs and outputs to the VIs, so it's critical to understand what they are if you are going to do any LabVIEW DAQ programming. One of the DAQ VIs with its inputs and outputs is shown for clarity.

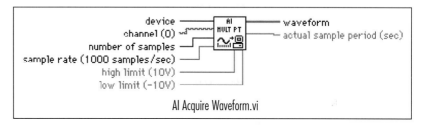

AI Acquire Waveform.vi

device is the "device number" you assigned the board in the NI-DAQ configuration utility (the same as the slot # on a Mac). This parameter tells LabVIEW which board you're using, keeping the DAQ VI itself independent of the board type (e.g., if you later used a different compatible board and assigned it the same device number, all your VIs would probably work without modification).

sample represents one A/D conversion: it's just one point, one numeric value corresponding to the real analog value at the time the measurement was taken.

channel specifies the physical source of the data sample(s). A board with 16 analog input channels, for example, means you can acquire 16 sets of data points at the same time. In the Lab-VIEW DAQ VIs, a channel or set of channels are specified as a string in a pretty intuitive manner. The reason it's a string and not a numeric type is that nonnumeric characters (": " and ", ") are needed to specify a set of more than one channel. For example:

Channels	Channel string
channel 5	5
channels 0 thru 4	0:4
channels 1, 8, and 10 thru 13	1,8,10:13

Sometimes arrays of strings are used in lower-level DAQ VIs. Each array element would contain a channel description.

A **scan** is a sample from each separate channel being used, assuming you are sampling multiple channels.

A **waveform** is a set of samples from ONE channel, collected over some period of time, and ordered by time. Usually, but not always, the time period between data points is constant for a given waveform.

Beginners often confuse scans and waveforms. A scan is a set of samples versus *channels* (one sample from each channel at just one instant in time); a waveform is a set of samples (from the same channel) versus *time*.

high limit and **low limit** are the voltage limits you expect to have on your signal. By changing these inputs from their defaults of 10 and -10 V, you can set the *gain* of your DAQ system. For example, on most boards, if you set them at 5 and -5 V, the gain is set at 2. If you set them at 1 and -1 V, the gain is 10. Thus, you wire these inputs if you know that the expected *range* of your input signal is different from the default. Use this formula to determine the gain applied:

$$\text{Gain} = \text{Board Input Range}[1] / (\,|\,\text{High Limit}\,| - |\,\text{Low Limit}\,|\,)$$

Be aware that many DAQ boards support only certain predetermined gain values. If you set a theoretical gain that the board does not support, LabVIEW will automatically adjust it to the nearest preset gain. A typical board might have available gains of 0.5, 1, 2, 5, 10, 20, 50, 100.

taskID is a 32-bit integer (I32 type numeric) that some DAQ VIs use to identify a specific I/O operation that's going on. Many of the DAQ VIs require a **taskID in** and return a **taskID out** to pass on to the next VI. By wiring this taskID, you don't have to provide every VI with information about your board, sampling rate, limits, etc. An initial VI can take all this and pass a taskID out, which will tell the other VIs what the configuration is. If you're confused by all this, just hold on and look at an example. It really isn't that hard.

■ Quick & Dirty Analog I/O: The Top Tier

If you can't wait to try out your DAQ board, try some of these easy VIs for starters. Even if you will move on to more complex data acquisition, it's a good idea to use the following examples

---

1. The Board Input Range for most boards is 20 V, since they can read inputs from −10 to 10 V.

as a means to test your board. All the VIs at the top tier of the **Analog Input** and **Analog Output** palette are "easy" VIs: they can be used as stand-alone VIs with minimal setup and configuration.

Here's what the "easy" DAQ VIs do:

■ Analog Input

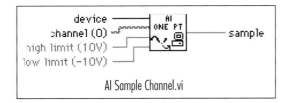

AI Sample Channel.vi

Obtains one sample from the specified channel.

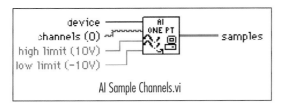

AI Sample Channels.vi

Obtains one sample from each of the specified channels in the **channels** string. The samples are returned in the **samples** array, ordered by channel number.

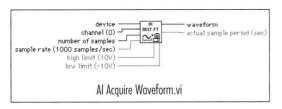

AI Acquire Waveform.vi

Obtains one waveform (a set of samples acquired over a period of time) from one channel, at the specified sampling rate. The samples are returned in the **waveform** array, each element being a sample one sampling period apart from the next.

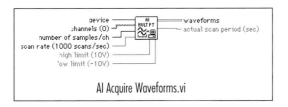

AI Acquire Waveforms.vi

Obtains a waveform from each channel in the **channel** string. The samples are returned in the **waveform's** 2D array, ordered by channel number and sample period. Each point of channel data is stored in a column; time increases as you go down the rows.[2]

---

2. The way this VI stores multichannel data can be confusing when it comes time to plot it. If you wired the output array of this VI directly to a graph indicator, you would get the data plotted on the x-axis, instead of the y-axis where you want them. This VI stores each channel's data in a *column*, but waveform graphs take each *row* of an array and plot it versus the array's index (indices are plotted on the x-axis, data is plotted on the y-axis). Most of the time you will want to pop up on the graph indicator and select **Transpose Array.**

◼ ANALOG OUTPUT

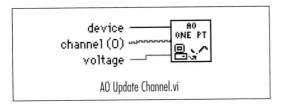

AO Update Channel.vi

Sets the specified **voltage** at the specified output **channel.** This voltage remains constant at the output channel until it is changed or the device is reset.

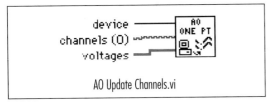

AO Update Channels.vi

Sets the specified **voltages** at the specified output **channels.** These voltages remain constant at the output channels until they are changed or the device is reset.

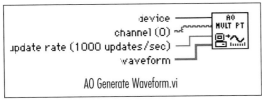

AO Generate Waveform.vi

Generates a waveform at the specified output **channel.** The waveform points, in volts, should be provided at the **waveform** array. The **update rate** specifies the time between points.

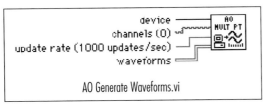

AO Generate Waveforms.vi

Same as the above, except multiple waveforms, one for each channel, can be generated simultaneously. Each waveform should be stored in one column of the 2D array.

All the previous analog input VIs are synchronous with the I/O on the DAQ board; that is, the VI does not finish executing until all the data is read or written.

Try this simple example.

Activity 11-1

1. Connect a voltage source, such as a function generator (or even a 1.5-V battery, if that's all you have) to channel 0 on your DAQ board. Be sure you know whether you've configured your board for differential or single-ended measurements.

2. Build the front panel and block diagram shown. **AI Sample Channel** is found on the **Analog Input** palette. If your DAQ board is not device #1 in the NI-DAQ utility, change the device constant to match your device number.

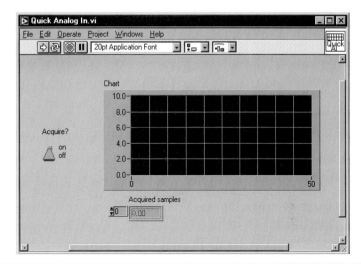

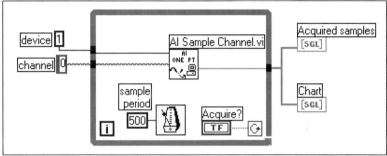

3. Save your VI as **Quick Analog In.vi**

4. Run the VI with <u>Acquire?</u> turned on, and turn it off after a few seconds.

5. Examine the data in the <u>Acquired samples</u> array.

The **While Loop** in the previous activity is the tradeoff for simplicity; you add unnecessary software overhead. Not a problem for a quick look at your data, but in a decent-sized block diagram, you really don't want the software trying to do all the sampling control; the DAQ boards and low-level drivers are designed to do that.

This method of DAQ works perfectly fine under many circumstances, provided that:

1. The sampling rate is slow (once per second or slower)

2. Other time-consuming operating system (OS) events are not occurring while the VI is running

3. Slight variations in the sampling times are acceptable.

For example, suppose you are running a VI using a LabVIEW timing function in a loop (as in the previous activity), sampling once per second, and you decide to drag the window around. Your VI may not acquire any data during the drag. Suppose you drag the window for five seconds. When you finally release the mouse button, the VI continues, but it gives *no indication* that the data was not sampled for a five-second period! When using these VIs that use the LabVIEW looping mechanism for timing, you should make sure that other programs are not running concurrently, as they may interfere with your timing, and you should avoid time-consuming OS events (mouse activity, disk activity, network activity, etc.) while the VI is running. To be extra cautious, you can use the LabVIEW timing functions such as **Tick Count (ms)** to gauge the accuracy of your loop timing.

Let's look at a second example for easy I/O. The following sample VI lets you acquire several scans and display them on the graph.

Activity 11-2

1. Connect four DC or low-frequency voltage sources to channels 0-3. If you don't have this many voltage sources, you can connect one source to all channels, or make a resistor network to vary the amplitude of the voltage at each channel.

2. Build the front panel and block diagram shown. On the graph, select **Transpose Array** from the pop-up menu of the graph. The **AI Sample Channel** function returns a 2D array, where each column contains voltages from one channel. The graph normally plots row versus columns, so you need to transpose the array to make the y-axis represent voltage data.

3. Set the scan rate, channels, and number of scans to acquire as shown.

4. Save your VI as **Acquire Multiple Channels.vi.**

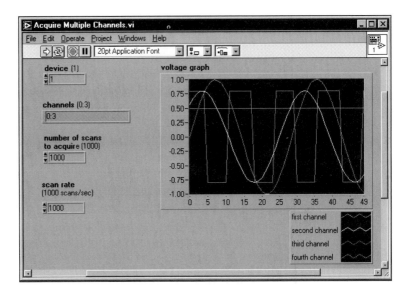

The block diagram is very simple. The **AI Acquire Waveforms** does everything for you. Notice the **Bundle** function we used for the graph. The **actual scan rate** output of the AI function is used for graphing the points on an accurate time scale. Note this VI performs a nonbuffered, software-triggered ADC.

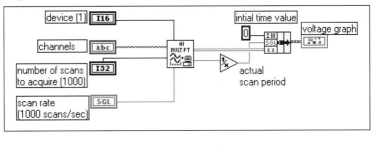

◆   ◆   ◆

An important limitation of multiple channel I/O needs to be mentioned. If you set a high scan rate for multiple channels, and observe the data from each channel over actual time (not over array index values) you'd notice a successive *phase delay* between each channel. Why is this? Most boards can only perform one A/D conversion at a time. That's why it's called scanning: the data at the input channels is digitized sequentially, one channel at a time. A delay, called the *interchannel delay*, arises between each channel sample. By default, the interchannel delay is as small as possible, but is highly dependent on the board.[3]

---

3. A few boards, such as the EISA A2000, do support simultaneous sampling of channels. Other boards support either *interval scanning* or *round-robin scanning*. Boards that support interval scanning have a much smaller interchannel delay than those that use round-robin scanning.

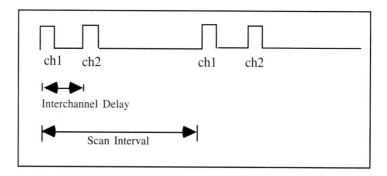

In DC and low-frequency signals, this phase delay in scans is generally not a problem. The interchannel delay may be so much smaller than the scan period that the board *appears* to virtually sample each channel simultaneously. For example, the interchannel delay may be in the microsecond range, and the sampling rate can be 1 scan/sec (as shown in the next figure). However, at higher frequencies, the delay can be very noticeable and may present measurement problems if you are depending on the signals being synchronized.

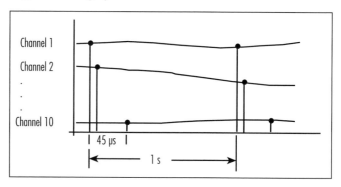

If you are not going to be doing any DAQ with LabVIEW, you may want to skip the rest of the DAQ section.

■ Better Analog I/O: The Middle Tier

One fundamental limitation of the easy VIs discussed previously is the redundancy of performing DAQ tasks. Every time you call **AI Sample Channel**, for example, you set up the hardware for a specific type of measurement, tell it the sampling rate, etc. Obviously, if you're going to take a lot of samples very often, you shouldn't need to "set up" the measurement on every iteration. The "top tier" VIs in the DAQ palette make for easy, quick programming, but can add a lot of unnecessary software overhead and have little flexibility in applications in which you are manipulating a lot of data.

The "middle tier" VIs offer more functionality, flexibility, and efficiency for developing your application. These VIs feature capabilities such as controlling intersampling rates, use of external triggering, and performing continuous I/O operations. The following table describes briefly each of the middle-tier analog I/O VIs. You'll notice some of these have a zillion inputs and outputs. Efficient use of the analog I/O VIs equates with only wiring the terminals you need. In most cases, you don't need to worry about the optional ("grayed out") terminals that show up in the Detailed Help window.

■ Analog Input

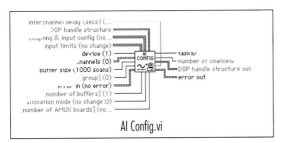

AI Config.vi

AI Config configures the analog input operation for a specified set of channels, configures the hardware, and allocates a buffer in computer memory. **Device** is the device number of the DAQ board. **Channel** is a *string array* that specifies the analog input channel numbers. **Input limits** specifies the range of the input signal and affects the gain applied by your hardware. **Buffer size** is specified in *scans* and controls how much computer memory **AI Config** reserves for the acquisition data. **Interchannel delay** sets the interchannel skew for interval scanning.

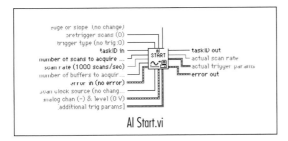

AI Start.vi

AI Start starts a buffered analog input operation. This VI controls the rate at which to acquire data, the number of points to acquire, and the use of any hardware trigger options. Two important inputs to **AI Start** are **scan rate (scans/sec),** how many scans per second to acquire on each channel; and **number of scans to acquire,** how many times to scan through the channel list.

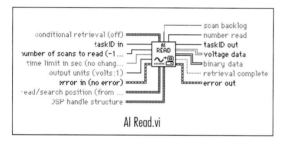

AI Read.vi

AI Read reads data from the buffer allocated by **AI Config**. This VI can control the number of points to read from the buffer, the location in the buffer to read from, and whether to return binary data or scaled voltage data. The output of this VI is a 2D array of data, where each column of data corresponds to one channel in the channel list.

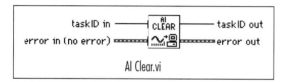

AI Single Scan.vi

AI Single Scan returns one scan of data. The **voltage data** output of this VI is the voltage value read from each channel in the channel list. You use this VI only in conjunction with **AI Config**; no need to use the **AI Start** and **AI Read** VIs when using **AI Single Scan.**

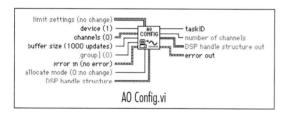

AI Clear.vi

AI Clear clears the analog input operation, deallocates the buffer from computer memory, and frees any DAQ board resources such as counters.

The first VI you always use when setting up an analog input application is **AI Config. AI Config** produces a **taskID** and an error cluster. All other Analog Input VIs accept the **taskID** as an input to identify the device and channels on which to operate, and output the **taskID** when they complete it. Because the **taskID** is an input and output to other Analog Input VIs, this parameter forms a data dependency between the DAQ VIs that controls the execution flow of the diagram, making sure the DAQ VIs execute in the correct order.

■ Analog Output

AO Config.vi

AO Config configures the analog output operation for a specified set of channels, configures the hardware, and allocates a buffer in computer memory for the operation. **Device** is the device number of the DAQ board. **Channel** is a *string array* that specifies the analog output channel numbers. **Limit settings** specifies the range of the output signals. The **taskID** output is used by all subsequent analog output VIs to identify the device and channels on which to operate.

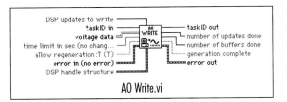

AO Write.vi

AO Write writes the data in **voltage data** into the buffer used for the analog output operation. The data must be a 2D array where each *column* of data is written to the corresponding channel in the channel list.

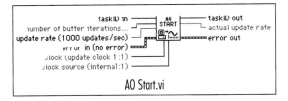

AO Start.vi

AO Start starts a buffered analog output operation. **Update rate** is the number of updates to generate per second. If you wire a 0 to **number of buffer iterations**, the board will continuously output the data buffer until you run the **AO Clear** function.

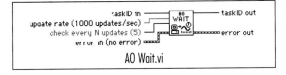

AO Wait.vi

AO Wait waits until the task's waveform generation is complete before returning. It checks the status of the task at regular intervals, and waits asynchronously between intervals to free the processor for other operations. The wait interval is calculated by dividing the optional **check every N updates** input (default = 5) by the **update rate**.

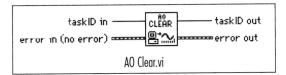

AO Clear.vi

AO Clear stops the analog output operation, deallocates the buffer from computer memory, and releases any DAQ board resources such as counters.

All the VIs shown in the previous two sections are meant to rely heavily on data flow to work together. As you probably already noticed, they follow a logical sequence. The analog functions are tied together by the taskID and error cluster nodes. The following table shows the intended sequence of VIs for analog waveform acquisition and generation.

Acquisition   Generation

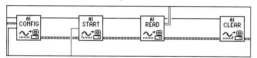   

1. **AI Config**: Configure the channels and buffer.

2. **AI Start**: Start the acquisition.

3. **AI Clear**: Clear the buffer and deallocate resources.

4. **AI Read**: Read from the buffer.

1. **AO Config**: Configure the channels and buffer.

2. **AO Write**: Write data to the buffer.

3. **AO Start**: Start the generation.

4. **AO Write**: Write new data to the buffer.

5. **AO Wait** [optional]: Wait for buffer to empty.

6. **AO Clear**: Clear the buffer and deallocate resources.

■ Buffered Analog Input

Remember buffered I/O, from the beginning of the chapter? Here is where you get to see it done. Previously, we used a While Loop to acquire multiple points. Now we'll use the "intermediate" VIs to see how we can avoid using a While Loop, by programming the hardware to acquire a set number of points at a certain sampling rate.

Activity 11-3

This is an example of buffered DAQ. First, let's build the following front panel, with a simple waveform graph for viewing the data.

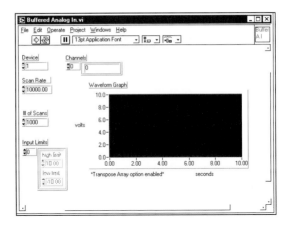

Note that Input Limits is an array of cluster data, with each cluster element containing two numeric controls. You can generate it by popping up on the **Input Limit** input to **AI Config** and selecting **Create Control.**

Next, connect a voltage source to channel 0. Then wire the block diagram using the analog input VIs, in the order we mentioned before.

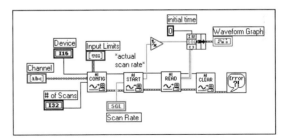

Save your VI as **Buffered Analog In.vi**

Voilà—you've just built a (very simple) oscilloscope! There are some details we should note about the block diagram you just built:

◆ The Input Limits control is used to adjust the gain of the board. For example, if you set each limit to 100 mV and –100 mV, respectively, the board will apply a gain of 100.

◆ The **AI Start** VI actually triggers the data acquisition. As soon as it's finished executing, the data has been loaded into the buffer. The buffer data is not "available" in LabVIEW until we decide to read it. Although we connected **AI Read** immediately after this VI, we could have read the data at any later time, provided the data in the buffer was not cleared or overwritten by some other process.

◆   The **actual scan rate** (the inverse of the actual scan period) output
      from **AI Start** turns out to be very useful for plotting the data, as
      we can use it to determine the "Δx" for the graph. If the actual
      scan rate doesn't match the theoretical scanned rate we pro-
      grammed, the graph will reflect that.

◆   **AI Clear** is important to use because it clears the buffer. Other-
      wise, a future **AI Read** might read some old data.

Activity 11-4

This is a modification of the previous exercise. Write a waveform display-and-store VI. This VI should have a front panel
similar to the one above, but with the added ability of storing all
the data in a spreadsheet file (see Chapter 9). The user should be
able to name the file. This same VI should be able to retrieve a
file, and display on a chart the old data from this file. You'll find
the solution in CH11.LLB/**Acquire and Save.vi**.

Hint: add a Boolean for "read/write" and a Case Structure so only one operation can be done at a time.

Activity 11-5

Write a "waveform synthesizer" analog output VI. This VI
should let you choose from a sine, triangle, square, and sawtooth waveform to output. The Full Development version of
LabVIEW includes some VIs for generating arrays whose values
represent sine, triangle, square waves, etc. You can find these in
the **Analysis▶Signal Generation** palette. Use the analog output
VIs in a similar fashion to the analog in VIs. As an optional feature, you could allow this VI to acquire a waveform from an external source, and have this waveform be one of the possible
analog output functions.

You'll find the solution to this activity in CH11.LLB/
Function Generator.vi.

◆   ◆   ◆

■ Nerd-level Analog I/O

Yep, this is where we throw you some fancy concepts such as circular buffering (continuous data acquisition), hardware triggering, and
disk streaming. Actually these concepts are pretty useful in practice.
We'll cover each one briefly with an example.

■ Continuous Data Acquisition

Continuous data acquisition, or real-time data acquisition, returns data from an acquisition in progress without interrupting the acquisition. This approach usually involves a circular buffer scheme as shown in the next figure. You specify the size of a large circular buffer. The DAQ board collects data and stores the data in this buffer. When the buffer is full, the board starts writing data at the beginning of the buffer (writing over the previously stored data, whether or not it has been read from LabVIEW). This process continues until the system acquires the specified number of samples, LabVIEW clears the operation, or an error occurs. Continuous data acquisition is useful for applications such as streaming data to disk and displaying data in real time.

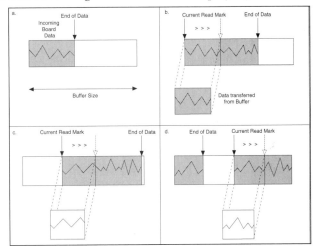

Activity 11-6

1. Build the front panel shown.

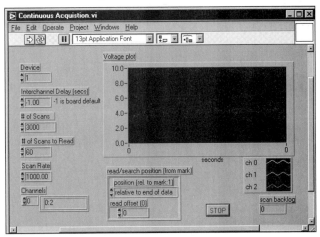

*Generate the <u>Read/Search Position</u> cluster by popping up on the similarly named input to **AI Read** and selecting **Create Indicator**.*

2. Save this VI as **Single Channel AI.vi**. Then wire the analog input VIs as shown in the next figure.

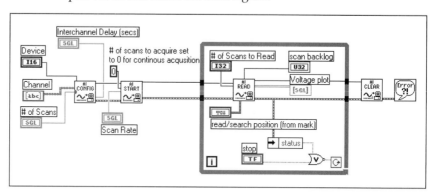

3. Configure LabVIEW for continuous data acquisition by instructing **AI Start** to acquire data indefinitely. To do this, set **number of scans to acquire** in **AI Start** to 0. This acquisition is *asynchronous*, meaning that other LabVIEW operations can execute during the acquisition. The following figure shows the block diagram for the previous front panel. **AI Read** is called in a looping structure to retrieve data from the buffer. From there, you can then send the data to the graph. **AI Clear** halts the acquisition, deallocates the buffers, and frees any board resources. Notice that the differences between this VI (which performs *continuous* acquisition) and the one for a waveform acquisition (the ordinary buffered DAQ VI in the previous section) are the While Loop, the 0 wired to the **number of scans to acquire** input of **AI Start**, and the **number of scans to read** parameter.

The previous VI is designed to run continuously, or "forever." In order to accomplish this without running out of memory, a fixed-length buffer is allocated, filled up with samples of data from beginning to end, and then the data at the beginning is overwritten as the buffer is filled up again. As an example, suppose this VI were run with a buffer size of 10 and a sampling rate of once per second. Here are snapshots of how the buffer is filled up with data.

Before the VI is run, the buffer is empty:

After one second, one sample is filled:

After nine seconds, the buffer is almost full:

After twelve seconds, the buffer has filled up and the data at the beginning is being overwritten:

This VI uses the "Intermediate" VIs to take advantage of Lab-VIEW's hardware timing and memory handling capabilities. If, while **Single Channel AI.vi** is running, an OS event ties up the microprocessor, the intermediate level VIs will use the DAQ board's buffer and DMA capability (if any) to continue collecting data without any microprocessor involvement. DMA allows the DAQ hardware to write directly to the computer's memory, even if the processor is tied up. Only if the OS event interrupts the microprocessor longer than the on-board FIFO buffer and DMA buffer can handle will LabVIEW lose data samples. This can best be understood by referring to the previous four snapshots of the buffer. First, assume the DAQ hardware has no on-board FIFO, but it has DMA capability. Ideally, the DAQ hardware will write data into the buffer continuously, and LabVIEW will be continuously reading a few samples behind the last sample written to the buffer. Suppose the microprocessor gets tied up after one second, and LabVIEW has read the first sample. When the microprocessor is tied up, the DAQ hardware device can write to this buffer, but LabVIEW can't read from the buffer. If after 12 seconds, the microprocessor is still tied up, then LabVIEW has missed the data in the second slot from the left when it contained good data (light gray). The four "intermediate" AI VIs used in this previous example will indicate such an error.

Why the Scan Backlog indicator from **AI Read**? It's pretty useful to know if LabVIEW is keeping up with reading the data. If the buffer fills up faster than your VI can retrieve the data in it, you will start losing some of this data, since the buffer will be overwritten, and Scan Backlog will reflect this.

■ Hardware Triggering

There are two ways to begin a DAQ operation: through a software or hardware trigger.

With software triggering, the DAQ operation begins when the software function that initiates the acquisition executes. For example, the "Easy" DAQ VIs use software triggering. As soon as LabVIEW executes the VI, the acquisition or generation starts. All DAQ boards support software triggering.

Another common method to begin a DAQ operation is to wait for a particular external event to occur. You usually begin the acquisition depending on the characteristics of a digital or analog signal such as the signal state, level, or amplitude. Hardware circuitry on the plug-in board uses this analog or digital event to start the board clocks that control the acquisition. Most DAQ boards support *digital* triggering to initiate an acquisition. Other boards also support *analog* triggering to initiate an acquisition. The input pin on your DAQ board that accepts the triggering signal is either EXTTRIG or START TRIG. All National Instruments MIO and Lab series boards support digital triggers.

You can also use an external analog trigger to initiate the *read* of the data from the buffer in LabVIEW, rather than just *initiate* the DAQ operation. This sort of triggering is called a *conditional retrieval*. When performing a conditional retrieval, your DAQ board acquires data and stores it in the acquisition buffer using software triggering. However, LabVIEW does not retrieve data until it acquires a sample that meets certain level and slope conditions. Careful here—it's easy to confuse conditional retrieval with hardware analog triggering. Systems using analog triggering do not store data in the acquisition buffer until the trigger condition occurs; systems using conditional retrieval do store the data in a buffer, but this data is not read into LabVIEW until the trigger occurs.

The following figure shows the conditional retrieval cluster that sets the criteria for retrieving data from the acquisition buffer. This cluster is an input to the **AI Read** VI. After the acquisition starts, the board continually samples the signal and compares it to the retrieval conditions. After the conditions are met, **AI Read** returns the amount of data specified in its **number of scans to read** input.

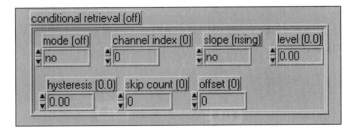

Activity 11-7

Build an analog input VI that uses hardware triggering to start and stop the acquisition.

1. Connect a switch with a TTL signal to the EXT TRIG (or similar) pin on your DAQ board, and connect a couple of analog input signals.

2. Build the front panel shown. The <u>Trigger Type</u> control is a ring control that lets you specify the type of trigger:

0   no triggering (default)

1   analog trigger

2   digital trigger A

3   digital trigger A and B

4   scan clock gating

You can use the one on the front panel of the example in CH11.LLB\ **Triggered Analog In**, or simply use a numeric control.

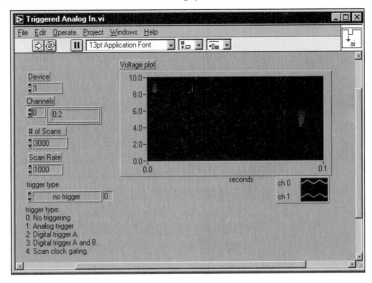

3. Make a block diagram like the one shown. Notice that the only difference between this one and the simple buffered analog input is the connection of the <u>trigger type</u> input on **AI Start.**

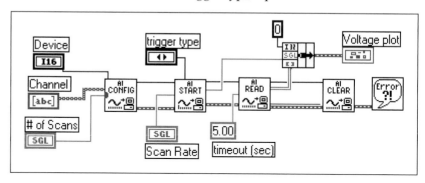

4. Save this VI as **Triggered Analog In.vi**

◆   ◆   ◆

■ Streaming Data to File

You have already written exercises that send data to a spreadsheet file. In these exercises, LabVIEW converts the data to a spreadsheet format and stores it in an ASCII file after the acquisition completes. A different and sometimes more efficient approach is to write small pieces of the data to the hard disk while the acquisition is still in progress. This type of file I/O is called streaming. An advantage of streaming data to file is that it's fast, so you can execute continuous acquisition applications, and still have a stored copy of all the sampled data.

With continuous applications, the speed at which LabVIEW can retrieve data from the acquisition buffer and then stream it to disk is crucial. You must be able to read and stream the data fast enough so that the board does not attempt to overwrite unread data in the circular buffer. To increase the efficiency of the data retrieval, you should avoid executing other functions, such as analysis functions, while the acquisition is in progress. Also, you can use the binary data output from **AI Read**, instead of the voltage data output, to increase the efficiency of the retrieval, as well as the streaming. When you configure **AI Read** to produce only binary data (specified by the **output units** input), it can return the data faster to the disk than if you used the **File I/O VIs** after acquiring the data into an array. One disadvantage to reading and streaming binary data is that users or other applications cannot easily read the file.

You can easily modify a continuous DAQ VI to incorporate the streaming feature. Although you may not have used binary files before, you can still build the VI in the following example. We will be discussing binary and other file types in Chapter 14, *Communications and Advanced File I/O*.

Activity 11-8

From the full version of LabVIEW, open the example **Cont Acq to File (binary)**, found under `LabVIEW\examples\daq\anlogin\strmdsk.llb`. You will observe how it streams to disk.

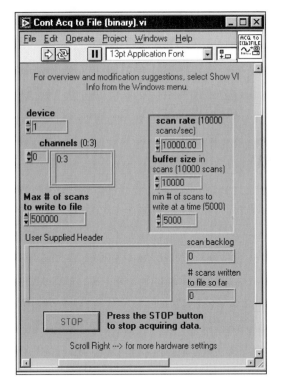

After you connect a signal to the DAQ board, run this VI for a few seconds—you may want to first enter header information in <u>User Supplied Header</u> that will go in your file.

To see the data written to disk, use the companion example VI **Display Acq'd File (binary)**, found in the same path as the previous VI.

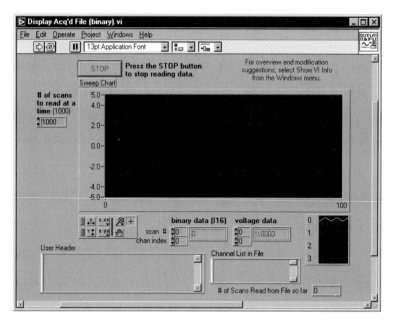

Run this VI, choosing the filename you used in the previous VI. Notice that the first VI, which streams data, doesn't attempt to graph the data. The point of streaming is to acquire data to the disk *fast*—and look at it later. Check out the block diagrams of both VIs to see how they work.

Digital I/O

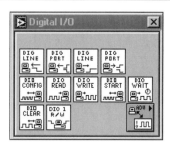

Ah, the digital world, where everything—for the most part—is much simpler than its analog counterpart. 1 or 0, high or low, on or off; that's mostly about it. Actually, some knowledge of binary representation and arithmetic is required, along with knowing the following LabVIEW definitions.

A digital **line** is the equivalent of an analog channel, a path where a single digital signal is set or retrieved. Digital lines are usually either **input lines** or **output lines**, but sometimes can be **bidirectional.** On some boards, digital lines must be configured as input or output; they can't act as both at the same time.

A **port** is a collection of digital lines that are configured in the same direction and can be used at the same time. The number of digital lines in each port depends on the board, but most ports consist of four or eight lines. For example, an AT-MIO-16E-10 board has eight digital lines, configurable as one eight-line port, two four-line ports, or even eight one-line ports. Ports are specified as string types, in a similar manner to channel strings.

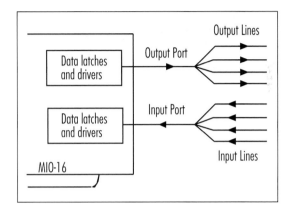

Port width is the number of lines in a port.

State refers to the one of two possible cases for a digital line: a Boolean TRUE (same as logical 1 or "on"), or a Boolean FALSE (same as logical 0 or "off").

A **pattern** is a sequence of digital states, often expressed as a binary number, which describes the states of each of the lines on a port. For example, a four-line port might be set with the pattern "1101," meaning the first, third, and fourth lines are TRUE, and the second line is FALSE. The first bit, or least significant bit (LSB) is the rightmost bit on the pattern. The last (fourth in this example), or most significant bit (MSB) is the leftmost bit in the pattern. This pattern can also be converted from its binary equivalent to the decimal number 13.

Incidentally, National Instruments DAQ boards use TTL positive logic, which means a logical low is somewhere in the 0-0.8V range; a logical high is between 2.2 V and 5.5 V.

■ Quick & Dirty Digital I/O: The Top Tier

For simple digital I/O, the top-tier VIs on the **Digital I/O** palette are great—very easy and intuitive to use.

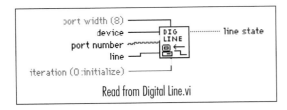

Read from Digital Line.vi

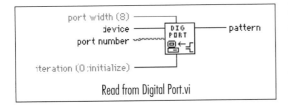

Read from Digital Port.vi

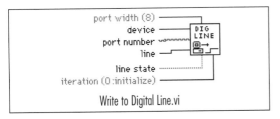

Write to Digital Line.vi

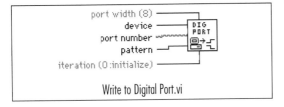

Write to Digital Port.vi

Read from Digital Line reads the logical state of a digital line. **Device** is the DAQ board device number and **port number** specifies the port containing the line. **Line** specifies the digital line to read. **Line state** returns the state of the line: high (TRUE) or low (FALSE).

Read from Digital Port reads the state of all lines in a port. **Port number** specifies the digital port to read. **Pattern** returns the digital line states as a decimal number, which if you convert to a binary representation, makes it easy to see the state of each individual line in the port.

Write to Digital Line sets a particular line on a port to a logical high or low state. **Device** is the DAQ board device number. **Port number** specifies the port containing the line. **Line** specifies the digital line to write to, and **line state** indicates the state to write to the line: high (TRUE) or low (FALSE).

Write to Digital Port outputs a digital pattern to the specified port. **Port number** specifies the digital port to update. **Pattern** is the decimal number equivalent of the digital or binary pattern to be written to the lines in the port.

All of the above VIs implement *immediate,* or *nonlatched,* digital I/O. This means that as soon as **Write to Digital Line** is executed, for example, that line is set to high or low immediately and remains in that state until another call to the VI changes it.

The **iteration** input on all these VIs needs some explanation. Since these are top-level digital I/O functions, by default they configure the digital ports for the appropriate direction and type when they are called. This configuration normally only needs to be done once for a series of reads or writes. If you use one of these functions repeatedly, such as in a loop, you can get rid of the extra overhead of configuring the ports over and over by wiring any nonzero number to the **iteration** input. Zero, the default value, tells the VI to initialize when it executes. The following example should help clarify the use of these VIs.

Suppose you had a high-voltage switch with a digital relay that let you know when it closed. The digital line from this relay is normally high (switch open). When the switch closes, the line goes to a logic low.[4] To create a simple VI that checks the status of digital line 0 on port 0 of your DAQ board (and thus determine the position of the switch), you might design something like this block diagram:

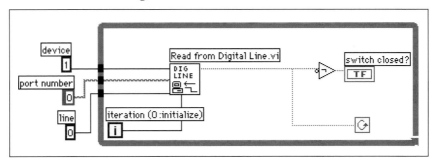

Iteration Terminal

Notice how the loop's iteration terminal is wired to the **iteration** input on the digital VI. The first time around the loop, when $i=0$, the **Read from Digital Line** will perform the configuration and initialization of the digital port. Subsequently, after $i>0$, the VI will only be reading from the line, allowing the loop to execute in less time. Once the digital line is set to FALSE, when the switch closes, the loop terminates and our <u>switch closed?</u> Boolean comes on.

Activity 11-9

Write a VI that has four Boolean LEDs on the front panel in an array. Make these controls (by popping up on them and selecting **Change to Control**). The objective of this panel is to turn on some real LEDs connected to the digital lines on your board (be sure and put a resistor in series with the LED). The user should be able to turn the real LEDs on and off with the "virtual" LEDs. Also, make this program write to the whole digital port, and not to each individual line, to reduce overhead. Save your VI as **Digital Port.vi**.

---

4. Logic low for "closed" and high for "open" may seem counterintuitive, since a closed switch usually means something has turned on, which we associate with a logic high. Nevertheless, most standard relays do use this "negative" logic.

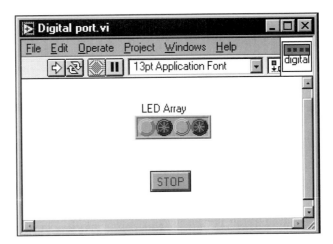

The solution is fairly simple—but is made easier if you find the **Boolean Array to Number** function on the **Boolean** palette. You can take the Boolean array and, using this function to convert the binary value of the Booleans in the array to a decimal number, wire the **pattern** input painlessly.

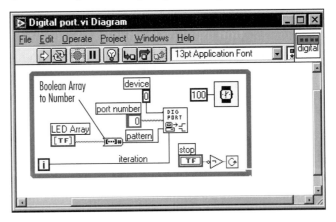

You'll notice there are many other functions in the **Digital I/O** palette. For more advanced applications, consult the LabVIEW manuals.

The GPIB VIs

We could devote a whole chapter or even a book to the subject of GPIB (remember, the GP stands for "general purpose"). Nonetheless, we will skim over the basics to help you get started.

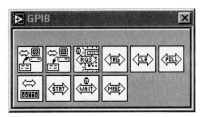

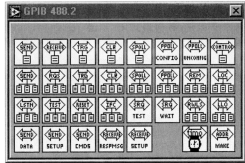

First of all, you'll notice there are *two* palettes for GPIB communication: one is called **GPIB** and the other, chock-full of more functions, **GPIB 488.2**. A word of explanation is due here: GPIB 488.2, also called IEEE 488.2 is the latest "version" of the standard. 488.2 establishes much tighter rules about the communications and command sets of instruments in an attempt to make programming easier and more uniform for different GPIB devices. Most GPIB devices will still work if you use the **GPIB** palette. If you need to adhere to the IEEE 488.2 standard, though, you should use the **GPIB 488.2** palette.

GPIB VIs will eventually be replaced by the functions in the VISA palette. VISA VIs are more general-purpose functions that can communicate with other instrument types, such as VXI. For more information on VISA VIs, consult the LabVIEW manuals.

Remember, all GPIB instruments have an *address,* which is a number between 0 and 30 uniquely identifying the instrument. Sometimes instruments will also have a *secondary address.* In all the GPIB VIs, you must provide the address of the instrument (in string format) in order to talk to it. This makes it easy to use the same VIs to talk to different devices on the same GPIB bus; you can enclose the addresses in a Case Structure, for example.

The majority of GPIB communication involves initializing, sending data commands, perhaps reading back a response, triggering the instrument, and closing the communication. With the following subset of simple GPIB VIs from the **GPIB** palette, you can do quite a bit.

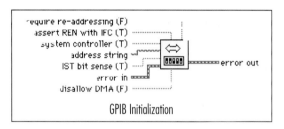

GPIB Initialization

Initializes the GPIB device whose address is given in **address string.** As in DAQ channels, the address is specified as a string because occasionally it may contain nonnumeric characters.

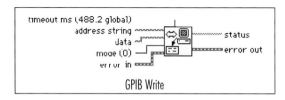

GPIB Read

Reads data from the device referenced by **address string**. The number of bytes read is determined by **byte count**. The data read is returned in the **data** string.

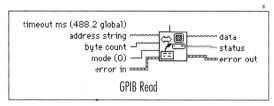

GPIB Write

Writes string **data** to the device.

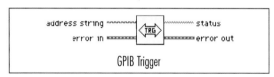

GPIB Trigger

Sends a trigger command to the device.

GPIB Clear

Clears the device.

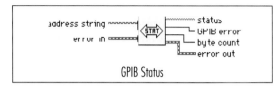

GPIB Status

Returns the status of the GPIB controller indicated by **address string** and number of bytes sent in the previous GPIB operation.

A good way to test if you are communicating over GPIB with your instrument is to use the built-in example, **LabVIEW<->GPIB.vi** (found in `examples\instr\smplgpib.llb`) in the full version of LabVIEW.

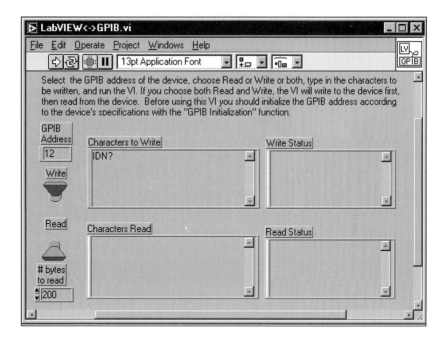

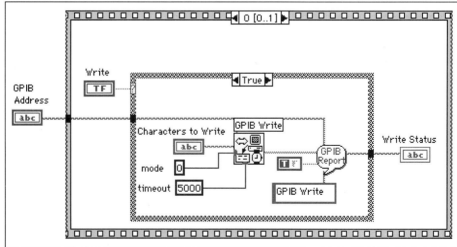

The **GPIB Report** VI in this block diagram is a useful utility VI that reports the status and errors (if any) of the previous GPIB operation. The Boolean input determines whether a dialog box is shown to the user in case of an error. You can find this VI in `examples\instr\smplgpib.llb` in the full version of LabVIEW.

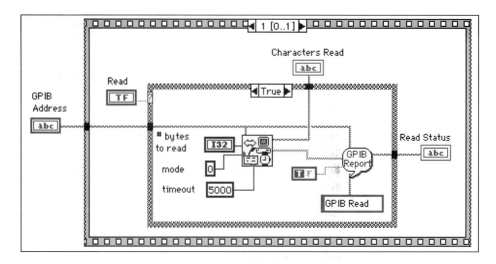

Sets of VIs for communicating specifically with hundreds of different instruments have already been written. These are the famous *instrument drivers* you hear about. Most of them are free of charge; you can get them on the LabVIEW Full Development CD from National Instruments or you can download them over the Internet. Other, more specialized or custom instrument drivers are available from various companies such as Alliance members (3rd party developers endorsed by National Instruments). Sometimes the instrument manufacturer will have the LabVIEW instrument driver available. Instrument drivers can save you hundreds of hours in programming time because the low-level GPIB code has already been taken care of, including specific commands and codes required by the particular instrument.

Often you will only use a few of the VIs included in an instrument driver, since you may only need to manipulate the functions pertinent to your application. One commercial instrument driver, **VI Strength** (available from VI Technology, Austin, Texas), is designed to control Instron Materials Testers through GPIB. The following diagram shows a simple application using just three of the VIs included with **VI Strength**.

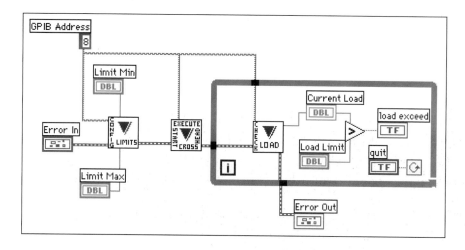

Serial VIs

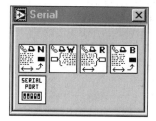

In one sense, serial communication is the simplest to program: after all, only five serial VIs exist on the **Serial** palette. On the other hand, serial communication suffers from abused and ignored hardware standards, obscure and complex programming protocols, and relatively slow data transfer speeds. The difficulties people encounter with writing an application for a serial instrument are rarely LabVIEW-related! We'll take a quick look at how to use the serial port VIs to communicate with a serial device.

You should become familiar with some basic concepts of how serial communication works if you've never used it before. If you have used a modem, and know what things like baud rate, stop bits, and parity roughly mean, then you should know enough to get started. Otherwise, it might be a good idea to read some documentation about the serial port (any good book on RS-232).

From LabVIEW's point of view, serial communication can occur at any specified *port* on the computer. The **port number** has different meanings depending on your platform:

| | | UNIX | |
Windows	MacOS	Solaris 1	Solaris 2
0: COM1	0: modem port	0: /dev/ttya	0: /dev/cua/a
1: COM2	1: printer port	1: /dev/ttyb, etc.	1: /dev/cua/b, etc.
2: COM3			
. . .			
8: COM9			
10: LPT1			
11: LPT2, etc.			

Following is a brief overview of what each serial communication VI does:

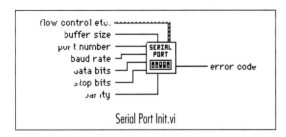

Serial Port Init.vi

Initializes the selected port to the specified settings. Often you can leave inputs unwired if you use the default values which are:

port number: 0

baud rate: 9600

data bits: 8

stop bits: 1

parity: none

The **flow control** input is a cluster of several different handshaking modes. By default, this VI uses no handshaking. Open the front panel of this VI and copy this complex cluster if you need to use it.

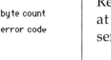

Bytes at Serial Port.vi

Returns the number of bytes currently at the input buffer of the designated serial port.

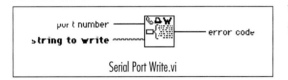

Serial Port Write.vi

Writes the data in the input string to the specified serial port.

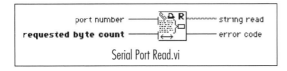

Serial Port Read.vi

Reads the number of characters specified by **requested byte count** from the specified serial port.

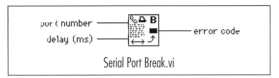

Serial Port Break.vi

Sends a break to the specified serial port for at least as long as **delay (ms)**.

One serial byte corresponds to one ASCII character, which is the way most serial communication is done. Another useful concept to remember is that your computer (and possibly your serial device as well) has a *serial port buffer*. The FIFO (first in, first out) buffer collects the characters as they come in over the serial line. This is important to know, because when you perform a serial read, you are reading the oldest data still in the buffer. Once the data is read, it is discarded from the buffer.

Looking at the above VIs, the sequence for serial communication should appear fairly obvious: initialize, write a command, read the response, repeat. The example included with the full version LabVIEW, **LabVIEW<->Serial Port.vi** (found in `examples\instr\smplserl.llb`) is a great starting place for writing a serial communication VI.

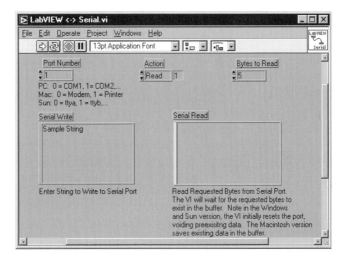

Notice that this VI can read or write to the serial port, but not at the same time. To read, it waits until it finds the expected number of bytes at the serial port buffer. It's a wise move to wait for the expected number of bytes to make sure all the data from the serial device has arrived—otherwise you'll get erroneous results. However, if you specify a read of a larger number of bytes than actually arrive, your VI will wait indefinitely.

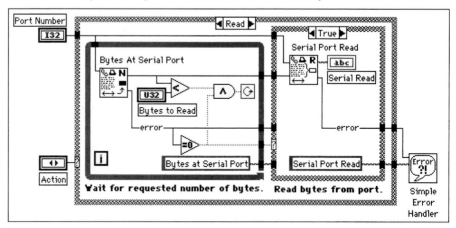

Writing is even easier; just send the string out the serial port:

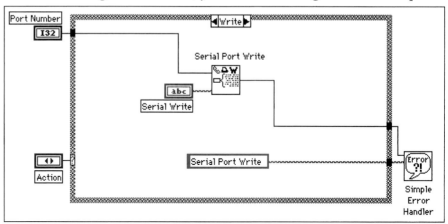

Some of the following guidelines may help you when writing an application that communicates with a serial instrument:

1.   Be absolutely certain that the configuration of the serial port on your instrument matches *exactly* the configuration in your **Serial Port Init** function (baud rate, handshaking, etc.). If even one of these parameters doesn't match, they won't communicate.

2.   Many serial devices expect a carriage return and/or a linefeed character after each string command. LabVIEW doesn't automatically append one, as many terminal emulators do. To add one or

both of these characters to your command string, pop up on the string terminal and choose **Enable '\' Codes**. Then type the carriage return (\r) and/or linefeed (\n).

3. Use **Bytes At Serial Port**! This VI is very useful. For example, if you didn't know ahead of time how many characters to expect from a serial device, you could use this VI to find out, so you can read the correct amount. You can wire the output of this VI directly to the **requested byte count** on **Serial Port Read.**

4. If in doubt, flush the buffer before issuing a command to which you expect a response. There is no inherent way to know if the serial data you read with **Serial Port Read** just came in or has been sitting there for hours. To flush the buffer, perform a read of all the bytes in the buffer, and discard the data.

5. There is a very useful example VI called **Serial Read with Timeout.**

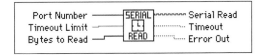

This VI can be found in `LabVIEW\examples\instr\ smplserl.llb` in the full version of LabVIEW. This function works just like **Serial Port Read,** except that it lets you specify a timeout if the serial instrument doesn't respond. Very useful when your program "hangs" waiting for a response at the serial port.

6. After LabVIEW has called any of the serial VIs, it does not normally "release" the serial port until you quit LabVIEW. This means that no other application (your modem, for example) can access the same serial port unless you first close the serial port (not the same as a serial port break). For some unexplained reason, you don't see a "serial port close" VI on the serial palette. Here's a secret: you can find this **Close Serial Driver** VI,

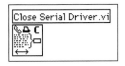

in the following location: `LabVIEW\vi.lib\_sersup.llb\ Close Serial Driver.vi` in the full version of LabVIEW. You have to call up this VI by using the **Select a VI...** We recommend adding it to the **Serial** palette if you find that you need to use it often.

Wrap It Up!

This chapter has helped get you started using LabVIEW to perform data acquisition and/or instrument control.

We took a look at the concepts of buffering and triggering. *Buffering* is a method for storing samples temporarily in a memory location (the buffer), and is used for fast and accurate sampling rates. *Triggering* is a method to initiate and/or end data acquisition dependent on a trigger signal. The trigger can come from the software or from an external signal.

The **Analog Input** and **Analog Output** palettes contain VIs for analog I/O at different levels of complexity. The top-tier VIs, or "easy" analog VIs, are useful for simple I/O. These easy VIs let you acquire one point (nonbuffered AI) or a whole waveform (buffered AI) from one or multiple channels. The middle-tier VIs, or "intermediate" VIs provide more control over data acquisition. With these VIs, you use them in a specific sequence to configure, start, read, and clear the operation. The intermediate VIs allow you to perform buffered I/O, hardware triggering, and disk streaming. The bottom-tier, advanced and utility, VIs, are not generally suitable for use by beginners.

The **Digital I/O** palette contains some simple VIs for reading and writing to the digital lines of a DAQ board. You can read or write to either a single *line* or a whole *port* (4 or 8 lines, depending on the board).

The **GPIB** and **GPIB 488.2** palettes contain about all the VIs necessary for communication with GPIB instruments. Generally, if you have a LabVIEW instrument driver available for your particular instrument, you won't need to use these. But some of the simple GPIB functions can be useful for testing and verifying communication with the instrument.

The **Serial** palette contains just 5 VIs, which cover the basic serial communication commands: initialize, read, write, etc. With these VIs, you can easily and quickly build a VI to communicate with any other RS-232 instrument.

Daq, Scxi, And LabVIEW Simulate And Test Power Systems

By Larry Park,
Senior Systems
Engineer, MicroCraft
Corporation

The Challenge

Developing a tool for simulating and testing semiconductor power systems.

The Solution

Creating a sophisticated system simulator using data acquisition (DAQ) boards, SCXI signal conditioning, and LabVIEW to initiate events and collect, analyze, and report real-time data.

Introduction

Applying semiconductor technology to power systems has emerged as a major advance in the electric power industry. In an industry where efficiency and optimization are key requirements, this state-of-the-art technology optimizes high-power transfer levels for more efficient use of transmission lines by increasing stability limits. The power system simulator at the ABB Power T&D Company's Transmission Technology Institute (ABB-TTI) in Raleigh, North Carolina, is an effective tool for developing and testing power systems that use this semiconductor technology.

The simulator at ABB-TTI is a physical model of one portion of a client's power transmission system—for example, a high-voltage direct current (HVDC) system or a system using other power electronic devices. ABB-TTI scales power system voltages and currents to appropriate simulation levels and configures the simulator with electrical models of components, such as transformers, breakers, controls, synchronous machines, and transmission lines. Naturally, this complex simulation requires a sophisticated DAQ system to initiate events and to collect, analyze, and report real-time data.

Configuration

Together, ABB-TTI and MicroCraft Corporation developed a Windows-based DAQ system using National Instruments AT-MIO-16F-5 boards, SCXI signal condi-

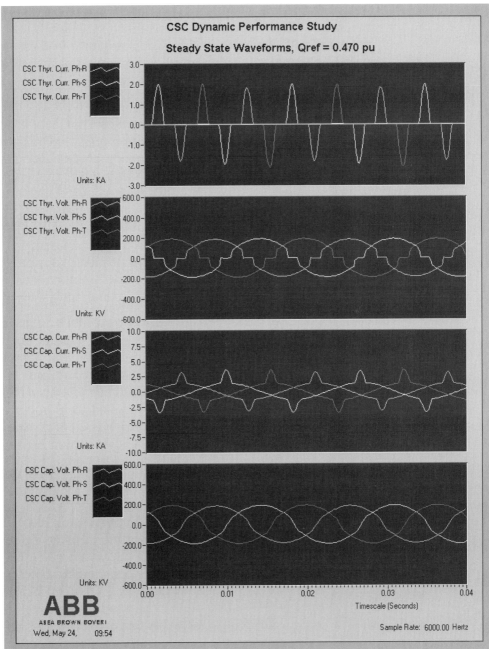

A sample data report showing four waveforms in various portions of the network under test. The user can determine which channels to graph, how many waveforms to display, and the number of graphs displayed on the page.

tioning hardware, and LabVIEW. This system triggers events in the simulator with a combination of analog, digital, relay, and timer outputs and records simulation response to these stimuli with 64 channels of analog input. In addition, the system is easy to use and flexible, so the user can focus on the tasks of power system analysis and design.

Meeting quantity and speed requirements for data collection presented the first challenge in developing ABB's DAQ system. The simulator is monitored on 64 analog channels that are simultaneously sampled at rates of up to 6,000 samples per second on each channel. To achieve this acquisition rate of more than 400,000 samples per second, we used two AT-MIO-16F boards. In addition, with eight SCXI-1140 signal conditioning modules connected to the front end of the AT-MIO boards, users can simultaneously sample all channels to prevent channel-to-channel time skew.

> This DAQ system is easy to use and flexible, so the engineer can focus on the tasks of power system analysis and design.

Analog and digital outputs, relay outputs, and precision timers initiate disturbances and other events (simulating faults and protective actions) as we collect response data. We selected the AT-DIO-32F for its 32-bit, double-buffered output rates of 5,000 patterns per second. Several output bits operate a National Instruments SC-2062 relay module attached to high-current or high-voltage equipment. The remaining bits control simulation equipment operating at standard 24 V logic. When higher resolution is necessary, we use a National Instruments PC-TIO-10 in its variable-duty-cycle mode to produce pulse outputs.

> To achieve the acquisition rate of more than 400,000 samples per second, we use two AT-MIO-16F boards. In addition, with eight SCXI-1140 signal conditioning modules, we can sample all channels simultaneously to prevent channel-to-channel time skew.

To achieve the necessary data bandwidth, we installed the input/output (I/O) boards in a Windows 486 66 MHz EISA-bus PC with expanded direct memory access (DMA) capabilities. During the simulation cycle, the PC bus transfers data at a rate up to 820,000 bytes per second with DMA operations. With 16 MB memory, the buffers are large enough to hold six seconds of acquired data and the entire digital output sequence without having to read from or write to disk.

The next challenge to ABB and MicroCraft was using the PC to synchronize event triggering and data collection operations. To synchronize the start of a cycle, a zero-crossing detector connects to each AT-MIO board, the AT-DIO, and the PC-TIO. The simulation cycle begins when the instruments detect a zero crossing in the 60 Hz base of the system. Achieving this precise synchronization between the collected data and the event triggering is necessary to enable "point on wave" analysis of events.

LabVIEW is the ideal platform for merging all aspects of the DAQ system. LabVIEW easily handles the quantity and speed of the required data collection. It also simplifies system development by providing instrument drivers and memory management utilities. And, most important, LabVIEW provides an easy-to-use graphical user interface (GUI) for the engineer performing the power system analysis.

MicroCraft Corporation is a member of the National Instruments Alliance Program.

For More Information Contact

MicroCraft Corp.

3209–154 Gresham Lake Rd.

Raleigh, NC 27615

tel (919) 872-2272

fax (919) 872-58221

OVERVIEW

This chapter will show you how to use some of the more advanced and powerful functions and structures of LabVIEW. LabVIEW has the capability to manipulate and store local and global variables—much like conventional programming languages. You will also see how you can make controls and indicators more flexible and useful by using their attribute nodes, which determine the behavior and appearance of front panel objects. In addition, this chapter covers some miscellaneous advanced functions such as system calls, calling external code written in another programming language, dialog boxes, etc. Finally, you'll take a look at advanced data conversions and why you might need them.

GOALS

- Understand local and global variables, and know how and when to use them
- Customize the appearance and behavior of your front panel objects using attribute nodes
- Learn about some miscellaneous advanced functions
- Get an overview of what LabVIEW can do to interface with external code
- Perform conversions between different data types

KEY TERMS

- local variable
- global variable
- read and write mode
- race condition

- attribute node
- occurrence
- CIN
- Call Library

- ASCII string
- binary string
- typecasting

Advanced LabVIEW Functions and Structures

12

Local and Global Variables

Local and global variables are, technically speaking, LabVIEW structures. If you've done any programming in conventional languages like C or Pascal, then you're already familiar with the concept of a local or global variable. Up until now, we have read data from or written to a front panel object via its terminal on the block diagram. However, a front panel object has only one terminal on the block diagram, and you may need to update or read it from several locations on the block diagram or from other VIs.

Local variables (locals, for short) provide a way to access front panel objects from several places in the block diagram of a VI in instances where you can't or don't want to connect a wire to the object's terminal.

Global variables allow you to access values of any data type (or several types at the same time if you wish) between several VIs in

cases where you can't wire the subVI nodes, or when several VIs
are running simultaneously. In many ways, global variables are
similar to local variables, but instead of being limited to use in a
single VI, global variables can pass values between several VIs.

This section will teach you some of the benefits of using locals
and globals, as well as show you some common pitfalls to watch
out for.

■ Local Variables

Undefined Local

Local variables in LabVIEW are built-in objects that are accessible from the **Structures** subpalette in the **Functions** palette. When
you select a local variable object, a node showing a "**?**" first appears to indicate the local is undefined. By clicking on this node
with the Operating tool, a list of all current labeled controls and
indicators will appear; selecting one of these will define the local.
Or you can pop up on the local variable and choose **Select Item➤**
to access this list. You can also create a local variable by popping
up on an object's terminal and selecting **Create➤Local Variable.**

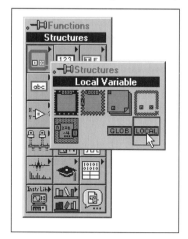

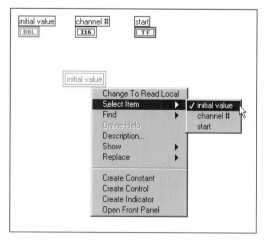

There are at least a couple of reasons why you might want to
use locals in your VI:

◆   You can do things, such as control parallel loops with a single
    variable, that you otherwise couldn't do

◆   Virtually any control can be used as an indicator, or any indicator
    as a control.

Controlling parallel loops

We've discussed previously how LabVIEW controls execution order through dataflow. The principle of dataflow is part of what makes LabVIEW so intuitive and easy to program. However, occasions may arise (and if you're going to develop any serious applications, the occasions *will* arise) when you will have to read from or write data to front panel controls and indicators without wiring directly to their corresponding terminals. A classical problem is shown here; we want to end the execution of two independent While Loops with a single Boolean stop control.

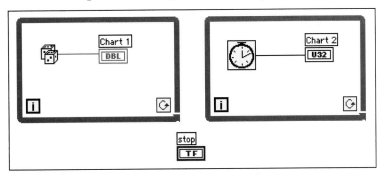

How can we do this? Some might say that we could simply wire the stop button to the loop terminals. However, think about how often the loops will check the value of the stop button if it is wired from outside the loops.

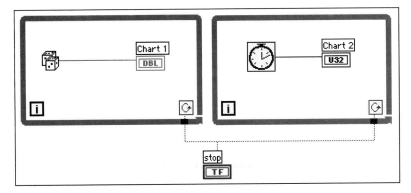

Wiring the stop button from outside the loops to both conditional terminals will not work, since controls outside the loops are not read again after execution of the loop begins. The loops in this case would execute only once if the stop button was FALSE when the loop starts, or execute forever if stop was TRUE.

So why not put the <u>stop</u> button inside one loop, as shown in the next picture? Will this scheme work?

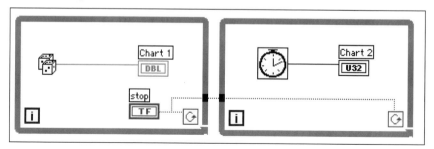

Putting the <u>stop</u> button inside one loop and stretching a wire to the other loop's conditional terminal won't do the trick either, for similar reasons. The second loop won't even begin until the first loop finishes executing—data dependency, remember?

The solution to this dilemma is—you guessed it—a local variable. Local variables create, in a sense, a "copy" of the data in another terminal on the diagram. The local variable always contains the up-to-date value of its associated front panel object. In this manner, you can access a control or indicator at more than one point in your diagram without having to connect its terminal with a wire.

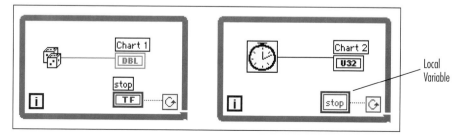

Local Variable

Referring to the previous example, we can now use one <u>stop</u> button for both loops by wiring the Boolean terminal to one conditional terminal, and its associated local variable to the other conditional terminal.

*There's one condition to creating a Boolean local variable: the front panel object can't be set to **Latch** mode (from the **Mechanical Action** option). Although it isn't obvious at first, a Boolean in **Latch** mode along with a local variable in read mode produce an ambiguous situation. Therefore, LabVIEW will give you the "broken arrow" if you create a local variable of a Boolean control set to **Latch** mode.*

Blurring the Control/Indicator Distinction

One of the really nice features about local variables is that they allow you to *write to* a control or *read from* an indicator, which is something you can't normally do with the regular terminals of an object. Locals have two modes: *read* and *write*. A local variable terminal can only be in one mode at a time, but you can create a second local terminal for the same variable in the other mode. Understanding the mode is pretty straightforward: in read mode, you can read the value from the local's terminal, just as you would from a normal control; in write mode, you can write data to the local's terminal, just as you would update a normal indicator. Just remember this formula for wiring locals:

READ mode = CONTROL
WRITE mode = INDICATOR

Another way to look at it is to consider a local in read mode the data "source," while a local in write mode is a data "sink."

You can set a local variable to either read or write mode by popping up on the local's terminal and selecting the **Change To...** option. A local variable in read mode has a heavier border around it than one in write mode (just like a control has a heavier border than an indicator), as shown in the following figure.

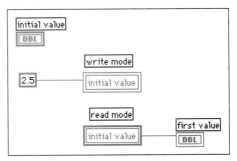

Pay close attention to these borders when you are wiring the locals, to make sure you are in the correct mode. If you attempt to write to a local variable in read mode, for example, you'll get a broken wire—and it may drive you crazy trying to figure out why.

Last but not least, you must label controls and indicators in order to make locals for them. That is, when creating a control or indicator, if you don't give it a name, you won't be able to create a local variable for it.

Activity 12-1

As a simple example, let's say you want to create a knob that represents a timer; the user can set the time, and watch the knob turn as it counts down the time, just like those old-fashioned timers used in the kitchen. Obviously, the front panel object is going to have to be a control since the user will set the time, but it must also be able to accept block diagram values and "turn" accordingly to indicate the passage of time.

1. Build a front panel with a knob labeled `Timer (seconds)` and select the **Show►Digital Display** option on it. Rescale the knob as shown in the following illustration.

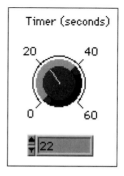

2. Create a local variable by selecting it from the **Structures** palette. You will get a local variable icon. Click on this icon with the Operating tool and select <u>Timer (seconds)</u>. The local should be in write mode by default

3. Build the following simple block diagram.

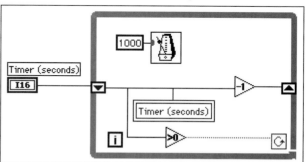

In this activity, the shift register is initialized with the value of <u>Timer</u> that was set at the front panel. Once loop execution begins, the shift register's value decreases once per second, and this value is passed to the <u>Timer</u> local variable, which is in write mode. The knob on the front panel rotates to reflect the changed value.

4. Set your knob to a desired time and run the VI. Watch the knob count down to zero!

5. Save your VI as **Kitchen Timer.vi**

A nice feature to add to this activity would be a sound to alert you that the timer has reached zero, just like the old-fashioned ones (Ding!). Later in this chapter you will see how we can play sounds with LabVIEW.

◆   ◆   ◆

Locals sound like a great structure to use, and they are. But you should watch out for a common pitfall when using locals: race conditions. A race condition occurs if two or more copies of a local variable in write mode can be written to in an unpredictable order.

There is a hazard to using locals and globals: accidentally creating a *race condition*. To demonstrate, build this simple example.

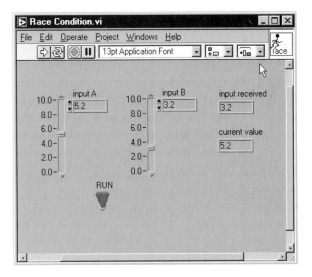

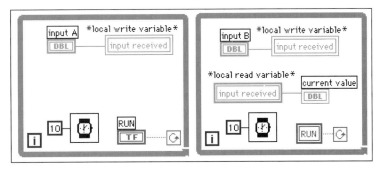

Notice the two **While Loops** controlled by Boolean <u>RUN</u> and a local variable <u>RUN.</u> However, the local variable <u>input received</u> is being written in the left loop as well as in the right loop. Now, execute this VI with <u>RUN</u> set to FALSE and different values on the two sliders. The loops will execute just once. What value appears at <u>current value</u>? We can't tell you, because it could be either value from <u>input A</u> or <u>input B!</u> There is nothing in LabVIEW that says execution order will be left to right, or top to bottom.

If you run the above VI with <u>RUN</u> turned on, you will likely see <u>current value</u> jump around between the two input values, as it should. To avoid race conditions, one must define the execution order by dataflow, Sequence structures, or more elaborate schemes.

Another fact to keep in mind is that every read or write of a local creates a copy of the data in memory. So, when using locals, remember to examine your diagram and the expected execution order to avoid race conditions, and use locals sparingly if you're trying to reduce your memory requirement.

Activity 12-2

Another case where locals are very useful is in an application where you want a "status" indicator to produce a more interactive VI. For example, you may want a string indicator that is updated with a comment, or requests an input every time something happens in the application.

1. Build a simple data acquisition VI similar to one you wrote in Chapter 11 (such as **Buffered Analog In.vi**), but modify it to have a string indicator that tells the user:

◆   *when the program is waiting for input*

◆   *when the program is acquiring data*

◆   *when the program is graphing the data*

◆   *when the program has stopped.*

To help you get started, the following figures show the front panel and the first two frames of a Sequence Structure.

2. Save your VI as **Status Indicator.vi**.

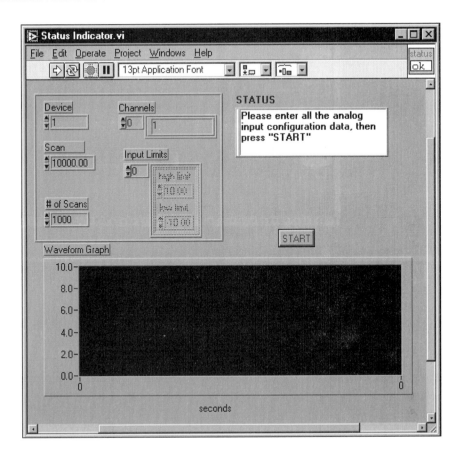

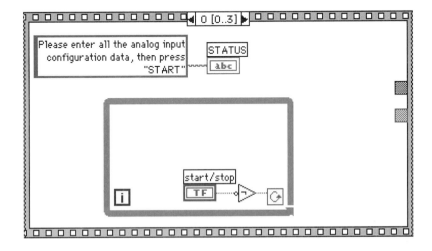

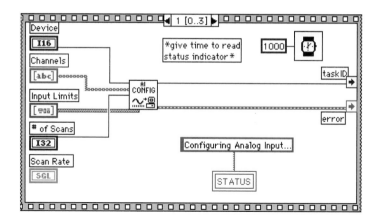

Activity 12-3

In many applications, you may want some type of "master" control that modifies the values on other controls. Suppose you wanted a simple panel to control your home stereo volumes. The computer presumably is connected to the stereo volume control in some way. In the VI shown next, a simulated sound control panel has three slide controls: left channel, right channel, and master. The left and right channel can be set independently; moving the master slide needs to cause increments or decrements in the left and right volumes proportionally.

1. Build the block diagram for the front panel shown below. The fun part about this is that by moving the master slide, you should be able to watch the other two slides move in response.

2. Save your VI as **Master and Slave.vi**.

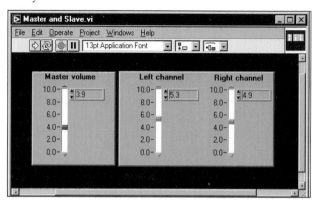

You will need to use shift registers for this activity.

■ Global Variables

If you've never used globals in your programs before, congratulations! Global variables are perhaps the most misused and abused structure in programming. Globals are more often than not the cause of mysterious bugs, unexpected behavior, and awkward structures. Having said this, there are still a few occasions when you might want to resort to globals.

Recall that you can use local variables to access front panel objects at various locations in your block diagram. Those local variables are accessible only in that single VI. Suppose you need to pass data between several VIs that run concurrently or whose subVI icons cannot be connected by wires in your diagram. You can use a global variable to accomplish this. In many ways, global variables are similar to local variables, but instead of being limited to use in a single VI, global variables can pass values between several VIs.

Consider the following example. Suppose you have two VIs running simultaneously. Each VI writes a data point from a signal to a waveform chart. The first VI also contains a Boolean <u>Power</u> button to terminate both VIs. Remember that when both loops were on a single diagram, we needed to use a local variable to terminate the loops. Now that each loop is in a separate VI, we must use a global variable to terminate the loops. Notice that the global terminal is similar to a local terminal, except that a global terminal has a "world" icon inside it.

First VI Front Panel   First VI Block Diagram

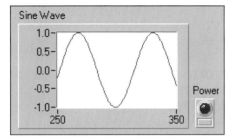
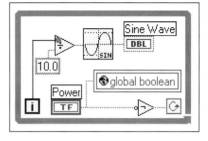

Second VI Front Panel

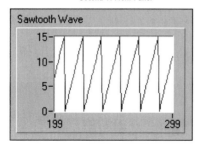

Second VI Block Diagram

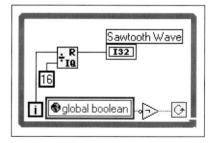

▉ Creating Globals

Like locals, globals are a LabVIEW structure accessible from the **Structures** palette. And like locals, a single global terminal can be in write or read mode. *Unlike* locals, different VIs can independently call the same global variable. Globals are effectively a way to share data among several VIs without any having to wire the data from one VI to the next; globals store their data independently of individual VIs. If one VI writes a value to a global, any VI or subVI that reads the global will contain the updated value.

Undefined Global

Once the global structure is selected from the palette, the icon shown at the left appears on the diagram. The icon symbolizes a global that has not been defined yet. By double-clicking on this icon, you will see a screen pop up that is virtually identical to a VI's front panel. You can think of globals as a special kind of VI—they can contain any type and any number of data structures on their front panels, but they have no corresponding diagram. Globals store variables without performing any kind of execution on these variables. Placing controls or indicators on a global's front panel is done in an identical fashion to a VI's front panel. An interesting tidbit about globals: It makes no difference whether you choose a control or an indicator for a given data type, since you can both read and write to globals. *Finally, be sure to give labels to each object in your global, or you won't be able to use them.*

A global might contain, as in the following example, a numeric variable, a stop button, and a string control.

Save a global just like you save a VI (many people use a ".gbl" extension when naming globals just to keep track of them). To use a saved global in a diagram, choose **Select a VI...** in the **Functions** palette. A terminal showing one of the global's variables will appear on the diagram. To select the variable you want to use, pop up on the terminal and choose **Select Item▶**, or simply click on the terminal using the Operating tool. You can select only one variable at a time on a global's terminal. To use another instance of that variable, or to use another element in the global, create another terminal for that global (cloning it by <control>-dragging or <option>-dragging is easiest, although you can always **Select A VI...**).

Operating Tool

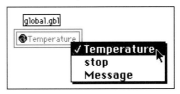

Just like locals, globals can be in a read or a write mode. To choose the mode, pop up on the terminal and select the **Change To...** option. Read globals have heavier borders than write globals. As with locals, globals in read mode behave like controls, and globals in write mode act like indicators. Again, a global in read mode is a data "source," while a global in write mode is a data "sink."

READ mode = CONTROL
WRITE mode = INDICATOR

Some important tips on using global variables:

1.   Always, always initialize your globals in your diagram. The initial value of a global should always be clear from your code. Globals don't preserve any of their default values unless you quit and restart LabVIEW.

2.   Never read from and write to global variables at the same time; i.e., where the order of events is undefined (this is the famous "race condition").

3.   Since one global can store several different data types, group global data together in one global instead of several globals.

It's important that you pay attention to the names you give the variables in your globals. All the VIs that call the global will reference the variable by the same name; therefore, be especially careful to avoid giving identical names to controls or indicators.

■ An example

Let's look at a problem similar to the two independent While Loops. Suppose that instead of having two independent While Loops on our diagram, we have two independent subVIs that need to run concurrently.[1]

The subsequent figure shows two subVIs and their respective front panels. These two VIs, **Generate Time** and **Plot,** are designed to be running at the same time. **Generate Time** just continuously obtains the tick count from the internal clock in milliseconds, starting at the time the VI is run. **Plot** generates random numbers once a second until a stop button is pressed, after which it takes all the tick count values from **Generate Time** and plots the random numbers versus the time at which the numbers were generated.

---

1. Of course, both subVIs can't *really* be executing at the same time on a single-processor machine! LabVIEW does a good job, however, of simulating parallel processes without you, the user, needing to worry about how it does this.

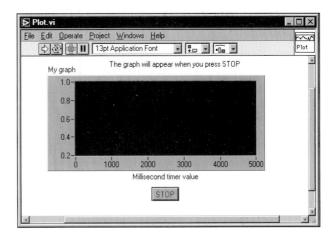

The way these two subVIs exchange data is through the use of a global. We want **Plot** to obtain an array of time values provided by **Generate Time**, and more importantly, we want both subVIs to be stopped by a single Boolean control.

First we create a global with the two necessary variables. Remember, to create a new global, select the **Global Variable** structure from the **Structures** palette, and double click on the "world" icon to define the global's components. In this case, we define a numeric component <u>Time (ms)</u> and a Boolean <u>stop</u>. The name of the global variable is **The Global.gbl**.

Then we use the global's variables at the appropriate places in both subVIs.

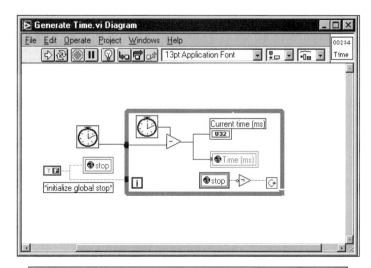

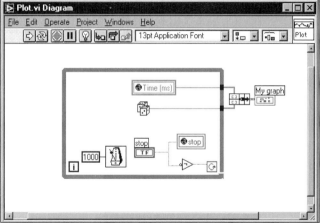

Notice how the stop Boolean variable is used: A stop button from **Plot** writes to the global variable stop, which is in turn used to stop the While Loop in **Generate Time**. Plot's front panel has been configured to open when the subVI is called so the user can see the plot and access the stop button. When the stop button is pressed in **Plot**, it will break the loop in **Generate Time** as well. Similarly, the time values from **Generate Time** are passed to the global variable Time, which is called by the **Plot** VI to build an array.

Hopefully, these two VIs have given you an example of how globals work. We could have avoided using globals for what we wanted to accomplish in this program, but the simple example is good for illustrative purposes.

If you look at the block diagram in the following figure, which calls the two subVIs, you'll see another problem with using globals: There are no wires anywhere! Globals obscure dataflow, since we can't see how the two subVIs are related. Even when you see a global variable on the block diagram, you don't know where else it is being written to. Fortunately, version 4 of Lab-VIEW addressed this inconvenience by including a feature that searches for the instances of a global.

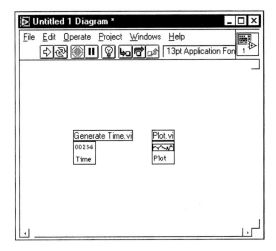

If you pop-up on a global's terminal, you can select **Find▶Global Definition**, which will take you to the front panel where the global is defined. The other option is **Find▶Global References**, which will provide you with a list of all the VIs that contain the global. For more information about LabVIEW's search capabilities, see the next chapter, *Advanced LabVIEW Features*.

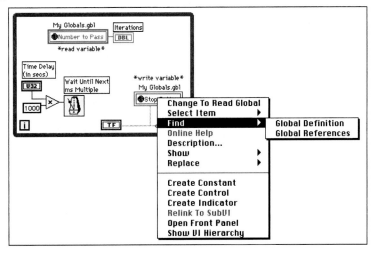

Attribute Nodes

With attribute nodes, first implemented in version 3 of Lab-VIEW, you can start making your program more powerful and a lot more fun. Attribute nodes allow you to programmatically control the attributes of a front panel object: things such as color, visibility, position, numerical format, etc. The key word here is *programmatically*, that is, changing the properties of a front panel object according to an algorithm in your diagram. For example, you could change the color of a dial to go through blue, green, and red as its numerical value increases. Or you could selectively present the user with different controls, each set of them appearing or disappearing according to what buttons were pressed. You could even animate your screen by having a custom control move around to symbolize some physical process.

To create an attribute node, pop up on either the front panel object or its terminal, and select **Create►Attribute Node**. A terminal with the same name as the variable will appear on the diagram. To see what options you can set in a variable's attribute node, click on the node with the Operating tool or pop up on the node and choose **Select Item►**. Now you have the choice of which attribute or attributes you wish to select. Each object has a set of *base attributes*, and sometimes, an additional set of attributes specific to that object.

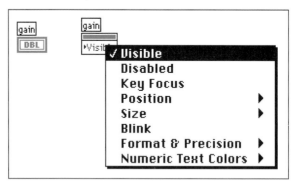

Just like with local variables, you can either read or write the attribute of an object. To change the mode of an attribute, pop up on it and select the **Change to...** option. The small arrow inside the attribute node's terminal tells you which mode it's in. An attribute node in write mode has the arrow on the left, indicating the data is flowing into the node, *writing* a new attribute. An attribute node in read mode has the arrow on the right, *reading* the current attribute and providing this data. The same analogy we used for locals—read mode works like a control and write mode works like an indicator—holds for attribute nodes.

An interesting feature of attribute nodes is that you can use one terminal on the block diagrams for several attributes (but always affecting the same control or indicator). To add an additional attribute, you can use the Positioning tool to *resize* the terminal and get the number of attributes you need, in the same way multiple inputs are added to functions like **Bundle, Build Array**, etc. The following figure shows two attributes on the same terminal for the numeric control <u>gain</u>.

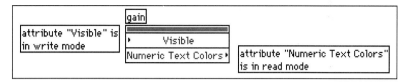

Let's look at a simple example. Suppose you wanted to have a front panel that would hide certain specialized controls except for those occasions when they were needed. In the following front panel, we see a tension gauge and a Boolean alarm switch. We include a button that says <u>Advanced . . .</u>, hinting at the possibility that if you pressed it, some really obscure and intricate options will pop up.

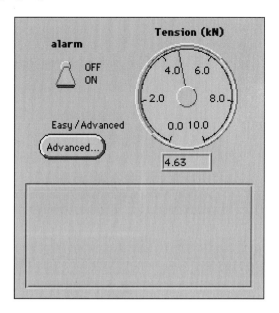

In this example, we've included two more controls, <u>gain</u> and <u>offset</u>, which are made invisible by setting their attribute nodes' option "Visible" to FALSE unless the button is pressed. If the <u>Advanced...</u> button is pressed, then ta-dah! The two knobs become visible.

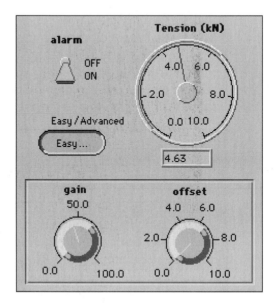

The entire block diagram is encompassed in a While Loop like the one shown next to make the button control the visibility of the two knobs and thus give the "pop-up" effect. For simplicity's sake, we've hidden most of the block diagram in the following figure.

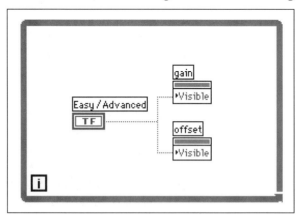

Often you will want to use more than one option in an object's attribute node. Remember, instead of creating another attribute node, you can select several options at a time by enlarging the terminal with the Positioning tool (much like you enlarge cluster and array terminals). You will see each new option appear in sequence; you can later change these if you like by clicking on any item with the Operating tool or popping up on them and choosing **Select Item▶**.

What do each of the base options in an attribute node refer to?

◆ **Visible**: Sets or reads the visibility status of the object. Visible when TRUE, hidden when FALSE. This is often a better choice than coloring an object transparent, since transparent objects can accidentally be selected.

◆ **Disabled**: Sets or reads the user access status of a control. A value of 0 enables the control so the user can access it; a value of 1 disables the control without any visible indication; and a value of 2 disables the control and "grays it out."

◆ **Key Focus**: When TRUE, the control is the currently selected key focus, which means that the cursor is active in this field. Key Focus is generally changed by tabbing through fields. Useful for building a mouseless application. See Chapter 13 for more information on Key Focus.

◆ **Position**: A cluster of two numbers that respectively define the top and left pixel position of the front panel object.

◆ **Size:** A cluster of two numbers that respectively define the height and width in pixels of the entire front panel object.

◆ **Blink:** When TRUE, the front panel object blinks.

◆ **Format and Precision**: Sets or reads the format and precision attributes for numeric controls and indicators. The input cluster contains two integers: one for format and one for precision. These are the same attributes you can set from the pop-up menu of the numeric object.

◆ **Color**: Depending on the type of object, you may have several color options. The input is one of those neat color boxes that sets the color of the text, background, etc., depending on the object.

The Help window really is helpful when using attribute nodes. If you move the cursor onto the terminal of an attribute node, the Help window will show you what the attribute means, and what kind of data it expects. You can also pop up on the attribute node terminal and choose **Create Constant** to get the correct data type wired up immediately—this comes in very handy when the input is a cluster.

Almost all controls or indicators have the base attribute options. Most of them have many more, especially tables and graphs, which have as a many as 73 options! We won't even begin to go into most of these options, because you may never care about many of them and you can always look up the details in the manuals. The best way to learn about attribute nodes is to create some to go with your application and to play around with them. You'll find that attribute nodes are very handy for making your program more dynamic, flexible, and user-friendly (always good for impressing your nontechnical manager).

■ Another example

Graphs and charts have zillions of options in their attribute nodes, as you can see by clicking with the Operating tool on the terminal of a chart's attribute node.

Some attribute nodes have several components (such as the Position attribute, which is a cluster of the left and top pixel coordinates). These attribute nodes, called compound attributes, are identified by the arrows indicating they have submenus. The fact to notice is that you can access the compound attribute as a whole, or just select one of its components.

This next example, which can be found in the attribute node examples in the full version of LabVIEW in `examples\general\attribute.llb`, shows just one of the many aspects of a graph you can control programmatically. **Chart Attribute Node** lets you programmatically select one of three display types for the chart: Strip, Scope, or Sweep (if you need to, review Chapter 8, *Charts and Graphs,* to see what they do).

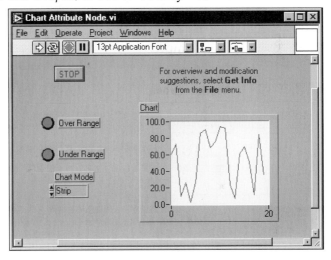

You can select the chart mode and watch it change even while the VI is running. The way this is done is through the **Update Mode** option on a chart's attribute node.

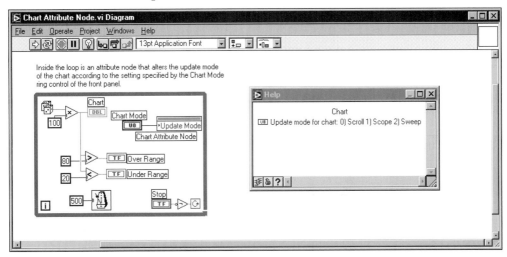

Activity 12-4

1. Write a VI that graphs three channels of data (either real-time data through a DAQ board, or random data). Let the user turn on or off any of the plots with three buttons. Since this is a chart and not a graph, the data will accumulate along with the old data.

2. Add a "CLEAR" button that clears the chart.

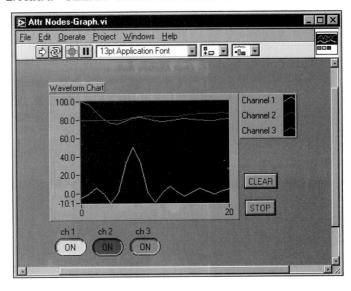

*LabVIEW includes a "transparent" color. To set a color attribute, you can create a cluster of **Color Box** constants. The **Color Box** is actually a numeric type, found buried deep in the **Numeric**➤**Additional Numeric Constants** palette. Pop up on a **Color Box** constant to select its color (remember the "T" is the transparent color).*

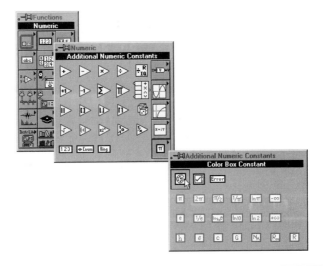

When using a multiplot graph or chart, you can only affect the attributes of one plot at a time. The plots are numbered 0,1, ...,n for attribute node purposes. A specific attribute, called **Active Plot**, is used to select the plot for the attributes you are modifying or reading. In this activity, you will need to build some case statements like the following:

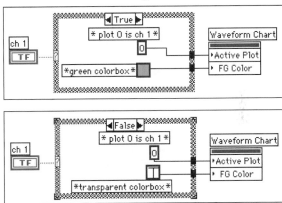

To clear a chart, use the **History Data** attribute of the chart. Then wire an empty array to this attribute node.

3. Save your VI as **Attr Nodes-Graph.vi**.

Activity 12-5

Ring controls, along with attribute nodes, can create some powerful interactive VIs. Examine the following example, called **Ring Control Menus,** which is in CH12.LLB.

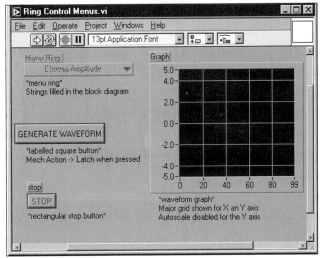

A pull-down menu ring, initially empty, will be used to query the user for the type of output. Once <u>GENERATE WAVEFORM</u> is pressed, the ring control will become enabled and say "Choose Waveform" and present several options: sine, square, or sawtooth wave.

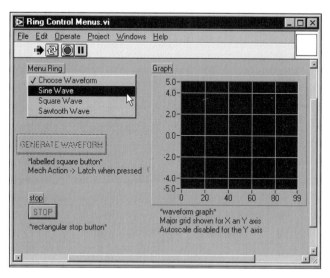

Next, the same ring control *will change* to let the user choose the output amplitude. Finally, the waveform is generated at the DAQ board and graphed.

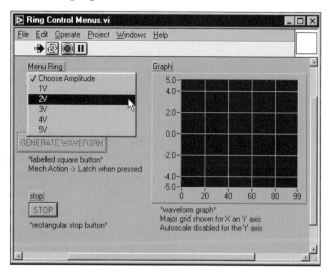

Here's a peek at part of the block diagram—check out the software solution for more detail.

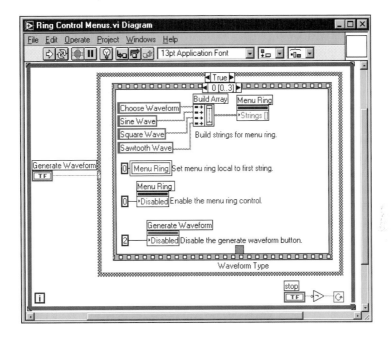

Other LabVIEW Goodies

A "miscellaneous" category is always hard to avoid—like the ever-growing one in our budgets. Many miscellaneous functions are located in the **Advanced** palette. We'll cover some of the functions in that menu as well as some others that don't quite fit anywhere else but nevertheless can be quite useful.

■ Dialogs

LabVIEW carrying on a real conversation with you? For now, you can make LabVIEW pop-up dialog windows that have a message and often some response buttons, like "OK" and "Cancel," just like in other applications. We've mentioned dialog functions before, but you may find the review and additional information useful.

The dialog functions are accessible from the **Time & Dialog** palette. LabVIEW provides you with two types of pop-up dialog boxes: one-button and two-button. In each case, you can write the message that will appear in the window, and specify labels on the buttons. The two-button dialog function returns a

Boolean value indicating which button was pressed. In any case, LabVIEW halts execution of the VI until the user responds to the dialog box. The dialog box is said to be *modal*, which means that even though windows (such as other VIs' front panels) will continue to run and be updated, the user can't select any of them or do anything else in LabVIEW via the mouse or keyboard until he or she has dealt with the dialog box.

As an example, suppose you wanted to add a dialog box to confirm a user's choice on critical selections. In the following example, a dialog box is presented when the user presses the computer's se

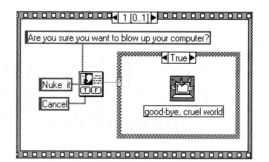

LabVIEW also provides dialog boxes for reading and saving files. These are discussed further in Chapter 14, *Communications and Advanced File I/O.*

Dialog boxes are a nifty feature to make your application have the look and feel of a true Windows or Mac executable, but don't overdo them. Remember, LabVIEW programs already provide the interactive graphical interface, so having unnecessary dialogs popping up can make your application more cumbersome and cluttered. It's best to reserve the dialogs for very important notifications and/or confirmations of a user's choice. Also, note that LabVIEW does not let you have more than two response buttons in a dialog box. For more elaborate "dialogs," you're best off writing a subVI whose front panel contains the dialog options you'd like. You can make this subVI pop open and even customize its window appearance (see the next chapter on how to do this).

■ Occurrences

The *occurrence* functions, **Generate Occurrence, Wait On Occurrence,** and **Set Occurrence**, all have kind of scary-sounding names—which is probably why we know hardly anybody who uses these functions. Nevertheless, occurrences are a powerful programming tool. You can use occurrences to force one or more parts of the block diagram to wait for something (an occurrence!) without using messy While Loops or globals. They can also speed up your program by saving processor time that would otherwise be spent examining conditions in loops. The occurrence functions are accessible from the **Occurrences** subpalette of the **Advanced** palette.

Occurrences are somewhat like Booleans—they're in one of two states: set or waiting to be set. This is what each of the functions do:

Generate Occurrence

Generate Occurrence creates an *occurrence refnum* (a magic number LabVIEW uses to keep track of the occurrence functions) that you pass on to the other occurrence functions.

Wait On Occurrence

Wait On Occurrence forces the part of the structure that contains it to halt execution until the occurrence is set. You can optionally wire a timeout to this function.

Set Occurrence

Set Occurrence triggers the occurrence; that is, it changes the "state" of the occurrence passed to it. When this function is executed, all other **Wait On Occurrence** functions that are wired to the *same* refnum detect it and allow the structure they're in to continue, regardless of what part of the block diagram these structures are in.

Occurrence functions operate on the part of the block diagram structure they're contained in, such as a particular frame of a Sequence Structure or a particular case of a Case Structure. In this next example, an occurrence is set if the <u>continue?</u> Boolean becomes TRUE (presumably a button available to the user). When the Sequence Structure at the right reaches frame 2, it does not continue to the next frames until the occurrence is set. Essentially, it allows just *part* of the Sequence Structure to run parallel to the While Loop, while the other part (after frame 2) must wait until the While Loop has concluded or the <u>continue?</u> button is pressed.

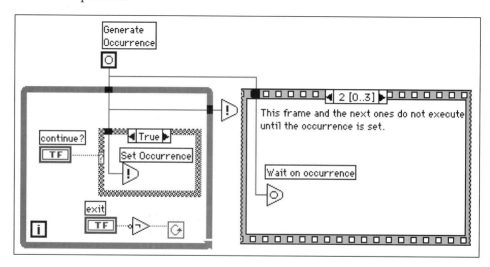

Note how both the **Set Occurrence** and **Wait on Occurrence** functions require an input from **Generate Occurrence**. You always need to provide this **Generate Occurrence** because in the case of multiple occurrences, the **Set Occurrence** and **Wait on Occurrence** need to know which one you're talking about. In other words, the **Generate Occurrence** doesn't really generate anything except the occurrence refnum used by the other functions. Another important detail you should notice in the above example is the second **Set Occurrence** wired to execute after the While Loop is finished. Think about what could happen if we left this out.

*When using occurrence functions, always remember to wire a **Set Occurrence** as an "escape" in your block diagram for all possible outcomes. Check your program to make sure it won't wait forever at a **Wait On Occurrence** event.*

Often you can achieve the same results using local variables as you would with occurrence functions. The advantage of occurrence functions is that they are slightly more memory- and time-efficient.

Saying "NO" harshly

LabVIEW gives you a couple of functions to abort the execution of your code immediately: **Stop** and **Quit LabVIEW**. Like the occurrence functions, these functions are also in the **Advanced** palette.

Stop Function

Stop has a Boolean input, that when TRUE (which is the unwired default input), halts execution of the VI and all its subVIs, just as if you had pressed the Abort button on the Toolbar.

Quit LabVIEW Function

Quit LabVIEW does just that, when its input is TRUE (also the default input). Be careful with this one!

Some of us reminisce about a third function that was included back when LabVIEW was in version 2.2: **Shutdown**. On certain Macintosh models, this function would effectively turn off the whole computer, monitor and all. We never understood how it could be useful, but it was fun to leave a colorful VI running with one big button that said "DO NOT PRESS THIS BUTTON" on someone else's Mac, and then just wait to see who would be too curious. The button was, of course, connected to the operating system's Shutdown function. It was a good way to amuse ourselves with the less experienced LabVIEW users.

On a serious note, however, you should use these functions with caution. It is considered bad programming etiquette (in any language) to include a "hard stop." Strive to always have a graceful way to exit your program.

*Make sure your VI completes all final tasks (closing files, finishing the execution of subVIs, etc.) before you call the **Stop** function. Otherwise, you may encounter unpredictable file behavior or I/O errors, and your results can be corrupted.*

Sound

A very nice or very obnoxious feature to have in many applications is sound, depending on how you use it. An audible alarm can be very useful in situations where an operator cannot look at the screen continuously during some test. A beep for every warning or notification, however, can be downright annoying. So use sounds with some thought to how often they will go off.

LabVIEW gives you access to the operating system's beep, through the **Beep** function. **Beep** is located in the **Advanced** palette.

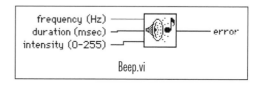

In the MacOS, you can set the frequency, intensity, and duration of the sound.

In Windows, these inputs are ignored and only the standard system beep is used. However, LabVIEW has the ability to call *.WAV sounds. The full version of LabVIEW includes an example VI called **Play Sound.vi** (you'll find it in `examples/dll/sound/plysnd.llb`). This VI calls external system code (see next section) that makes use of a Windows multimedia sound DLL.

Calling Code from Other Languages

What happens if you already have some code written in a conventional language (such as C, Pascal, FORTRAN, Basic) that you'd like to use? Or if for some reason you just miss typing in all those semicolons in your familiar text-based code? LabVIEW does give you some options for interfacing with code from other languages. If you are thinking about writing *all* your code in C or C++, you should check out LabWindows/CVI® (available from National Instruments), a programming environment very similar to LabVIEW; the main difference is that C code replaces the graphical block diagram. But if you'd like to (dare we say it?) actually have *fun* programming, then stick to LabVIEW and use conventional code only when you have to.

You may want to skip this section if you aren't familiar with writing code in other programming languages such as C, or if you don't expect to need to interface LabVIEW and external code.

You have two options for calling external code in Windows: you can call a Dynamic Link Library (DLL) using the *Call Library* function, or you can call the externally-generated executable code directly by using the *Code Interface Node (CIN)*. Both of these functions are accessible from the **Advanced** palette. Under Windows 3.1, you can call 16-bit DLLs, and under Windows 95, you can call 32-bit DLLs. LabVIEW 4 now also allows you to call code from LabWindows/CVI front panels by converting the C functions into subVIs.

In the MacOS, you can also use the Code Interface Node (CIN) to call executable source code. On Power Macintosh systems, you can use the Call Library function to communicate with Code Fragment libraries. On the older 680x0 Mac systems, unfortunately you cannot communicate with the Apple Shared Library (ASL), like you can with DLLs on Windows systems.

Using a CIN involves basically the following steps:

1. Place the CIN icon on the block diagram (from the **Advanced** palette)

2. The CIN has terminals for passing the inputs and outputs from the code. By default, the CIN has only one pair of terminals. You can resize the node to include the number of parameters you need.

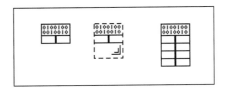

By default, each pair of terminals is an input-output: the left terminal is an input, and the right terminal is an output. However, if a function returns more outputs than inputs (or has no input parameters at all), you can change the terminal type to **Output-Only** by popping up on that terminal and choosing this option.

3. Wire the inputs and outputs to the CIN. You can use any LabVIEW data type when wiring the CIN terminals (of course, it *must* correspond to the C parameter data type for the function you're calling). The order of the terminal pairs on the CIN corresponds to the order of the parameters in the code. In the following example, we call a CIN that filters the <u>Raw Data</u> input array and passes the

output into the <u>Filtered Data</u> array. Notice also how the terminals, once wired, indicate the type of data you are passing.

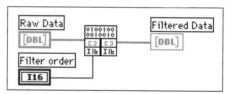

4.   Create a .c file by selecting this option from the pop-up menu. The .c file created by LabVIEW, in the style of C programming language, is a template in which you write the C code. With a little effort, you can paste in your existing code if you have it.

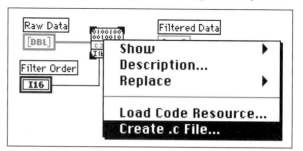

5.   Compile the CIN source code. This step could be tricky, depending on what platform, compiler, and tools you are using. You have to first compile the code with a compiler that LabVIEW supports, and then, using a LabVIEW utility, modify the object code so LabVIEW can use it.

6.   Load the object code into memory by selecting **Load Code Resource...** from the pop-up menu. Select the .lsb file you created when you compiled the source code.

   Once all these steps are completed successfully, you're calling C code in the CIN as if it were a subVI, with one major exception: calls to CINs execute synchronously, meaning that, unlike most other tasks in LabVIEW, they do not share processor time with other LabVIEW tasks. For example, if you have a CIN and a For Loop at the same level in the block diagram, the For Loop halts execution until the CIN finishes executing. This is an important consideration if your timing requirements are tight.

   One important final fact about CINs: They are most definitely NOT portable across platform. If you compile your LabVIEW code with a CIN on a Power Mac and then try to run it under Windows, it won't work. The way around this is to rebuild the CIN using a C compiler for the platform that LabVIEW is running on. At press time of this book, the C compilers compatible with LabVIEW are:

Windows 3.1	Watcom C
Windows 95/NT	Microsoft Visual C++, Microsoft Win32 SDK C/C++ compiler
MacOS	THINK C (ver. 5 or above), Symantec C++ (ver. 8 or above), Metrowerks CodeWarrior, MPW from Apple
Solaris	Sun ANSI C compiler
HP-UX	HP-UX C/ANSI C compiler

There are many complex issues involved in LabVIEW communicating with external code. Since many of these are highly dependent on the processor, operating system, and compiler you're using, we won't attempt to go any further into discussing CINs or Call Library functions. If you'd like more details, contact National Instruments and request their application notes on this subject, or look at the *Code Interface Reference Manual* included in the LabVIEW manual set.

Fitting Square Pegs into Round Holes: Advanced Conversions and Type-casting

Remember *polymorphism*, discussed early in this book? It's one of LabVIEW's best features, allowing you to mix data types in most functions without even thinking about it (any compiler for a traditional programming language would scream at you if you tried something like adding a constant directly to an array—but not LabVIEW). LabVIEW normally takes care of doing conversions internally when it encounters an input of a different but compatible data type than it expected at a function.

If you're going to develop an application that incorporates instrument control, interapplication communication, or networking, chances are you'll be using mostly string data. Often you'll need to convert your numeric data, such as an array of floating-point numbers, to a string. We talked about this a little in Chapter 9. It's important to distinguish now between two kinds of data strings: *ASCII strings* and *binary strings*.

ASCII strings use a separate character to represent each digit in a number. Thus, the number 145, converted to an ASCII string, consists of the characters "1," "4," and "5." For multiple numbers (as in arrays), a delimiter such as the <space> character is also used. This kind of string representation for numbers is very common in GPIB instrument control, for example.

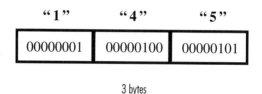

3 bytes

Binary strings are a bit more subtle—for one thing, you can't tell what data they represent just by reading them as ASCII characters. The actual bit pattern (binary representation) of a number is used to represent it. Thus, the number 145 with **I8** representation is just a single byte in a binary string (which, incidentally, corresponds on my computer's ASCII set to the "ë" character—not that most humans could tell that means 145). Binary strings are common in applications where a minimal overhead is desired, because generally it is much faster to convert data to binary strings and they take up less memory.

1 byte

Ultimately, all data in computers is stored as binary numbers. So how does LabVIEW know if the data is a string, Boolean, double-precision floating-point number, or an array of integers? All data in LabVIEW has two components: the *data* itself, and the data's *type descriptor*. The data type descriptor is an array of I16 integers that constitute code that identifies the representation of the data (integer, string, double-precision, Boolean, etc.). This type descriptor contains information about the length (in bytes) of the data, plus additional information about the type and structure of the data. For a list of data type codes, see Appendix A in the *LabVIEW User's Manual.*

In a typical conversion function, such as changing a number to a decimal string, the type descriptor is changed and the data is modified in some way. The conversion functions you use most of the time convert numbers to ASCII strings, such as in this example.

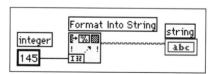

The topics in the rest of this section can be very confusing to beginners. If you do not have any need to use binary string conversions in your application, you can safely skip the rest of this section.

In some cases, you may want to convert data to a *binary* string. Binary strings take up less memory, are sometimes faster to work with, and may be required by your system (such as a TCP/IP command, or an instrument command). LabVIEW provides a way to convert data to binary strings using the **Flatten To String** function. You can access this function from the **Data Manipulation** subpalette under the **Advanced** palette.

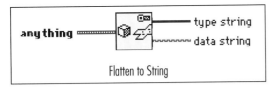

Flatten to String

The input to **Flatten To String** can be literally any LabVIEW data type, including complex types such as clusters. The function returns two outputs: the binary **data string** representing the data, and **type string** (which is not a string, but actually an array of **I16**, yet called a string because of nerdy programmer's terminology). The type string gives you the information in the data type descriptor.

A flattened binary string contains not only the data in compact form, but four bytes of *header* information as well. The header information included at the beginning of the binary string contains information about the length, type, structure, etc. of the data. For example, the following diagram shows an array of DBL numeric types flattened to a string. The **type string**, which is always an array of I16, contains this header information.

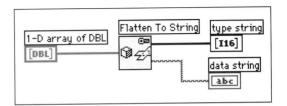

The key to using these confusing binary flattened strings is that most of the time you don't need manipulate or view the strings directly—you just pass these strings to a file, network, instrument, etc., for little overhead. Later, to read the data, you will need to unflatten the string.

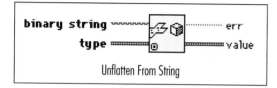

Unflatten From String

To read back binary strings, use the inverse function, **Unflatten from String**. The **type** input is a dummy LabVIEW data type of the same kind that the **binary string** represents. The **value** output will contain the data in the **binary string**, but will now be of the same data type as **type. err** is TRUE if the conversion was unsuccessful.

Here's the example of how we convert back our flattened binary string to an DBL array.

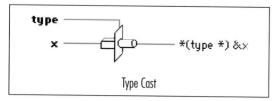

We wire the binary string input, as well as a dummy (empty) 1D array of DBL to specify the data type, and we get back our original array of DBL. Notice how we flattened and unflattened without ever needing to "peek" or "manipulate" the binary strings directly.

For fast, efficient conversions let's take a final look at a very powerful LabVIEW function: **Type Cast.**

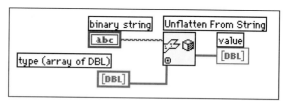

Type Cast

This function allows you to change the data type descriptor, without changing the data at all. There is no conversion of the data itself in any way, just of its type descriptor. You can take almost any type of data (strings, Booleans, numbers, clusters, and arrays), and call it anything else. One advantage of using this function is that it saves memory since it doesn't create another copy of the data in memory, like the other conversion functions do.

The following figure shows the comparative results of converting an array to a string through flattening, typecasting, and numerical formatting.

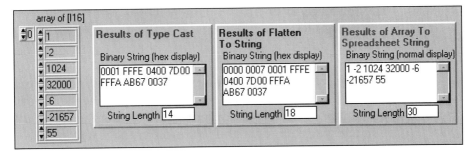

A common use of *typecasting* is shown in the following figure, where some instrument returns a series of readings as a binary string, and they need to be manipulated as an array of numbers. We are assuming that we know ahead of time that the binary string is representing an array of **I16** integers.

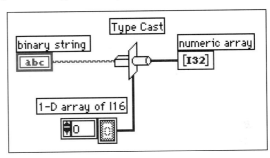

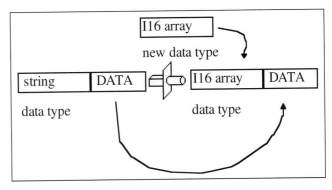

Type casting from scalars or arrays of scalars to a string works like the **Flatten To String** function, but it has the distinct feature that it doesn't add or remove any header information, unlike the four bytes of header strings created by the Flatten and Unflatten functions. The **type** input on the **Type Cast** function is strictly a "dummy" variable used to define the type—any actual data in this variable is ignored.

You need to be very careful with this function, however, because you have to understand exactly how the data is represented so your type casting yields a meaningful result. As mentioned previously, binary strings often contain header information (such as the output from the **Flatten to String** function). **Type Cast** does not add or remove any header information in binary strings. If you don't remove headers yourself, they will be interpreted as data and you'll get garbage as a result.

Make sure you understand how data to be type cast is represented. Type casting does not do any kind of conversion or error-checking on the data. Byte ordering (MSB first or last) is platform-dependent, so understand what order your data is in.

You might try experimenting with the **Type Cast** function to see what kind of results you get. To show you how bizarre you can get with typecasting, we took a cluster containing two numeric controls (type **I16**) and cast the data to a string, a numerical array, and a Boolean.

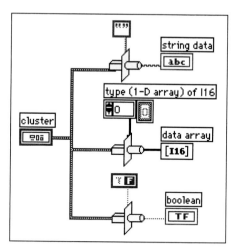

The front panel shows you what results we obtained.

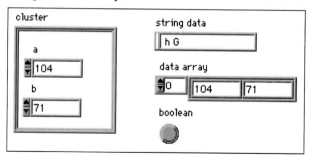

The string returned two characters corresponding to the ASCII values of 104 and 71. The array simply reorganized the numbers into array elements instead of cluster elements. And the Boolean? How would a Boolean variable interpret cluster data? Since we didn't know, we tried this, and found out that the Boolean is TRUE if the numerical value is negative; otherwise it's FALSE. In the case of a cluster, it apparently ORs the interpretation of the cluster elements, so if any element had a negative number, the Boolean interprets it as TRUE. Pretty strange, eh?

Wrap It Up!

In this chapter, we mined LabVIEW's gemstones: local variables, global variables, and attribute nodes. We also looked at some of the advanced functions: occurrences, dialogs, sound, calling external code, binary string conversion, and typecasting. The power and versatility offered by these structures and functions allow you to go a level deeper into programming applications in LabVIEW.

Local variables allow you to create a block diagram "copy" of a front panel object, which you can read from or write to in different places. *Global variables* are similar to locals, but they store their data independently of any particular VI, allowing you to share variables between separate VIs without wiring them together. Locals are very useful for controlling parallel loops or updating a front panel object from multiple locations in the diagram. Globals are a powerful structure, but care is required in their use because they are prone to cause problems.

Attribute nodes give you immense control over the appearance and behavior of indicators and controls, allowing you to change their attributes programmatically. Every control and indicator has a set of base attributes (such as color, visibility, etc.). Many objects, such as graphs, have numerous attributes that you can set or read accordingly.

With LabVIEW, you can use external C routines by using a *CIN (code interface node)*. A CIN lets you call C routines compiled with external compilers. Additionally, you can directly call DLLs (Dynamic Link Libraries) in Windows.

The functions **Flatten to String, Unflatten from String,** and **Type Cast** are powerful conversion utilities for working with different data types. These functions allow you to convert Lab-VIEW data types to and from binary strings. Binary strings are often needed in many applications.

OVERVIEW

*It's now time to examine the potpourri of advanced tools you can use in Lab-VIEW to make your programming environment easier and more powerful. By using the numerous preference options, you can customize LabVIEW so it works best for you. You will also learn about the VI Setup options and how LabVIEW's numeric types give you the option of representing them as binary, octal, or hexadecimal, as well as decimal numbers. In addition, you can include built-in units as part of a numeric variable and allow LabVIEW to perform automatic conversions. You will see how to instantly transform a section of your diagram into a subVI and explore some useful development tools in the **Project** menu.*

GOALS

- Use **Preferences** to customize your LabVIEW environment
- Become familiar with **VI Setup** options that allow you to make choices on the appearance and execution of your VI
- Be able to set up keyboard access of controls
- Learn how to represent numbers with other radices and assign units to them
- Create a subVI from a selection of the block diagram
- Discover the functions under the **Project** menu

KEY TERMS

- Preferences
- VI Setup
- Priority
- Reentrant Execution
- Key Focus
- Radix
- Unit
- Find
- Profile window
- Hierarchy window

Advanced LabVIEW Features

13

Preferences, Preferences...

Maybe you've already used the **Preferences...** command under the **Edit** menu. In any case, it's a good idea to skim through all the options available, since often here is where you will find the solution to that persistent LabVIEW problem you've had for a month. You can specify things such as search paths, default front panel colors and fonts, time and date formats, etc.

The following figure shows some of the preference categories.

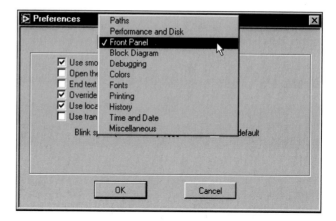

Each of the categories shown in the previous figure brings up several options when selected. Although a comprehensive preference list would be too long to show here, some preference options are more commonly used and turn out to be quite useful:

◆   In the **Paths** preferences, you can specify LabVIEW's default directory. Normally, this is just the `LabVIEW` directory. However, if you keep your project VIs somewhere else, you can change the default path so that when you start LabVIEW, it's easier to open your project VIs.

◆   The **Front Panel** preferences have some important options. For example, "Use smooth updates during drawing" turns off the annoying flickering you see on a graph or chart that is updated regularly. "Use transparent name labels" is another popular option because many people don't like the 3D box on the control and indicator labels. Finally, you can set the blink speed here (for the blink attribute of an object)—you'd think this feature would be a programmatic option, but it's buried here instead.

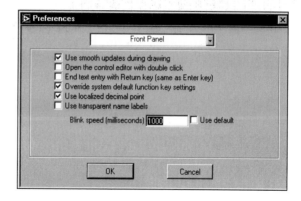

◆ The **Colors** preferences let you set the default colors for the front panel, block diagram, blink colors, etc. If you just must have that magenta background all the time, this is the place to set it.

◆ The **Printing** preferences allow you to choose between bitmap, standard, and Postscript printing. Depending on the printer you have, you may need to choose a different option to get the best print results.

Take a look around at the rest of the categories. You can specify font styles, time and date styles, levels of resource sharing between LabVIEW and other applications, etc.

Configuring Your VI

Often you may want your program to bring up a new window when a certain button is pressed or a certain event occurs. For example, you may have a "main" front panel with several buttons that present options. Pressing one of the buttons would lead the user to a new screen, which may in turn contain an "Exit" button that would close the window and return to the main front panel. Since this pop-up window will usually be the front panel of a subVI, you also may want to customize what features of this window should appear, such as the **Toolbar**, or the clickable box that closes the window. You can set several options that govern the window appearance and the execution of your VIs in two different places: **VI Setup...** (from the pop-up menu of the VI's icon pane, which you will find in the upper righthand corner of the front panel window) and **SubVI Node Setup...** (from the pop-up menu on a subVI's icon in the block diagram). A very important distinction should be made at this point: The setup options under **SubVI Node Setup** for a subVI affect *only that particular instance of the called subVI*, while the **VI Setup** options *always* take effect, whether it's run as a top-level VI or called as a subVI. We'll start first with the few and simple **SubVI Node Setup** options.

■ SubVI Node Setup Options

When you select this setup option from the pop-up menu on a subVI, you get the following dialog box.

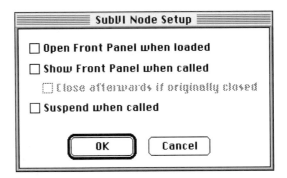

You can select any of the following options:

◆   **Open Front Panel when loaded**—The subVI's front panel pops open when the VI is loaded into memory (such as when you open a VI that calls this subVI).

◆   **Show Front Panel when called**—The VI front panel pops open when the subVI is executed. You'll find this option and the next to be pretty useful for creating interactive VIs.

◆   **Close afterward if originally closed**—Causes the subVI front panel to close when its execution is complete, giving that "pop-up" window effect. This option is only available if the previous option is selected.

◆   **Suspend when called**—Same effect as setting a breakpoint, that is, causing the VI to suspend execution when the subVI is called. You can use this as a debugging tool to examine the inputs to your subVI when it is called but before it executes.

Remember, all the subVI setup options apply only to the particular subVI node you set them on. They do not affect any other nodes of the same subVI anywhere else.

Activity 13-1

This activity will let you use the **SubVI Setup Options** to create a "login" shell that can be used with any application.

1. Make a simple front panel, as shown in the following figure, that will call a "pop-up" subVI when the button <u>Change User</u> is pressed. Call this top-level VI **Shell.vi**

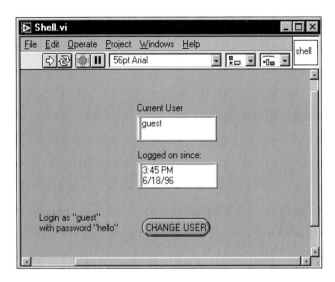

2. Use the subVI **Login.vi** (which you will find in CH13.LLB) to build the block diagram as shown.

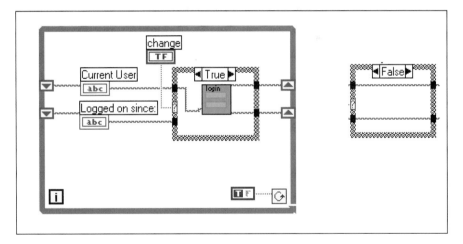

3. Pop up on the **Login** subVI, and choose **SubVI Node Setup** as follows.

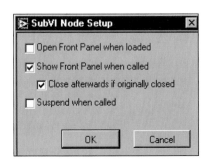

When this VI is run, it will display the last value of the <u>Current user</u> (from the uninitialized shift register), until the **Login** VI is called and changes it.

Login must be closed before you run this example. Otherwise it will not close after it finishes executing—hence the phrase "Close afterwards" if originally closed."

■ VI Setup Options

VI Setup options are a bit more numerous. The dialog box that appears when you choose **VI Setup...** by popping up on the icon pane gives you three option sets: **Window Options, Execution Options,** and **Documentation**.

■ Window Options

The **Window Options**, shown in the following figure, let you control many details of your VI's window appearance.

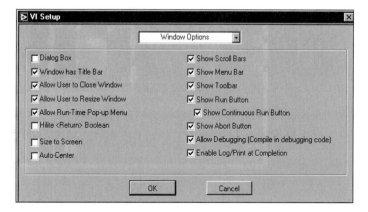

Most of the options are self-explanatory; however, a few comments are in order:

◆ Be careful with options such as disabling the **Show Abort Button**. A VI with this option can't be forced to stop, even with the keyboard shortcut, once it's running!

◆ The **Dialog Box** option gives the VI an appearance of the standard OS dialog box, and prevents accessing other LabVIEW windows while the window is active.

◆ **Hilite <return> Boolean**. OK, so not all of these options are self-explanatory. This option will highlight (put a black border around) the Boolean control, which has been assigned the <return> or <enter> key, as discussed in the next section.

◆ **Allow Debugging,** checked by default, gives you access to Lab-VIEW's debugging tools when the VI is running. You can uncheck it to obtain a slight 1-2% increase in speed and decrease in memory usage of the VI.

▆ Execution Options

The **Execution Options** window contains the same options as the **SubVI Node Setup** options, plus a few more. Two potentially confusing options are worth explaining: *Priority* and *Reentrant Execution*. These options are difficult to understand and are rarely used, so attention all nerds, read on!

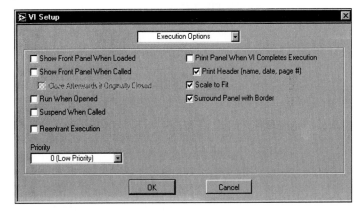

When more than one VI is supposed to execute in parallel, LabVIEW gives each VI (or task) a small amount of time to execute in turn, lining the VIs up in a queue. Each VI executes for the allotted time, then rotates to the end of the queue, and the next VI or task in line executes. *Priority* assigns a level of CPU time priority to the VI on the scale 0, 1, 2, 3, or Subroutine, where 0 is the lowest (default) and Subroutine is the highest.

Higher priority VIs execute before lower priority VIs. For example, say you had three VIs running simultaneously, and you assigned two of them priority 3 while the other one was assigned priority 2. The two level 3 VIs would execute "simultaneously" until their completion; only then would the level 2 VI begin execution. But if the level 3 VIs were in some kind of repeating loop, the level 2 VI might never get a chance to run.

The subroutine priority is somewhat different than the rest. When a VI runs at subroutine priority, no other VI can execute until the subroutine VI is finished, even if the other VI is also a subroutine VI. In addition, front panel controls and indicators are not updated while the VI runs, so the front panel will not give you any information while the VI is running. Subroutine VIs are ideal for simple computations with minimal overhead and no user interaction.

Normally, you shouldn't have to fiddle with the scheduling of multiple, simultaneous VIs executing. If you need to allot more time to a certain VI, you should probably try using the **Wait** functions before resorting to the **Priority** option. Refer to the *LabVIEW User Manual* for a more detailed discussion.

Using priorities to control execution order may give you unexpected results, such as low priority tasks being ignored altogether.

Reentrant execution is another important concept. Normally, when you have two or more instances of a call to the same subVI that occur in parallel, LabVIEW assigns the same data storage space to both VIs, which they share. In some cases, though, you need separate data storage blocks for each subVI node. In the following block diagram example, we are using the subVI **Running Average** on each data channel.

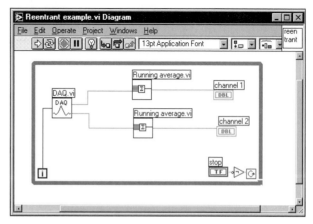

Running Average uses *uninitialized* shift registers as memory elements. Remember that uninitialized shift registers keep their last value even after the VI is stopped and run again (see Chapter 6 for a discussion of uninitialized shift registers). If this VI were left in the default (non-reentrant) mode, it would give unexpected results because the call to each subVI node would contain the shift register data from the last call to the other node (since the subVIs will normally take turns executing). By choosing the "Reentrant execution" option, each subVI node is allocated an independent data storage space, just as if they were two completely different subVIs.

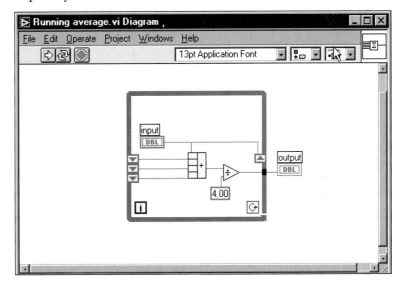

Activity 13-2

As an interesting activity, run this VI (called **Running Average.vi** in `CH13.LLB`) first to see how it works. Then run the **Reentrant** VI (also in `CH13.LLB`), which calls **Running Average.vi** in two places. Next, select the "Reentrant Execution" option for the **Running Average** VI and run **Reentrant** again to compare the differences.

◆   ◆   ◆

■ Documentation Options

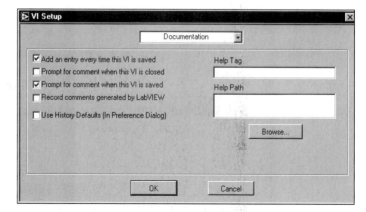

The **Documentation** options of a VI are a more advanced development tool that give you options related to the **VI History** window (an option available from the **Windows** menu). The VI History records the developer's comments as the programming progresses.

Keyboard Navigation

If you are one of those people who thinks mice are unfriendly (or your end users think so), there's good news—you can set up your VI to allow users to "navigate" the controls with the <tab> key. Without any special setup, you can always use <tab> to pick the control you want to receive input (you can't pick indicators this way, since indicators don't accept inputs). A "selected" control—called the *key focus*—has a rectangular border enclosing it. Once a control is the key focus, you can use the appropriate keys to enter its value. The following tips may be useful:

◆ If you directly type the value into the selected control, you must hit the <enter> key when you are done to make your entry valid.

◆ For numerical and ring controls, you can also use the arrow keys to advance to the desired number. Pressing <shift> with the <up> or <down> arrow key advances the value faster. You can set the minimum increment from the **Data Range…**option in the pop-up menu of the control.

◆ For Boolean controls, the <return> key toggles the Boolean value.

◆ Tabbing from control to control normally follows the order in which you created the controls

For VIs with several controls, you may wish to set your own tabbing navigation order; that is, once a control is selected, determine which control will be selected next when the <tab> key is pressed. This sequence is known as *panel order* in LabVIEW. To change the panel order, choose **Panel Order...** from the **Edit** menu. This works in the same way as the **Cluster Order...** option, discussed in Chapter 7. The following illustration shows what the front panel looks like when **Panel Order** is selected.

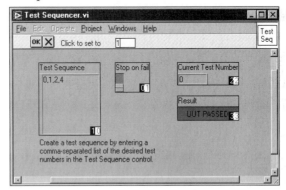

On your front panel, all controls will be boxed in with two numbers in the lower-case corner of each box. The number in the white background represents the previous panel order, the number in the black background represents the new panel order you are assigning to your diagram. To create the new panel order, type the number in the menu bar, then click on an object to assign it. When you are finished changing panel order, click OK. To cancel all changes, click the X button.

You can also assign certain "special" keys to a control. Choosing the **Key Navigation...** option from the pop-up menu of a control brings up a dialog box like the following.

You can assign function keys (<F1>, <F2>, etc.) to a control, as well as function keys with modifiers (such as the <control> key). Pressing the selected key will set the key focus on that control without having to "tab" to it. You will find Key Navigation useful if you have many controls but have a few you use more often.

Finally, you can also programmatically set or disable a control's key focus. The Key Focus attribute of a control is a Boolean that, when TRUE, means the control is "selected" and ready for keyboard input.

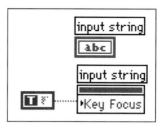

Activity 13-3   Build the **Login** VI you used in Activity 13-1, with a nice front panel as shown. This VI should make the Key Focus first appear in the <u>LOGIN</u> box. The VI should detect when the user has pressed <return> or <enter> and move the Key Focus to the <u>PASSWORD</u> box. The password characters should not be readable. Finally, the VI should appear in the middle of the screen with the same appearance as a dialog box from the OS.

1. *You will need to use the key focus attribute.*
2. *Pop up on the PASSWORD control and select the "Password" option for the string.*
3. *Pop up on the LOGIN control and select the "Limit to single line" option*
4. *How will this VI know when a user has finished the input string to the LOGIN? You will likely need a local variable to periodically check whether the input contains a carriage return character yet.*

As you know, the solution (called **Login.vi**) is found in CH13.LLB. Try not to cheat!

Radices and Units

A useful feature of LabVIEW numeric types is that you can manipulate them as more than just pure numbers. LabVIEW numeric controls and indicators have a base representation (called a *radix*) and an optional associated *unit*. Normally, you represent numbers in decimal format. But for some applications, you may need to view numbers in a binary, octal, or hexadecimal representation. Similarly, you may want to treat certain numerical variables as having an associated unit (such as feet, calories, or degrees Fahrenheit)—especially if you plan to do unit conversions.

■ Radices

LabVIEW's numeric displays are very flexible. You can choose to display numbers in decimal, hexadecimal, octal, or binary format by selecting **Show►Radix** from the numeric's pop-up menu. Click with the Operating tool (do NOT pop up) on the tiny letter that appears, and you'll pull down a *radix* selection menu.

*If the numeric control or indicator does not have integer representation, all options except **Decimal** are disabled.*

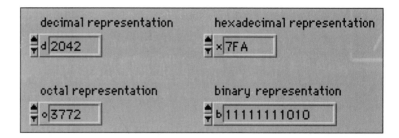

A user who is working with digital I/O, for example, might find it very useful to represent a numeric control in binary, since it would be easy to see the one-to-one correspondence of each digit to each digital line.

■ Units

Any floating-point numeric control or indicator can have physical *units*, such as meters or kilometers/second, associated with it. You can access the unit label by selecting **Show►Unit Label** from an object's pop-up menu.

Once the unit label is displayed, you can enter a unit using standard abbreviations such as m for meters, ft for feet, etc. If you enter an invalid unit, LabVIEW flags it with a ? in the label. If you want to know which units LabVIEW recognizes, enter a simple unit such as m, then pop up on the unit label and select **Unit...** A dialog box will pop up showing you information about LabVIEW's units.

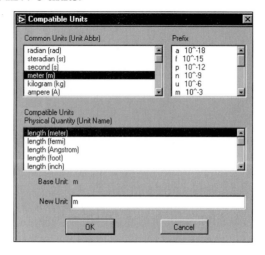

Any numeric value with an associated unit is restricted to a floating-point data type.

LabVIEW preserves the associated unit throughout any operations performed on the data. A wire connected to a source with a unit can only connect to a destination with a compatible unit, as shown.

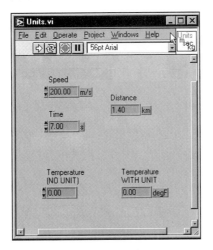

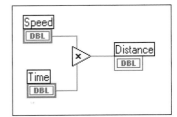

When we say "compatible," we mean that the destination unit has to make sense from a physical or mathematical point of view. The destination unit does not have to belong to the same classification system (e.g., SI versus English units), as shown in this next figure. Thus, you can transparently convert compatible units from different systems in LabVIEW.

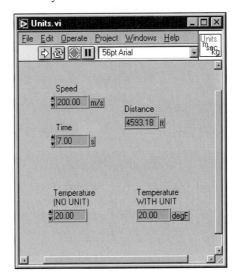

*Some functions are ambiguous with respect to units and cannot be used if a unit is assigned to the data you are trying to operate on. For example, if you are using distance units, the **Add One** function cannot tell whether to add one meter, one kilometer, or one foot.*

You cannot connect signals with incompatible units, as shown in the following illustration. If you try, the error window (discussed soon) will contain a `Wire: unit conflict` error.

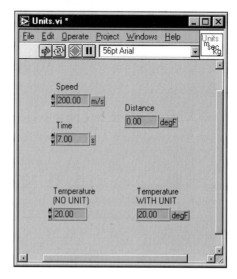

Finally, you need to be wary of doing arithmetic that combines numbers with and without units. For example, you can multiply a unitless constant times a number with a unit, but you cannot add a number with a unit to a number without a unit. LabVIEW allows you to "add" a unit to a unitless number using the **Convert Unit** function (from the **Conversion** palette in the **Numeric** palette).

 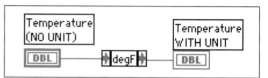

To specify the unit, just type the unit inside the middle box of the terminal, or as with controls and indicators, pop up on the terminal and select **Unit…** for a list of valid units. The **Convert Unit** function also allows you to "delete" a unit from a number.

Creating a SubVI from a Section of the Block Diagram

One key to writing your program right the first time with the least amount of bugs is *modularity.* The modules in LabVIEW are, of course, subVIs. When you're writing a large application, it's essential to break up the tasks and assign them to subVIs. The **Edit▶SubVI From Selection** menu option makes converting sections of your block diagram into subVIs much simpler. If you're working on some code and decide you should make part of it a subVI, no problem! Select the section of a VI's block diagram and choose **Edit▶SubVI From Selection**. LabVIEW converts your selection into a subVI and automatically creates controls and indicators for the new subVI. The subVI replaces the selected portion of the block diagram in your existing VI, and LabVIEW automatically wires the subVI to the existing wires. Choosing this feature is also an excellent option to use, for example, if you need to repeat part of your block diagram in another VI.

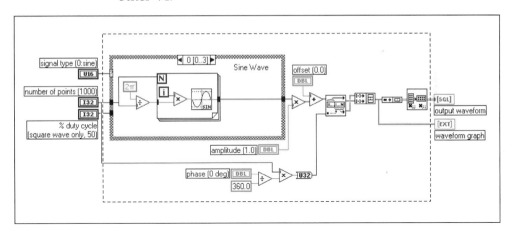

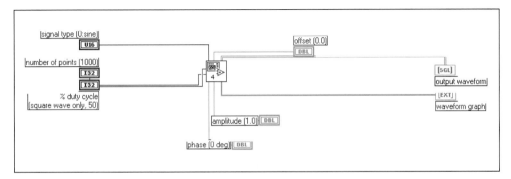

Although we've already mentioned this feature in the beginning of this book, it's time to look at some rules for using the **SubVI From Selection** option.

Because converting some block diagrams into subVIs would change the behavior of your program, there are some instances where you cannot use this feature. These potential problems are listed next.

◆   You cannot convert selections that would create a subVI with more than 28 inputs and outputs, because that is the maximum number of inputs and outputs on a connector pane. We'd feel sorry for the user who would have to wire even 15 inputs! In a case like this, select a smaller section or group data into arrays and clusters before selecting a region of the block diagram to convert.

◆   You cannot convert selections in which items inside and outside a structure are selected, yet the structure itself has not been selected.

◆   Because attribute nodes refer to controls in the same VI by definition, attribute nodes are not copied into a subVI. Instead, the attribute node remains in the calling VI and the attribute data is passed to or from the subVI by new controls of the same type, as required by the attribute nodes. Yeah, we know, this is confusing. Just try one for fun and you'll see what we mean.

◆   Despite the previous statement, you can't convert a section of the diagram to a subVI if an attribute node is contained inside a structure (such as a While Loop). This attribute node will not be updated if it's inside a subVI and inside a structure, because the attribute data is passed only once to the subVI. Trust us on this one.

◆   When your selection includes local variables but does not include the corresponding control, one local variable for the control remains in the calling VI and passes the data to or from the subVI. Within the subVI, the first local variable for a control actually becomes a new control and subsequent local variables refer to the new control in the subVI. Yikes! What does that mean? Bottom line: be *very* careful when you create a subVI from the selection if it includes locals.

◆   When your selection includes local variables or front panel terminals inside a loop, the value that they are measuring may be changed elsewhere on the block diagram as the loop runs. Thus, when you convert the loop into a subVI, there is a possibility that the functionality of the selected code changes. If you have selected some but not all of the local variables or front panel terminals, LabVIEW displays a warning that allows you to choose between continuing and canceling the operation.

*There is no way to "undo" a SubVI from **Selection**. Save your work and make a backup copy to be safe.*

The **SubVI from Selection** can be a really handy tool in cases where you start with what you thought was a small project (hah!) and you soon realize your program is going to be more complex than you originally thought. However, don't just squeeze a bunch of diagrams into a subVI to increase your workspace—subVIs should always have a clear, well-defined task. When considering whether to make part of your diagram a subVI, you might ask yourself, "Could I ever use this piece of code or function somewhere else?" We'll come back to subVIs and modular programming again in Chapter 15, where we look at good programming techniques.

The Project Menu

The **Project** menu provides a variety of features that are useful for debugging and managing large and complex VIs. The commands from the **Project** menu let you:

◆ Graphically display the hierarchy structure of a VI and its subVIs (**Hierarchy Window**)

◆ Search for controls, indicators, functions, text, and more throughout a VI or a set of VIs (**Find** command)

◆ Obtain information about the calls, timing, and memory usage of each subVI while running (**Profile Window**)

■ Hierarchy Window

Generally, you will find the *Hierarchy window* useful when you are working on a VI of fair to high complexity—such as one that contains 10 or more subVIs. The Hierarchy window can help keep track of where subVIs are called and who calls them.

With a VI's front panel open, select **Show VI Hierarchy** to bring up the Hierarchy window for that VI. The Hierarchy window displays a graphical representation of the calling hierarchy for all VIs in memory, including type definitions and globals.

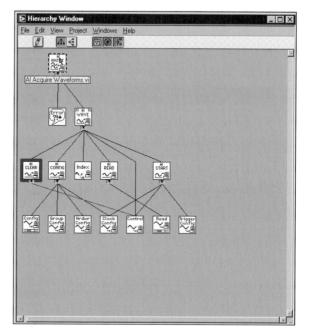

The buttons in the Hierarchy window Toolbar let you configure a number of aspects of the display. For example, the layout can be displayed horizontally or vertically; you can hide or show as many levels of hierarchy as you wish; and you can include or exclude globals or type definitions. A nice feature in this window is that you can double-click on a subVI icon to open its front panel.

■ Searching for Objects in the Virtual Haystack

A powerful search engine was introduced with version 4 of LabVIEW. LabVIEW's **Find…** function, under the **Project** menu, can quickly locate any LabVIEW function, subVI, global variable, attribute node, front panel terminal, or text in your block diagram. You can limit your search scope to a single VI, a set of VIs, or include all VIs in memory.

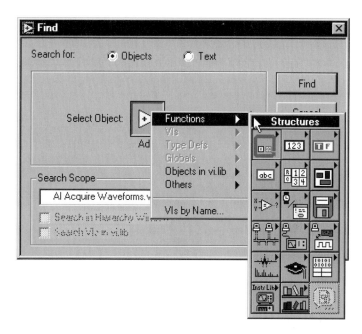

You can choose to search for objects or text by clicking one of the **Search for** buttons. If you select objects, click on the **Select Object** button to access a pop-up menu that allows you to choose the type of object you want to search for. If you select text, you can type in the text, and click on the **More Options...** button to further limit the scope of the search to sections of VIs and object labels.

If LabVIEW only finds a single match, it opens the corresponding VI and highlights the match. If LabVIEW finds multiple matches, it displays a Search Results dialog box, as shown.

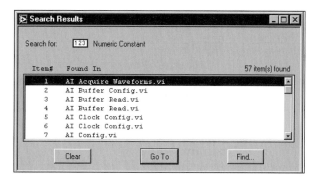

When multiple items are found, you can double-click on any item displayed in the **Search Results** window to have LabVIEW show you where that item resides.

■ Profile Window

LabVIEW includes an advanced development tool called the *Profile window.* The Profile window will show you where your application is spending its time, as well as how it is using memory. This information can be useful on large projects that require timing and/or memory optimization.

To access the Profile window, choose **Show Profile Window** from the **Project** menu with the front panel open. You get a bunch of columns with titles like VI Time, SubVIs Time, #Runs, etc. Each column category represents a statistic that the profiler measures. For example, memory usage is broken down into minimum, maximum, and average memory use for each VI. The rows correspond to each subVI in the caller VI. The top-level VI is shown in bold.

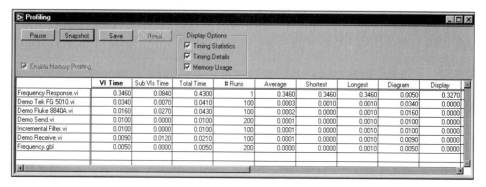

To use the Profile window, do the following:

1. Choose whether to collect memory and timing statistics, by checking the appropriate boxes. Memory analysis in particular will slow down your VI's execution speed considerably.

2. Press the Start button.

3. Run your VI (if it is not already running).

4. At any time in which your VI is running, press the Snapshot button to see the statistics that are available. You can press the Snapshot button as often as you wish to see how the statistics change.

5. The Save button gives you the option of saving the Profile window data to a text file.

6. Press Reset to start over.

Using the Profile window information is an advanced topic related to improving performance and memory usage. Memory management and performance issues will be covered briefly in Chapter 15, *The Art of LabVIEW Programming.*

Wrap It Up!

This chapter covered a potpourri of interesting and useful LabVIEW features. We covered aspects such as the **Preferences** options, **VI Setup,** and **SubVI Node Setup** options, key focus, radices, units, creating subVIs from selections, and the **Project** menu.

The **Preferences** command allows you to control a myriad of options under several categories. These options give you control over the LabVIEW environment and let you personalize aspects such as fonts, colors, printing types, paths, etc.

VI Setup and **SubVI Node Setup** contain useful selections to customize the aspects of a particular VI at the top level or as a subVI. With **VI Setup**, you can hide or show the Toolbar buttons, center the window, control execution options, etc. **SubVI Node Setup** allows you to "pop up" the window of a subVI as it is called and close it afterwards.

Key focus is a Boolean attribute of a control that decides whether the control is "selected" and ready for input from the keyboard. You can set the key focus programmatically, or simply tab through all the controls. You can also assign special keys, such as function keys to toggle a particular Boolean control.

Radices and *units* are special built-in features of LabVIEW numerical types. A numerical type whose radix is shown can be changed to a decimal, binary, octal, or hexadecimal representation. The optional built-in units on numeric types allow you to perform mathematical operations on numbers with units and provide automatic conversion for different unit systems.

The modularity of LabVIEW is highlighted by the ability to select any part of the block diagram, and under certain conditions, turn it into a subVI with the appropriate inputs and outputs.

The **Project** menu provides some advanced programming tools, such as a **Hierarchy** window, a **Find** utility, and a **Profiling** window to examine memory usage and performance of VIs.

Accelerating The Development Of Medical Diagnostic Instruments

By David G. Edwards,
President, FemtoTek, Inc.

The Challenge
Efficiently developing a comprehensive PC-based system for testing blood coagulation timing instruments.

The Solution
Using an AT-MIO-16X board to collect the data along with LabVIEW for its integrated acquisition, analysis, and presentation capabilities.

Electra 1600C Coagulation Analyzer with MLA-DAQ Data Acquisition System

Introduction

When Medical Laboratory Automation, Inc. (MLA) decided to accelerate the development of its medical diagnostic equipment, the company turned to National Instruments for DAQ boards and LabVIEW software and to Alliance Program member FemtoTek, Inc. for application software development. MLA had previously used PC-based hardware and single-function software packages. Functions that previously required data transfer between separate data acquisition, spreadsheet, and mathematics software packages are now integrated into a single LabVIEW program.

MLA, an industry leader in the design and manufacture of blood coagulation timers, features diagnostic equipment designed for fast and accurate processing of medical samples. The equipment combines physical measurements with advanced mathematical analysis. To provide effective development and checkout tools for their latest instrument, the Electra 1600C, MLA needed to integrate the functions of data acquisition, user interface and data display, advanced mathematical analysis, and data storage and retrieval into one PC-based application. FemtoTek developed this application, MLA-DAQ, using National Instruments DAQ boards and LabVIEW.

The Coagulation Analyzer

Blood coagulation timers are used for both analysis and screening of clotting disorders. There are a number of methods of measuring the time it takes blood to coagulate. MLA equipment measures changes in optical density of the sample. Automatic measurement in the Electra 1600C starts when an automatic pipette system places test samples of plasma in disposable cuvettes. A linear belt transport mechanism slowly steps the samples through a series of heated locations and finally to the optical photometers. Heating brings the samples to a controlled body temperature. Before the samples reach the optical detectors, clotting is initiated by adding a start reagent from one of six reagent pumps. In the Electra 1600C, cuvettes move in batches of four, one immediately after the other. The operator measures four samples at the same time using four separate optical channels.

The time a sample spends at any station, such as the optical detectors, varies from 40 to 200 seconds. Moving a new batch of samples into the optical detector station takes between 0.5 and 10 seconds, depending on the demands on the various pumps in the reagent delivery system. The instrument analyzes the optical detector signal to give the coagulation time result.

Connecting to the electrical signals of the Electra 1600C coagulation analyzer requires a DAQ system that can measure eight analog inputs for the optical detector and second derivative; measure four digital inputs for the trigger signals; and generate four digital outputs to drive the derivative clamp signals. The hardware for MLA-DAQ was a National Instruments AT-MIO-16X board installed in a 486-DX2/66-MHz personal computer with 8 MB of RAM. The AT-MIO-16X board has the necessary 16-bit analog accuracy as well as sufficient digital input/output (I/O) lines, while the 486 PC provides enough processing power to run the combined MLA-DAQ data acquisition and analysis tasks.

> **F**unctions that previously required data transfer between separate data acquisition, spreadsheet, and mathematics software packages are now integrated into a single Lab-VIEW program.

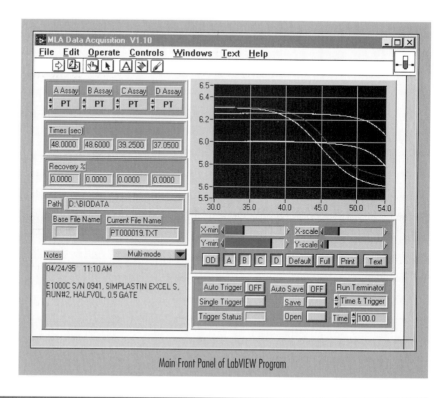

Main Front Panel of LabVIEW Program

Eliminating the Processing Bottleneck

The MLA-DAQ application measures the analog optical detector signal and its analog second derivative. A key requirement for the MLA-DAQ application is to keep up with the sequence of samples in the coagulation analyzer—recording and analyzing the signals for each sample as it moves through the analyzer. Accurate analysis of the signals demands that the operator record the full length of the signals, leaving little time between samples for data analysis and storage. FemtoTek removed the processing bottleneck by using the multiprocessing inherent in LabVIEW; MLA-DAQ captures current signals while it is analyzing the previous signals.

FemtoTek designed MLA-DAQ so that the different functions that run the data acquisition, data analysis, signal display, and RS-232 connection are independent LabVIEW loops. These functions can run independently and asynchronously yet still exchange data, using first-in-first-out (FIFO) buffers, which remove the need to program interlock features into the independently running loops.

An important part of MLA-DAQ is the mathematical analysis algorithms implemented by MLA in their coagulation analyzer. The extensive library of analysis routines in LabVIEW simplified the transfer of the algorithms to MLA-DAQ from the mathematical packages where they were developed. Integrating the analysis algorithms into the LabVIEW program is a higher performance solution than linking MLA-DAQ to a mathematical package through dynamic data exchange (DDE) or file transfer.

Using the System

To keep the user interface as simple as possible and still meet the needs of the users for complete information, FemtoTek consolidated the main user interface on one front panel. For many users, the most important feature on the front panel is the graph showing the signal output of the optical detectors in the instrument. To achieve a compact but flexible user interface, FemtoTek made extensive use of the powerful local variable and attribute node features of LabVIEW.

FemtoTek has designed MLA-DAQ to monitor a coagulation analyzer during long periods of use. The operator can record many thousands of samples. An automatic save feature generates a data file for each set of four samples. The operator can save the corresponding assay type and sample identification number to an information file with a matching file name. In addition to facilities for capturing and saving the optical density data, FemtoTek has provided the user with a zoom function to analyze specific portions of the signal. The user can also print out complete signal or zoom portions for later analysis. The user can reload old data files into the system for comparison, analysis, and printout.

> FemtoTek removed the processing bottleneck by using the multiprocessing inherent in LabVIEW; MLA-DAQ captures current signals while it is analyzing the previous signals.

Summary

MLA-DAQ is a complex application. However, thanks to a clear definition from Medical Laboratory Automation and FemtoTek's experience developing PC-based systems, the

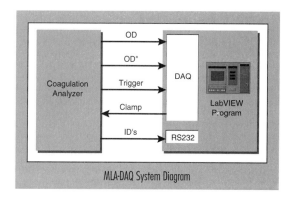

MLA-DAQ System Diagram

development time for MLA-DAQ was short and the resulting product went efficiently into use. Originally intended for internal R&D use, the MLA-DAQ application has been dubbed an unqualified success and is now used extensively in the final product qualification and field trial analysis of MLA's latest instrument. Several installations of the package are in use at MLA, with further installations planned for the laboratories of reagent suppliers and key end users of the Electra 1600C in the U.S. and Europe.

For More Information Contact

FemtoTek, Inc.

560 Fellowship Rd.

Mt. Laurel, NJ 08054

tel (609) 235-4435

fax (609) 722-0153

OVERVIEW

In this chapter, you'll learn about some advanced file and communication functions in LabVIEW. Knowing how to use the three kinds of files in LabVIEW (text, datalog, and binary) will help you pick the best file type for your application. LabVIEW also lets you interact with other programs using OS features such as DDE or OLE (Windows) and AppleEvents (MacOS). Communication with the Internet and other TCP/IP networks is possible with built-in LabVIEW networking functions. Finally, this chapter teaches you how to programmatically print from a VI.

GOALS

- Learn the differences between text, datalog, and binary files
- Use LabVIEW's low-level file I/O functions
- Learn to share data between LabVIEW and other applications by using the interapplication VIs specific to each platform, such as DDE or OLE
- Become familiar with LabVIEW's networking capabilities, such as TCP/IP
- Know how to print programmatically

KEY TERMS

- binary file
- datalog file
- ASCII file
- refnum
- DDE
- OLE
- AppleEvents
- TCP/IP

Communications and Advanced File I/O

14

Advanced File I/O

In Chapter 9, you saw how to save your data in a text file, whether it was plain old text or a spreadsheet format, using the VIs from the **File I/O** palette. These files, stored in *ASCII*, or text, format, have the advantage of being the most easily portable type of file. Virtually any computer running on any operating system can read or write a text file. However, text files do have some drawbacks: They are the least space-efficient (most bytes per piece of information), and they may require a lot of conversion and processor time if the data you want to store is not text (e.g., a graph). LabVIEW provides you with the ability to store and retrieve two other kinds of files: *datalog files* and *binary files*.

Datalog files are a special kind of binary file used by LabVIEW to store front panel information or any LabVIEW data you may want to save. For storing all front panel information, you can think of datalog files as a sort of screen dump of your

VI. When you create a datalog file, it records the values in all the controls and indicators at the time you saved it. You can later load this datalog file into your VI to see the stored values displayed in your front panel. You can even save several "sets," or *records*, of values from the same front panel in one datalog file. Datalog files can only be made and read by LabVIEW. They are fairly easy to use because you can manipulate datalog files from the menus in LabVIEW without writing code. You can also create datalog files that record some specific LabVIEW data types, such as a cluster or a string.

Binary files (also called *byte stream* files) normally contain a byte-for-byte image of the data as it was stored in memory. You can't just "read" a binary file with a text editor, or with any other program, unless you know exactly how the file is formatted—just as you need to know the data type when using binary strings (see Chapter 12). The advantages of a binary file are that you have the least overhead, since no conversion is required, and you save a lot of disk space compared to ASCII files. For example, storing an array of 100 eight-bit integers in a binary file takes up about 100 bytes, while a text file might require over 400 bytes. That's because each eight-bit integer takes up only one byte (in binary format), but the same number in text format can take up three or four bytes (one byte for each ASCII digit, then one space delimiter character to separate each number).

Summary of File Types

ASCII	Datalog	Binary
Easiest to use	Easy to use interactively; requires more elaborate programming to use it in applications	Require strict programming protocols
Compatible with other applications, easy to view and manipulate		Most efficient use of disk space and processor time
Require the most disk space and conversions	Can only be used within LabVIEW	Fast (disk streaming)
Suitable for small- to medium-size data sets that will be used in other applications (e.g., spreadsheets)	Best for storing front panel data or LabVIEW data types	With care, they can be read by other programs
		Useful for applications that will record large files in real time

■ Giving Directions to Find Your File

To locate a file within a file system, LabVIEW uses a special data type called a *path*. A path control or indicator, available from the **Path & Refnum** palette, looks and feels similar to a string control or indicator. You can specify the absolute or relative path name to a file.

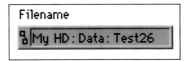

Unlike a string type, however, in a path control, you can only enter a path or filename according to the syntax of your operating system. If you are not familiar with the syntax and terminology of your file system, consult your operating system manuals. A full path name for the file `sample.txt` might look something like the following, for each operating system:

Windows	C:\TRAVIS\DATA\SAMPLE.TXT
MacOS	PowerHD:Travis:Data:Sample Text file
Sun and HP	usr/travis/data/sample_text_file

Whenever you open or create a file, you need to specify the path for that file. If you don't wire a path to the appropriate file function, LabVIEW will pop up a dialog window prompting you to find the file when functions like **Open File** or **New File** are called. You can also make LabVIEW prompt the user for a filename with a custom prompt, using the **File Dialog** function.

■ The Three-Step Process

When you write file I/O functionality into your programs, you should always follow the three-step process: *open, read* or *write*, and *close*. The old filing cabinet analogy works well here. Whether you are pulling out a folder or putting some papers into a folder, you need to open the file cabinet first. Then you do whatever you need to do with your files. Finally, you could leave your filing cabinet drawer open—but why? Coffee might spill into it, someone else might take some files or disorganize them, and it just looks plain messy! Closing your files on the computer as well will ensure your data integrity.

The following diagram illustrates this process. The functions **Open/Create/Replace File, Write File,** and **Close File** are all accessible from the **File I/O** palette. Notice the data dependency from each function to the next. After the **Open/Create/Replace File** operation is completed, LabVIEW generates a file *refnum* (file reference number) that it passes on to **Write File.** When the string is written to the file, **Write File** in turn passes a refnum to **Close File.** Finally, an **Error Handler** VI at the end is almost always mandatory so that you will know if something went wrong.

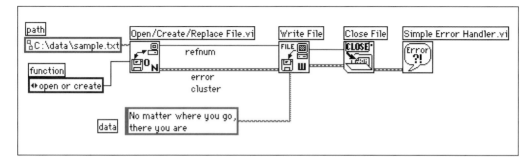

LabVIEW uses the refnum to keep track of what file your VI is referring to. You generally don't need to care what these refnums are or how they work—just be sure to wire them between file I/O functions. Wiring the refnum terminals between these VIs has the added advantage of enforcing data dependency; things will automatically happen in the right order.

A few file functions that you will probably use often are described here:

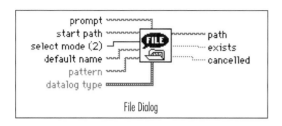

File Dialog

The **File Dialog** function (**File I/O►Advanced File Functions** palette) displays a file dialog box for file selection. This dialog is used for the selection of new or existing files or directories. You can specify the **prompt** message that will appear in the dialog box. We will discuss the use of the **datalog type** input later in this chapter.

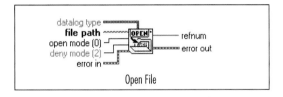

Open File

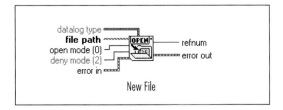

New File

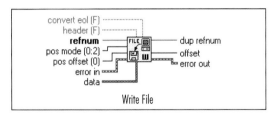

Write File

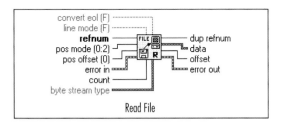

Read File

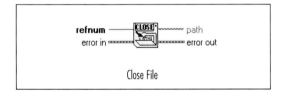

Close File

The **Open File** function (**File I/O►Advanced File Functions** palette) opens an existing file. You can wire a valid path to the **file path** input. This function is *not* capable of creating or replacing files. It opens only *existing* files. The **datalog type** input is used only when opening LabVIEW datalog files.

The **New File** function (**File I/O► Advanced File Functions** palette) creates and opens a new file for reading or writing. You must connect a path to the **file path** input of this function, and it must be a path to a *nonexistent* file. The **datalog type** input is used only when creating new LabVIEW datalog files.

The **Write File** function (**File I/O** palette) writes data to an open file. The behavior of this function varies slightly depending on if you are writing data to a byte stream file or a LabVIEW datalog file. The **header** input is used with binary file types and is ignored for ASCII files.

The **Read File** function (**File I/O** palette) reads data from an open file. When reading byte stream files, you can use the **byte stream type** input to indicate how LabVIEW should interpret data in the file. We'll discuss byte stream files in more detail later in this chapter.

The **Close File** function (**File I/O** palette) closes the file associated with **refnum**.

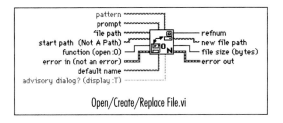

Open/Create/Replace File.vi

Open/Create/Replace File (File I/O palette) is a utility file function that programmatically lets you open a file, create a new one, or replace an old one of the same name. You can specify things such as a dialog prompt string, start path, etc. This VI calls several of the previously mentioned file I/O functions.

■ Writing and Reading ASCII Files

If you're going to write text files to save your data, and high-speed disk streaming is not required, save yourself some time by using one of the higher-level file I/O functions shown here (these were discussed in Chapter 9). These VIs do all the opening, closing, and error checking for you.

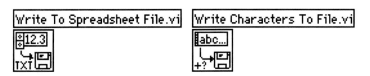

However, there may be times when you want to write your own customized file I/O routine. The heart of this process is the **Write File** function.

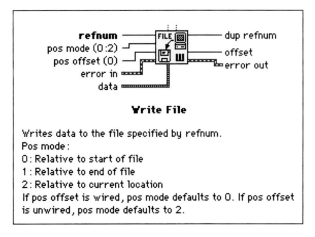

The **data** input to **Write File** is polymorphic; you can wire numeric, string, or cluster data types. The type of data you wire to

this function determines what file format you'll be using. Obviously, for text files, you should only wire a string to this input.

■ The Basics

The actual file management for text files is pretty easy (open, write, close!), but you should ask yourself a couple of questions:

1. Where will you get your path and filename from? Will LabVIEW create it automatically? Will the user input it from the front panel or a dialog box?

2. How will your program handle a file error (disk full, invalid path, etc.)? Should the program stop? Should it continue and simply notify the user?

And, of course, you need to know how to convert your data to a string. Fortunately, LabVIEW has all kinds of string conversion and manipulation functions (just browse the **String** palette or see Chapter 9).

Let's suppose you have written a simple VI, called **Acquire Data,** that reads one data point and returns it as a floating-point number. You want to periodically acquire the data and stream it to a file, until a stop button is pressed. The user should be able to choose the file name at runtime. We might start by building a diagram like this:

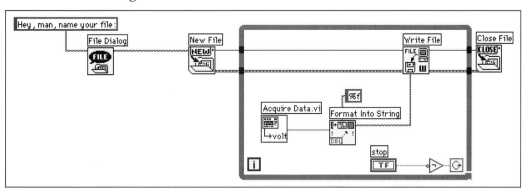

The user is prompted to select a filename, a new file is created, and then the While Loop begins. Each time the loop goes around, a data point is converted into string format with **Format Into String** and is written to the file. When the stop button is pressed, the loop ends and the file is closed. Simple enough, eh? Well, this VI *will* work as it stands. But, frankly, it's a pretty poor file management VI! It can be improved quite a bit. For starters, notice that nothing is being done to separate each string that represents a data point. The file is going to be just one long jumbled string! Also, what would happen if the user pressed the CANCEL button at the File Dialog?

The next diagram makes our program a little more robust. Notice a carriage return and linefeed are appended to each number (the "\r\n" characters)—this ensures that each number will be on a new line of text. We took the possibility of the user pressing the CANCEL button into account by adding a case statement based on the output of **File Dialog**. We also used the error clusters to check for a file I/O error; if one occurs, the loop stops.

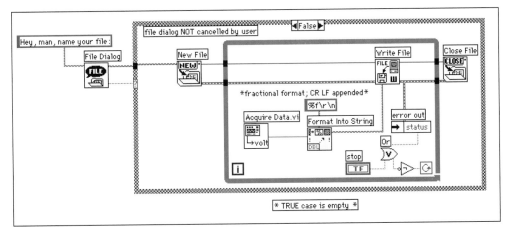

There's still another problem with this code that's easy to overlook. It has to do with the **Write File** inside the While Loop. Can you guess what it is? Unless the **Acquire Data** VI has a **Wait** function, this program will write to the file as often as it can push the microprocessor—maybe thousands of times per second! You need to put a **Wait** function inside the loop to limit how often you want to write to your file, say, once a second or something reasonable. Or better yet, acquire all the data into an array using auto-indexing, and write to the file just *once, outside* the While Loop.

When writing your file I/O routines, you may have runtime problems if you don't take into account worst-case scenarios. Try to implement graceful escapes if a file I/O error or anomaly occurs.

As you see, a lot of things can go wrong or at least get messy when you're dealing with file I/O. Don't be discouraged, however. Most of the time you can probably use LabVIEW high-level VIs for reading and writing text files. If you do need to do specialized text file I/O, be sure you understand what you're doing—and always consider the "what ifs"!

Reading back a text file is very similar to writing one. You use the **Read File** function in a complementary fashion to **Write File**.

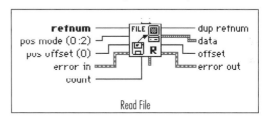

Read File

If you leave **count** unwired, **Read File** reads the entire file. Wiring a number to **count** specifies how many bytes (characters) to read. You can also, of course, use some of the simpler text file functions mentioned in Chapter 9.

Nerd-level File I/O Stuff

Pos mode and **pos offset** are related to your computer's file system and can be quite confusing, to say the least. To keep track of file I/O, an invisible variable called the *file mark* is used by your operating system. The file mark normally points to the current location of where data was last stored inside the file, measured in bytes from some reference point. Whenever you write data to an existing file, LabVIEW writes data at **pos offset** bytes from the reference point determined by **pos mode**.

This reference point can be the beginning of the file (**pos mode** = 0), the end of the file (**pos mode** = 1), or the current location of the last write or read (**pos mode** = 2). With **Write File**, you can manipulate the new data insertion point and move it around using **pos mode** and **pos offset**. This powerful feature allows random access to the data in a file with **Read File.** If you leave both of these unwired, by default, new data is appended after the most recent data for a write, or read from the beginning of the file for a read—which is what you likely will want to do most of the time.

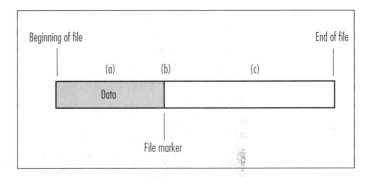

To help you understand these concepts, consider the previous figure. The file marker points to the place where data was last written. The labels (a),(b), and (c) indicate three of the possible places to which you can write to in the file. Here's one possible way you would access each one:

(a)   Setting **pos mode** to 0 (relative to start of file) and setting **pos offset** to a positive number

(b)   Leaving **pos mode** and **pos offset** unwired

(c)   Setting **pos mode** to 2 (which specifies a write relative to the current location of file marker) and setting **pos offset** to a positive number

Many possibilities other than those just mentioned exist, of course. If all this "pos" stuff confuses you, don't worry about it—you can get by fine without it if you don't need rapid random-access to files. Leaving both unwired will let you read from the beginning of a file and write to the position following the last place data was written, which is the way you might expect to normally handle files.

Activity 14-1

Build a VI that has the ability to both read and write ASCII files. The data type input should be a 2D array of numbers. A Boolean switch determines whether the VI will read a file or write a new one. The front panel of the VI is shown.

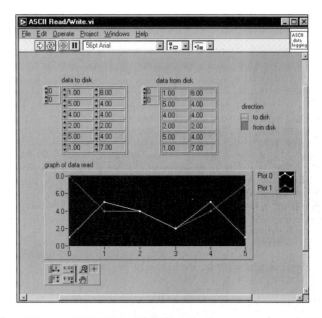

Save this VI as **ASCII Read/Write.vi**. One simple solution is given in `CH14.LLB`.

*To read the file, use an easy file I/O function, such as **Read Characters from File**.*

■ Writing and Reading Datalog Files

Datalog files store data from a LabVIEW object (such as cluster, string, Boolean array, etc.) in a special binary format. Each time you write to a datalog file, LabVIEW normally appends a *record* of the data. A record is sort of like a file within a file; the operating system only sees one file, but from within LabVIEW you can see several separate records in a datalog file. The nice thing about this structure is that you can randomly access any record in the file. Another advantage of using records is that you don't have to know how many bytes you need to "skip" to get to a particular set of data, just the record number. Datalog files are especially useful for storing mixed data types, such as a Boolean and an array. There are two ways to create a datalog file:

1. Front panel datalogging: use the built-in datalogging features found under the **Operate** menu. These commands allow you to log the whole front panel into a file, without writing any file I/O

code. The data is logged either at the completion of the VI, or by a user command in the **Operate Menu**.

2. Read and write datalog files with the **Read File** and **Write File** functions. In this case, you can specify the stored data to be whatever you want (instead of the whole front panel) and store it whenever you want.

■ Front Panel Datalogging

Front panel datalogging is very easy to use and doesn't involve any block diagram programming on your part. When you enable datalogging, LabVIEW saves the data in all front panel controls/indicators and a timestamp to a datalog file. You can have several separate files, each filled with logged data from different tests. You can later retrieve this data with the same VI you saved data from, or in another VI using the file I/O functions.

To have your VI log front panel data, choose **Data Logging▶Log...** from the **Operate** menu. The first time you log data, you will be asked for a log file—give it any name you like.

You can also enable your VI to log data automatically every time it completes its execution by selecting **Operate▶Log at Completion**. Every time you log data to the same log file, you create a new record in that file.

To view the logged data interactively, select **Operate▶Data Logging▶Retrieve...** The Toolbar will change to become a data retrieval toolbar as shown below. All the front panel controls and indicators will suddenly change their values to display the saved data.

The highlighted number tells you which record you're currently viewing. The number range in brackets to the right tells you how many records exist. You can switch records using the arrow buttons. As you switch records, the front panel objects will show the data corresponding to the record. To the right of the record number, a timestamp shows you when this record was logged. You can delete an individual record by choosing that record number and clicking on the trash can icon. Click OK to exit the data retrieval mode.

An easy way to programmatically retrieve information in a datalog file is to use the VI that logged the data as a subVI. Popping up on the subVI, you can choose the **Enable Database Access** option.

When you do this, a funny-looking yellow frame appears around your VI (they actually call this a *halo* in the LabVIEW manuals. It makes us wonder if we could start classifying VIs into angelic and demonic types).

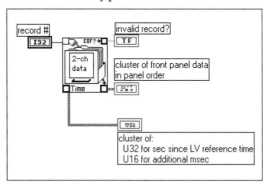

If you run this database-access subVI, it does not execute. Instead, it returns the saved data from its front panel as a cluster, according to the record number wired at the input. This cluster contains all the data on the front panel. Each item in the cluster appears in the same order as the front panel order of objects.

▥ Programmatic Datalog File I/O

Besides creating and reading datalog files using the built-in LabVIEW datalogging features, you can do a little more advanced datalog file I/O using the **Read File** and **Write File** VIs.

With *programmatic* datalog files, you don't need to store *all* the front panel data. You can store just some of it. In fact, the data needn't even be on the front panel; it could be generated at the block diagram. The purpose of datalog files is to let you store multiple records of one LabVIEW data type in one file.

Even though you can only store one type of data in a datalog file, you can combine several different variables into one cluster, which is a valid data type. Or the data type can be an array, a string, a numeric, or a Boolean. The importance of datalog files lies in the fact that you can read and write directly into LabVIEW variables without any concern for conversion to text, headers, etc.

Although you use the **Write File** and **Read File** to store datalog files in a similar manner as you do for text files, the meaning of the **pos mode** and **pos offset** inputs changes. Instead of these inputs relating to the file mark, they relate to the record number. Thus, in **Read File**, for example, you can wire the record number to the **pos offset** input to retrieve a specific record. The **count** input specifies how many records you want to read, not how many bytes.

*The **data, byte stream type,** and **datalog type** inputs on the different file I/O VIs all have a similar function: to specify either the data or the data type. In the Help window, this input always appears as a thick brown wire, indicating a polymorphic input. For simplicity's sake, we'll refer to any of these inputs as "data type" right here.*

You should always wire the data type input to all the auxiliary file functions (such as **New File** or **Open File**—but not the **Read File** VI) when using datalog files. Doing so is crucial because it tells the read and write VIs that you're dealing with datalog files (as opposed to binary files), which is important because the **pos mode** and **pos offset** input change their behavior. This fact is possibly one of the most confusing aspects we've ever seen about LabVIEW. It's easier to remember what to do with this rule:

For datalog files, ALWAYS wire the "data type" on all the relevant file I/O VIs, EXCEPT **Read File**. You must leave the "data type" input unwired on **Read File** to specify a datalog file.

For binary files, wire the "data type" input ONLY on the **Read File** or **Write File.** You must leave the "data type" unwired on the **New File, File Dialog,** etc., to specify a binary file.

Intuitive, isn't it? We thought so.

A simple datalog VI example included with the full version of LabVIEW (examples\file\datalog.llb\Simple Temp Datalogger.vi) is shown.

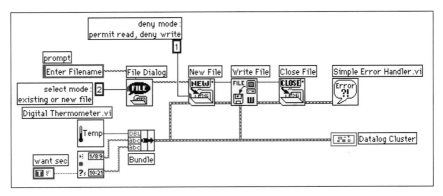

Simple Temp Datalogger stores datalog files that include a time and date stamp with a numeric type representing temperature. This VI incorporates some error handling, always a good idea when you're messing with file I/O. Notice again how the datalog type is wired to all the file I/O functions (except **Close File**, where the type is irrelevant).

The VI for retrieving the datalog files is **Simple Temp Datalog Reader**.

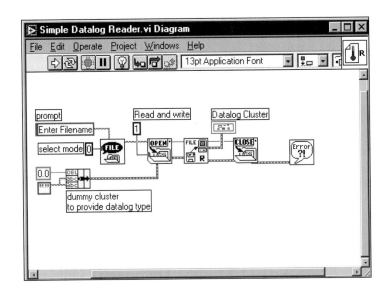

Notice how a block diagram constant of the appropriate cluster type was created to provide the **data type** input. Care is required here, because you need to specify the exact data type that was originally stored. For example, if the numeric type were an **I32** type instead of **DBL** type, attempting to read the data with the above VI would not work. The cluster order is also important to remember, to make sure you access the right variable.

■ Writing and Reading Binary Files

Binary files, also called *byte stream* files, are to text files what a Porsche might be to an 18-wheeler: They're much smaller and much faster. The disadvantage is that if you want to read a binary file, you'd better know every detail about how the file was written, or else you can forget about it. All the raw data may be there, but it's up to you, the programmer, to provide a way to interpret the data when it's time to read it. There's no explicit information in the binary file about the data types or convenient things like headers, so you have to know what's in there.

To summarize: binary files are not very easy to use, compared to other file types. However, LabVIEW includes some higher-level binary file VIs that let you read and write numeric data. These functions are found in the **Binary File VIs** in the **File I/O** palette.

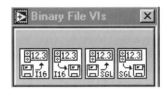

The first two VIs let you read and write **I16** numeric data; the second pair lets you read and write **SGL** numeric types. All VIs assume the data is in array format (either one or two dimensions). The following is a description of their functionality:

Write To I16 File.vi

Writes a 1D or 2D array of unsigned integers (**I16**) to a binary file specified by **file path**. You can either create a new file or append to an old one with the Boolean input **append to file?**.

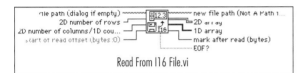

Read From I16 File.vi

Reads a binary file whose data type is an array of **I16** integers. If you know the number of rows and columns that you want in the 2D array output, you can wire these inputs to get the correct array size returned at **2D array.** Otherwise, leave the inputs unwired to read the entire data file into a 1D array returned at **1D array.** You can optionally specify a byte offset at **start of read offset** for random access to the file.

The other two VIs in the palette, **Write To SGL File** and **Read From SGL File** have identical functionality except for the numeric data type.

As a simple example, look at **Binary Read/Write.vi** (in CH14.LLB), which acquires a waveform and stores it in a file, or allows you to retrieve this data from the file. In either case, the data is passed to a graph.

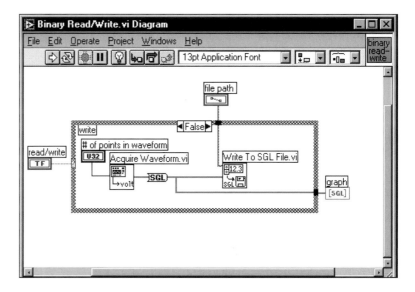

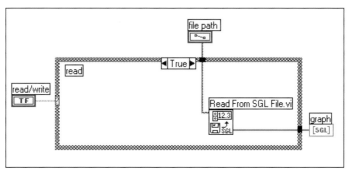

Notice the conversion function to an **SGL** type in the write feature. With these binary files, it's very important to wire the correct data type, or else you may not be able to retrieve the data correctly.

■ For the curious nerds . . .

If you'd like to take a peek on how to use binary files with the lower-level file I/O VIs, read on.

The **Write File** function, as used in the text file section examples, skips over any header information that LabVIEW uses to store the string in memory and simply writes the contents of the string data to the file. In fact, the **Write File** function does not distinguish an ASCII string from a binary string when writing the data to the file; it simply places the data in the file. In other words, there's no *functional* distinction between

ASCII text files and binary files—whether the string is meaningful as text or as binary information depends on how you interpret it.

But remember, the **data** terminal of the **Write File** function is *polymorphic*, meaning that it will adapt to any kind of data you wire into it. For example, you can wire a 2D array of numbers into the **data** input, just as you do with datalog files.

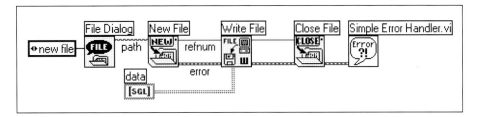

However, datalog files and binary files are very different in structure and behavior, even though you use the same LabVIEW functions to work with them. Because it's so confusing, we'll repeat the rule for working with binary vs. datalog files:

> For datalog files, ALWAYS wire the "data type" on all the relevant file I/O VIs, EXCEPT **Read File**. You must leave the "data type" input unwired on **Read File** to specify a datalog file.

> For binary files, wire the "data type" input ONLY on the **Read File** or **Write File.** You must leave the "data type" unwired on the **New File, File Dialog,** etc., to specify a binary file.

The data saved by the **Write File** function is the same as that obtained by removing the header information from the output of the **Flatten To String** function and then writing the resulting string to the file. Recall that the **Flatten To String** function collects all of the data in its input terminal and places it in a binary string. It precedes the data in the output string with header information necessary to decode the string back into the original data type.

This example exposes a very important aspect of binary files. If you do not store some kind of header information in the file, it will be virtually impossible to successfully interpret the file. In this example, the VI saved 24 bytes of data to a file. Even if you knew that the data was originally single-precision floating-point numbers, you cannot successfully reconstruct the 2D array. How's that, you say? You know that you have a total of six values in the array (24/4 = 6), but how do you know whether the original data was stored in a 1D array of six elements, an array with one row and six columns, or a table with three columns and two rows?

When you work with binary files, you will find that storing header information is vital if your data type is of variable length, which is the case with arrays, strings, and clusters. So, as a general rule:

When working with binary files, always wire the **header** input to TRUE on **Write File** if your data type is of variable length (a string, array, or cluster).

Fortunately, the **Write File** function has a very simple way for you to generate headers to include in the file. If you wire a Boolean value of TRUE to the **header** input of the **Write File** function, it will write exactly the same data to the file as if you had created a binary string with the **Flatten To String** function and written *that* string to the file. In other words, a standard header describing the data is appended to the beginning of the binary string.

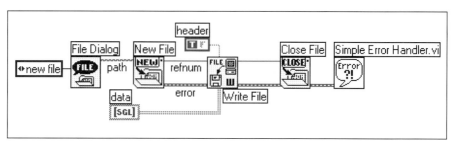

Writing binary data to a file with header information using the **header** feature.

You can use the following code to read the file. Note that the **byte stream type** (*data type*) input of the **Read File** function has a dummy two-dimensional array of single precision numbers wired to it. The **Read File** function uses only the *data type* of this input. In essence, when you use the **Read File** function in this manner, it operates on file data in the same way that the **Unflatten From String** function acts on an input string. In other words, once **Read File** knows what the data type is (from the *data type* input), it can use the header information stored previously to re-build the data into the correct format (in this case, a 2D array).

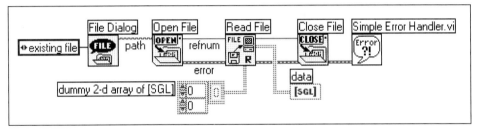

Reading the file created in the previous examples.

Read File always assumes header information was stored if the data type is a type that can have variable length, such as strings, arrays, or clusters. If a header was not included in this type of file (e.g,. the **header** input on **Write File** was set to FALSE), the data will be read incorrectly and the result will be garbage!

One big plus of using binary files is that you have random access to the file. For example, if you store arrays of numeric data in a file, you may find it necessary to access data at random locations in the file. Random access in ASCII files is hampered by the presence of negative signs, a varying number of digits in individual data points, and other factors. Such obstacles do not occur binary files.

The count input, for binary files, specifies how many elements of the data type to read.

Let's look at an example.

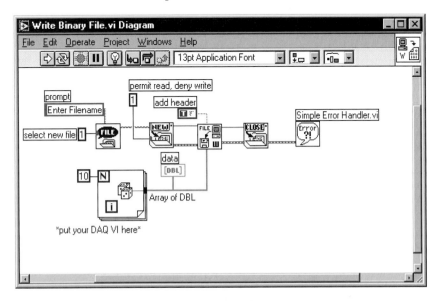

This **Write Binary File** VI takes an array of random numeric (**DBL**) data and writes it to a binary file. Notice how the **data type** input is only wired to the **Write File** VI. To retrieve the binary data, we use **Read Binary File**:

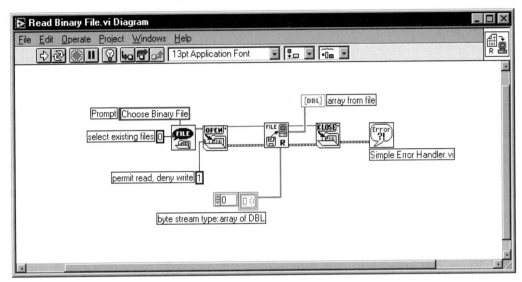

Reading the file is pretty straightforward. All we needed to do is wire the correct data type to **Read File** (an array of DBL). Since we set the **header** to TRUE when we wrote the file, **Read File** can figure out the array length and interpret the data correctly.

If some or all of this file I/O business is confusing, don't feel bad—you're not alone. Even seasoned LabVIEW programmers have difficulties grasping the subtleties of file markers, headers, etc. The best advice is to use good examples such as those in this book and those included with LabVIEW to do your file management.

Activity 14-2

Using some of your own DAQ VIs, write a VI that performs three file I/O functions: one that reads/writes to a binary file, another that reads/writes to an ASCII (text) file, and finally, a third that is a datalogger (datalog file types). Then, use the **File/Directory Info** VI (from the **File I/O➤Advanced** palette) to measure the file size for each of the three generated files. It should prove to be an interesting comparison. As an optional feature, incorporate a way to measure the time it takes to write or read the file.

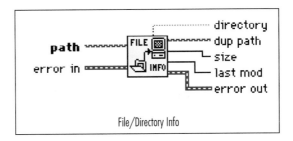

File/Directory Info

Save your VI as **File Type Comparison**.

Communications I: Talking to Other Programs

Wouldn't it be nice if you could acquire data directly into your spreadsheet program? Or if you could make your VI serve as a top-level graphical interface for running other applications such as utility software? The good news is that you *can* do this and more with LabVIEW. The bad news is that writing your own software to perform interapplication communication (IAC) requires a bit of study and detailed knowledge of the appropriate data formats and protocols in order to successfully exchange data between LabVIEW and other programs. Fortunately, the good folks at National Instruments include some example VIs (such as one that communicates in real time with Microsoft Excel) with the full version of LabVIEW that you can use right away, or modify for other uses.

This section will give you an overview of how you can use LabVIEW to communicate with other applications. The next section deals with how LabVIEW can work over a network connection such as Ethernet. The two subjects have a lot of overlap, however, because on many networks, you can call an application on a remote computer as if it were simply on another drive on your own machine (mounted servers on the Mac or network drives on Windows for Workgroups, Windows NT, or Windows 95). So when we talk about communicating with other programs in this section, they don't necessarily have to reside on your computer. For this section, we'll just assume the operating system takes care of making any network connections transparent to LabVIEW.

■ A Word About Protocols

A protocol is a common language used by computer communication processes. A communication protocol is a defined method that lets you specify what data you want to send where, without having to concern yourself about how the data gets there. Several protocols have become established standards for communications. Generally speaking, one protocol is totally incompatible with another, so choose wisely which protocol you are going to use beforehand, since it won't be easy to modify your VI to accept a new protocol. The following protocols are available in LabVIEW:

Protocol	Win 95/NT	Win 3.1	Mac	Sun & HP
OLE (Object Linking and Embedding)	✓			
DDE (Dynamic Data Exchange)	✓	✓		
AppleEvents			✓	
PPC (Program-to-Program Communication)			✓	
TCP/IP (Transmission Control Protocol/ Internet Protocol)	✓	✓	✓	✓
UDP (User Datagram Protocol)	✓	✓	✓	✓

■ The Client/Server Model

Models or paradigms help describe how applications communicate with each other. The most common model is the client/server model. In this model, one application or process (the client) requests services from another application or process (the server). The process or application that initiates the requests is called the client. For example, if you write a VI to acquire real-world data and record the data directly to Excel, your VI is acting as a client and your spreadsheet as the server because Lab-VIEW requested a service from Excel, namely, to record its data. In networked applications, it is very common to have the server on one computer and the clients on remote computers.

Most of the time VIs are written as clients, since they will often initiate a request for a process from another application. A VI can also be configured as a server, although it would probably only be accessed by another LabVIEW program.

■ Using OLE

Object Linking and Embedding (OLE) is a standard for a high level of communication between Windows applications that are "OLE-compliant." Data is exported from one application as an *object*, which can be accessed by another application. Linking and embedding are two of the methods used to access OLE objects. The most common example of OLE you may have seen is to embed a MS Excel spreadsheet into an MS Word document. The object is the spreadsheet data, which although it does not normally "belong" to Word, can be accessed and modified from within Word.

OLE is not a simple topic and is beyond the scope of this book. However, if you are familiar with the basic concepts in OLE automation, such as functions and methods, you will find the LabVIEW OLE VIs (in the **Communications➤OLE** palette) very useful.

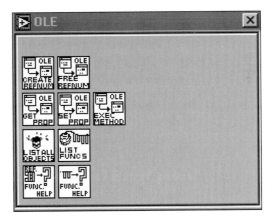

In the full version of LabVIEW, you can see some examples of OLE Automation VIs in the `examples\comm\OLE-xxx.llb` library.

One final note: currently, LabVIEW can only serve as an OLE client and not a server. This means that you can control (from LabVIEW) another application, but other applications can't control LabVIEW.

■ Communicating with DDE

Dynamic Data Exchange (DDE) is an older protocol for exchanging data between Windows applications. You can use DDE to communicate with applications on the same computer or over a network with applications on other computers. To use networked DDE (netDDE), you have to be running Windows 95, Windows for Workgroups, or Windows NT. Also, not all Windows applications have built-in DDE support. Check the documentation of your other applications to see if they support DDE.

You cannot use DDE or OLE VIs to communicate with computers that aren't running Windows (such as Macintoshes). If you need to communicate with other platforms, use a protocol common to all of them, such as TCP/IP.

In DDE, applications send messages to each other to exchange information. For example, a LabVIEW VI could send a connect message to Excel, then send some commands to request information from the spreadsheet, and finish by sending a "close" command.

You can write a VI that acts as a simple client by using a few DDE VIs, which are located in the **DDE** subpalette of the **Communications** palette. LabVIEW includes a very useful pair of example VIs that send and read a table to/from Excel. You can find them in `LabVIEW\examples\comm\DDE examples.llb`.

■ Sending System Commands to Your Computer from LabVIEW

Let's say you wanted to make a VI that could open and close other applications, without controlling them. Aside from all the complexity DDEs involve, LabVIEW provides you with a very simple VI for executing a system command: **System Exec**, located in the **Advanced** palette.

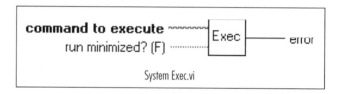

System Exec.vi

This VI simply executes the command string you provide, just as if you typed it at the DOS prompt. In a networked environment, under Windows 95 or Windows NT, you can provide commands to remote computers, by using the Universal Naming Convention (UNC). UNC filenames allow you to specify the location of a file or directory in a networked environment. They use the following path forms.

```
\\<machine>\<share name>\<dir>\...\<dir>\<file or dir>
```

When using UNC filenames, `machine` is the name of the computer that you want to access on the network. `share name` is the name of a shared drive on that machine. The remainder of the path is identical to the way you would work with a file if it was on your system.

■ Communicating with AppleEvents

AppleEvents are a MacOS-specific protocol that applications use to communicate with each other. AppleEvents send messages to other applications or to the operating system itself to open a document, request data, print, etc. An application can send a message to itself, to another application on the same computer, or to another application on a remote computer.

> You cannot use AppleEvent VIs to communicate with computers that aren't running MacOS (such as a PC running Windows). If you need to communicate with other platforms, use a protocol common to all of them, such as TCP/IP.

The actual low-level AppleEvent messages are quite intricate—you can use them with LabVIEW, but you'd better know the MacOS very well and have a good reference handy. For simpler stuff, though, LabVIEW comes with higher-level VIs for sending some of the popular commands to applications—telling the Finder to open a document, for example. The AppleEvent VIs are located in the **AppleEvent** subpalette of the **Communications** palette.

You can also send system commands with AppleEvents. For example, you might want a VI that could open and close a user-selectable application for viewing the text data file you just saved.

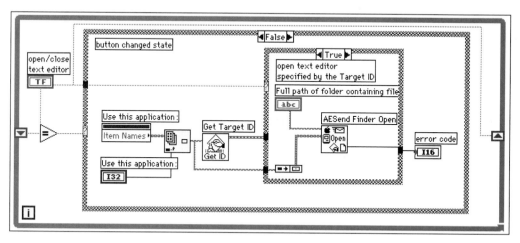

Using DDE and AppleEvent VIs will make your computer more liable to crash and possibly lose data, since these VIs use system commands. Make backups as you test your application thoroughly!

You can find a neat example of how LabVIEW can communicate with Microsoft Excel in the full version of LabVIEW under `examples:comm:AE examples.llb`.

■ Communicating with PPC

Program-to-Program Communication (PPC) is a low-level form of Apple interapplication communication (IAC) that allows applications to send and receive blocks of data. LabVIEW provides you with VIs to perform PPC, found on the **PPC** sub-palette of the **Communications** palette. PPC is a more complicated and advanced protocol than AppleEvents, so we won't say more about it here. If you know how to use PPC, then you can easily learn how to use LabVIEW's PPC VIs by looking at some examples or perusing the manuals.

Communications II: Talking to Other Computers

Ready to plug LabVIEW into the Internet? LabVIEW provides you with an assortment of communication VIs that allow you to use your programs in networked applications using *TCP/IP*. Choosing TCP/IP has several advantages over other protocols:

♦ You can connect computers from multiple platforms (Windows, Mac, Sun, HP).

♦ You can communicate with multiple computers simultaneously.

♦ The networks can be separated by great geographical distances.

In other words, you could have your data acquisition VI running on a PC-AT in the Arctic, passing the results to your analysis VI running on a Power Macintosh in Texas, while someone else observes the whole process in real time on a Sun machine in France.

Your networked computers don't have to be far from each other, of course. All they need is the hardware and software to connect to a network system (such as Ethernet) that can use the TCP/IP protocol. TCP/IP is built-in on the Sun, HP-UX, Windows NT, Windows 95, and MacOS System 7.5 operating systems. For some earlier operating systems, you should be able to get the necessary system software to plug in. A discussion of *how* to get hooked up with TCP/IP is beyond the scope of this book. The rest of this section will assume you're somewhat familiar with TCP/IP connections and have the necessary setup on your computers to do so.

Incidentally, TCP stands for *Transmission Control Protocol* and IP stands for *Internet Protocol*. IP divides your data into manageable packets called *datagrams* and figures out how to get them from A to B. Problem is, IP isn't polite and won't do any handshaking with the remote computer, which can cause problems. And like ordinary mail with the U.S. postal service, IP won't guarantee delivery of the datagrams. So they then came up with TCP, which added to IP, provides handshaking and guarantees delivery of the datagrams in the right order (more like Federal Express, to follow the analogy).

TCP is a connection-based protocol, which means you must establish a connection using a strict protocol before transferring data. When you connect to a site, you have to specify its IP address and a port at that address. The IP address is a 32-bit number that is often represented as a string of four numbers separated by dots, like 128.39.0.119. The port is a number between zero and 65535. You can open more than one connection simultaneously. If you're familiar with UNIX or have used Internet applications, then this should all be old hat to you.

LabVIEW has a set of VIs, found under the **TCP** subpalette of the **Communications** palette, that let you perform TCP-related commands, such as opening a connection at a specified IP address, listening for a TCP connection, reading and writing data, etc. They are all fairly easy to use if your network is configured properly. A good example for getting started on building your networked VIs are the examples found in the full version of Lab-VIEW (examples\comm\tcpex.llb): **Simple Data Client.vi** and **Simple Data Server.vi** Both their diagrams are shown here.

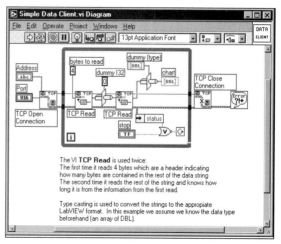

Simple Data Client

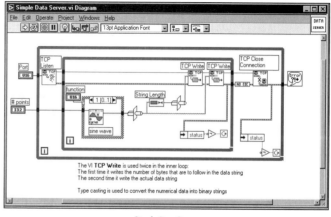

Simple Data Server

You can learn quite a bit about writing client/server VIs by examining these diagrams. The basic process for the client is:

1. Request a TCP connection. You can set a timeout to avoid hanging your VI if the server doesn't respond.

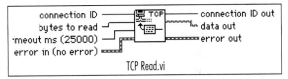

TCP Open Connection.vi

2. Read (or write, in other cases) data. Data is always passed as a string type.

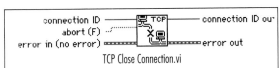

TCP Read.vi

3. Close the TCP connection.

TCP Close Connection.vi

The basic process for a server is:

1. Wait for a connection.

TCP Listen.vi

2. Write (or read, in other cases) data. Data is always passed as a string type.

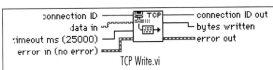

TCP Write.vi

3. Close the connection.

Because all data over a TCP/IP network has to be passed as a string, you will need to convert your data to the LabVIEW string type. The easiest way to do this, as in the previous examples, is to use the **Type Cast** function. But you need to ensure that both the server and client know exactly what kind of data they're passing. If the server, for example, typecasts extended-precision floats to a string, and the client tries to typecast the string into a double-precision number, the result will be garbage!

Networked applications are ideal when you need to write a program to control a large distributed system, such as in process control applications. You can find a lot more information about TCP/IP from many books at the bookstore, and you can get more information on the LabVIEW TCP/IP VIs from the *Networking Reference Manual*.

There is one more protocol that LabVIEW supports that we haven't talked about: *User Datagram Protocol* (UDP). UDP is similar to IP, does no handshaking, can transmit data to multiple recipients, and is somewhat faster. We personally don't know anybody who uses it, but it's there in the **Communications** palette.

Put It in Writing: Printing from LabVIEW

As we've seen before in Chapter 5, you can always print the active window in LabVIEW by choosing **Print Window...** from the **File** menu. This method isn't exactly thrilling if you want your application to automatically generate a large number of printouts. You may want an application that has a "Print" button on the front panel, or that prints a plot when a certain pattern in the data occurs. Fortunately, you can print programmatically in LabVIEW.

LabVIEW gives you the option for automatically printing the front panel whenever the VI finishes executing. Choose **Print at Completion** from the **Operate** menu to do this. This is a piece of cake if all you needed was a printout of your front panel each time your VI ran. But what if you only wanted to print part of your front panel, or something different? Or if you wanted to just print on certain occasions, not necessarily when your VI finishes executing? It's still easy to do this using the same **Print at Completion** option, although it may not be obvious how at first.

Just create a subVI whose front panel has the graphs, Booleans, or whatever you wanted to print! Then select the **Print at Completion** option for this subVI. You don't need to put anything in the block diagram. Finally, place the subVI in your main VI and wire it with the data you wanted to print. Anytime this subVI is called, it will finish "executing" almost immediately, because it has nothing in the block diagram, and will then send its front panel to the printer while your main VI continues along merrily. An example is shown in the following illustration, where a user can print the graph by pressing a "Print" button at any time.

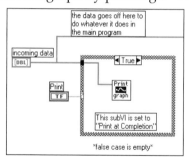

In this case, the **Print graph.vi** has a graph on its front panel and nothing else, and is configured to **Print at Completion.** Remember that the graph must be changed to a *control* for you to wire data into it.

*If you have a PostScript printer, turn on PostScript printing from the **Print** menu of the **Preferences...** dialog (under the **Edit** menu). You should get better-looking graphics and fonts. LabVIEW also supports QuickDraw GX printing for the MacOS.*

You can also direct other applications to print documents through the use of the DDE or AppleEvent functions mentioned earlier.

Wrap It Up!

In this chapter, we have examined some advanced I/O topics; specifically, file I/O, interapplication communication, networking, and programmatic printing.

We've seen how LabVIEW works with three kinds of files: text (or ASCII), datalog, and binary. Text files are the easiest to use but take up more disk space and time. Datalog files are useful for keeping *records* of a LabVIEW data type or of the whole front panel. Binary files are the most difficult to use, but provide the least overhead in terms of both speed and space. All 3 file types can use the basic file functions in the **File I/O** palette: **Open File** or **New File, Read File, Write File, Close**.

The capabilities of LabVIEW to communicate with other applications enhance the power of your VIs. Built-in functions allow you to use protocols such as OLE and DDE (Windows) or AppleEvents and PPC (MacOS) to interact with external applications such as spreadsheet programs or text editors.

With LabVIEW's TCP/IP functions, you can also access remote computers over the Internet or another network that uses these protocols. A VI can act either as a server or as a client under TCP/IP.

Finally, you learned a "trick" for printing programmatically from the block diagram: using a dummy subVI that is set to **Print at Completion**.

OVERVIEW

This chapter focuses on techniques, tips, and suggestions for making better Lab-VIEW applications. We start with the appearance of the front panel of your application—making it look as cool as possible. The whole concept of an intuitive, appealing graphical user interface (GUI) is embedded in LabVIEW. We'll show you a few tips and techniques for taking it a bit further with suggestions on panel arrangement, decorations, customized controls, dynamic help windows, and more. Some common programming problems and their solutions will be presented. We will also cover issues such as performance, memory management, and platform compatibility to help you write better VIs. Finally, we'll examine some general aspects of good programming techniques and style, which you will find especially useful in managing large LabVIEW projects.

GOALS

- Become familiar with general guidelines and recommendations for creating an aesthetically pleasing, professional-looking graphical interface
- Import external pictures onto your front panel and into picture rings
- Build custom controls using the Control Editor
- Dynamically open and close the Help window
- Become familiar with some nifty solutions to common LabVIEW programming challenges (the kind that will make you go, "a-ha!")
- Boost performance, gobble less memory, and make your VIs platform-independent when necessary
- Know some proposed guidelines for writing an awesome application
- Become the coolest LabVIEW programmer in town

KEY TERMS

- decorations
- custom controls
- Control Editor
- importing pictures
- custom help
- memory usage
- performance
- platform dependency

The Art of LabVIEW Programming

15

Why Worry About the Graphical Interface Appearance?

In our culture, image counts for a lot— even in software. People are often more impressed by what a program looks like on the screen than how it actually works. We can think of at least a couple of reasons why you would want to polish and improve your graphical user interface (GUI): 1) To impress and convince someone else (your supervisor, customers, or spouse) about the quality of your software and its goals; 2) To make your software more intuitive and easier to use for the end user; 3) It's easier than doing actual work. And therein lies the beauty of Lab-VIEW—even if you're a novice user you can put together an impressive graphical interface on the front panel (never mind that it doesn't do anything yet) before a guru C programmer can blink.

LabVIEW's front panel objects are already pretty cool-looking: knobs, slides, LEDs, etc. But this chapter will teach you a few more tricks for organizing and putting your objects on the front panel in such a way as to create an even better interface. You will also see how to add online help so that the end user can figure out what each control does without a manual. Making a good GUI is not just about aesthetics—it's also about saving the user time and effort.

Look at this next VI, a Temperature System Monitor. It doesn't look too bad...

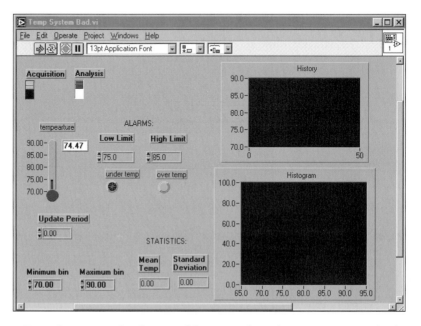

But almost anybody would agree that the same VI with *this* interface would be easier to work with and would certainly have greater visual appeal.

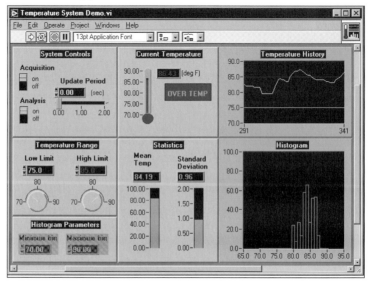

Arranging and Decorating

One way to improve the appearance of your front panel is to organize the objects by deciding where to physically place them on the panel. You can do this by aligning and distributing your objects evenly, and by grouping sets of objects that are logically related onto or inside a decoration.

LabVIEW provides you with the ability to *align* and *distribute* objects, much like a drawing program, as we saw in Chapter 4. You have a variety of choices for how to align your objects. Distribution of objects refers to the spacing arrangement between objects. Remember that the label of an object can be placed anywhere you want to, but it remains attached to that object and can be used as a reference point for alignment and distribution.

A subpalette in the **Controls** palette that you may have noticed by now is the **Decorations** set.

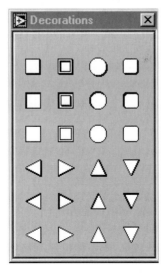

These front panel decorations are just that—they don't do anything, and they have no corresponding block diagram terminal. You can use these boxes, circles, frames, etc. to encapsulate and group your front panel controls and indicators into logical units. You might also use them to approximate the front panel of an instrument more closely.

Often you'll have to make use of the functions under the **Edit** menu to place the decorations "behind" your controls, such as **Move To Back**. The reason for these commands is obvious once you create a decoration to enclose a set of controls; decorations aren't normally transparent, so if you place a decoration on the front panel *after* you created controls, the decoration will obscure them, since it is at the "front."

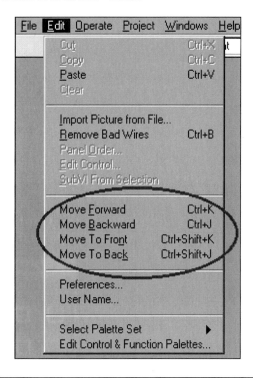

Clicking on a decoration is not the same as clicking on a blank part of the front panel. When using decorations behind controls, it's easy to unintentionally select the decoration with the cursors in the edit mode.

■ Vive l'Art: Importing Pictures

You can paste pictures right into LabVIEW and include them on your front panel. For example, you might want to make a block on your instrument that has your company logo. Or you might want to be a little more elaborate, like add piping and valve pictures to represent some process control loop (you can actually make your valve pictures be Boolean controls—see the next section).

To import a picture into LabVIEW under Windows or MacOS, simply copy the picture from the application in which it's open, and paste it onto the front panel. LabVIEW then treats the picture in the same way as a decoration; you can resize it, move it behind or in front of other objects, etc. However, you can't edit it from LabVIEW.

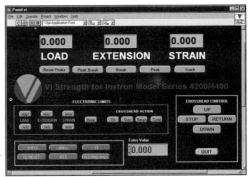

On this VI, the logo of VI Technology was pasted into the front panel as a picture.

On a Mac system, LabVIEW converts your graphic to the PICT format, which preserves a decent amount of resolution.

On Windows systems, LabVIEW normally converts the graphic in the clipboard to a bitmap image. The disadvantage of a bitmap image over another popular Windows graphic file format, a metafile, is that you lose some resolution if you resize the graphic. However, LabVIEW does support enhanced metafile graphics, a newer graphic format for Windows. An enhanced metafile can be stretched without "distortion" and can also have transparent sections, so you don't need to worry about matching the background color of your front panel. If you import an enhanced metafile graphic into LabVIEW, LabVIEW will accept the graphic without converting it to bitmap format.

Another place where you can put pictures is in a **Picture Ring** or a **Picture and Text Ring**, available from the **List & Ring** palette. Two examples are shown in the following illustration. Using the picture rings allows you to get very creative in presenting the user with a customized *graphical* set of options that you can index and track.

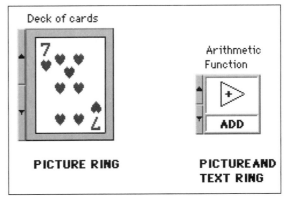

To add a picture to a picture ring, first copy your picture into the clipboard. Then choose **Import Picture** from the pop-up menu on the ring control. To add more pictures, choose either **Import Picture After** or **Import Picture Before** from the pop-up menu. Choosing **Import Picture** overwrites the current picture.

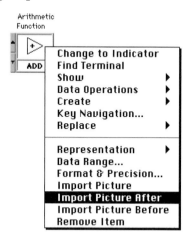

■ Custom Controls and Indicators

Q: What do all the following objects have in common?

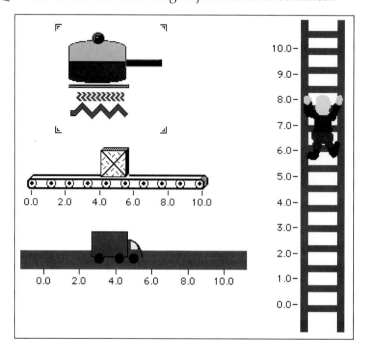

A: They are all LabVIEW controls!

You can customize controls and indicators to make them better suited for your application while displaying a more impressive graphical interface. In the previous examples, you might want to display the position of a box on a conveyor belt to represent the production stage of some product. Or you might want to turn on and off a furnace by clicking on the "fire" in the stove picture control shown on th previous page.

You can save a custom control or indicator (or "control" from now on, for simplicity) in a directory, or VI library, just as you do with VIs and globals. The standard convention is to name custom control files with a .ctl extension. You can use the saved control in other VIs as well as in the one in which it was created. It's possible to even create a master copy of the control if you need to use it in many places in different VIs, by saving the control as a *type definition*. When you make changes to a type definition, LabVIEW automatically updates all the VIs that use it.

You may want also to place a frequently-used custom control in the **Controls** palette. Do this by choosing **Edit Controls & Function Palettes** from the **Edit** menu.

OK, so how do you create a custom control?

Activity 15-1

Usually, you will want to import pictures for your custom control, so first of all, have your picture files available, maybe along with a graphics program. LabVIEW doesn't have any sort of picture editor. When you create a custom control, you always base it on the form of an existing control, such as a Boolean LED or a Numeric Slide control. In this activity, we'll create a custom Boolean that looks like a color-coded valve that is open or closed.

1. Place any Boolean control on a front panel, and select it (in this case, a switch).

2. Launch the Control Editor by selecting **Edit Control...** from the **Edit** menu.

3. The Control Editor window will show the Boolean.

4. In Edit mode, the Control Editor window works just like the front panel. You can pop up on the control to manipulate its settings—things like scale, precision, etc.

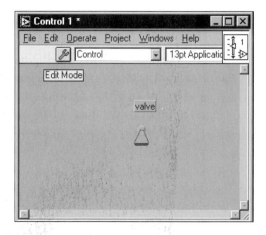

In the Customize mode, which you access by clicking on the Tool button, you can resize, color, and replace the various picture components of the control.

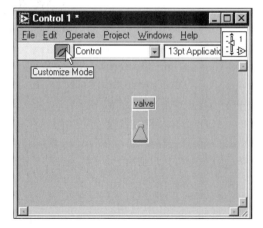

6. Copy the picture file of a closed valve or use the one provided on the CD (named `closed.bmp` for Windows, or `closed.pct` for MacOS). In the Customize mode, pop up on the Boolean and select **Import Picture** with the Boolean in the FALSE position.

7. Repeat step 6 for the TRUE case of the Boolean, using a different valve picture. The TRUE, or open valve picture is also on the CD (`open.bmp` for Windows, or `open.pct` for MacOS).

FALSE

 TRUE

8. Save the custom control by selecting **Save** from the **File** menu. By convention, controls are named with the `.ctl` extension

9. The following front panel is found in the examples of the full version of LabVIEW (`examples\apps\demos.llb\Control Mixer Process.vi`). This front panel uses the same Boolean controls you just created, as well as some others. Open it and take a look.

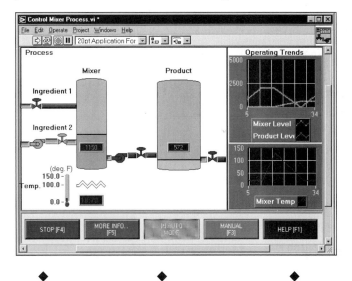

◆   ◆   ◆

An important concept to understand is that you cannot change the behavior of a control, you can only modify its appearance. That also means that generally you can't change the way a control displays its data (a custom control based on a slide will always have something that slides in one dimension, for example). Despite these limitations, you can produce some fancy graphical interfaces, especially if you're artistically inclined and willing to experiment.

If you want to create a master copy of your control so that all the VIs that contain it are automatically updated when you make a change to the master copy, select **Type Def.** from the ring at the Control Editor's Toolbar.

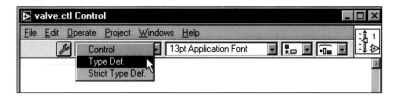

A type definition forces the control data type (I16, string, etc.) to be the same everywhere it is used, but different copies of the type definition can have their own name, color, size, etc. This is useful because only the master copy can be modified in behavior, thus preventing accidental changes. A *strict type definition* forces almost everything (size, color, etc.) about the control to be identical everywhere it is used.

Adding Online Help

LabVIEW's Help window is a godsend when you need to learn something in a hurry; most people come to depend on it for wiring their diagram—a good habit to develop. You can make your own application as easy to learn by adding your own entries for the Help window, as well as links to a hypertext Help document. Two levels of customized help are available:

1.   Window Help: the comments that appear in the Help window that describe controls and indicators as you move the cursor over them.

2.   Online help to a Help (hypertext) document: you can programmatically create a link to bring up an external help file.

Providing customized help in the Help window is fairly easy. We covered this briefly in Chapter 5. To do this, choose **Data Operations➤Description...** from the pop-up menu of a control or indicator. Enter the description in the dialog box, and click **OK** to save it. LabVIEW displays this description whenever you have the Help window open and move the cursor over the front panel object.

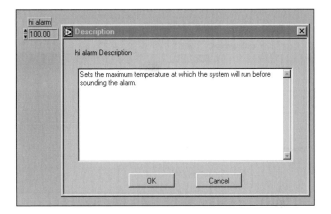

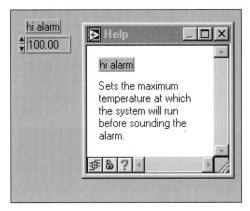

If you document all your front panel controls and indicators by entering their descriptions, the end user can open the Help window and quickly peruse the meaning of your front panel objects.

To provide a description for entire VIs, enter information in the text box that appears when you choose **Show VI Info...** from the **Windows** menu. These comments will show up in the Help window, along with the wiring diagram, whenever the cursor is positioned over the VI's icon in the block diagram of another VI.

You can also bring up, position, and close the Help window programmatically. Use the **Control Help Window** and the **Get Help Window Status** functions, available from the **Help** sub-palette of the **Advanced** palette.

Control Help Window

The Boolean **Show** input closes or opens the Help window; the cluster input consists of two numerical indicators that determine the top and left pixel position of the window.

Get Help Window Status

Returns the state and position of the Help window.

The more advanced help method, calling a link to an external help file, is more elaborate. To create the source document for the external help file, you will need a help compiler for your specific platform. Windows help compilers are available from Microsoft as well as from a number of third-party companies that provide compilers, such as *RoboHelp* from Blue Sky Software and *Doc-to-Help* from WexTech Systems. On the Macintosh, you can use *QuickHelp* from Altura Software. With UNIX, you can use *HyperHelp* from Bristol Technologies. All the help compilers include tools to create help documents. To call these help files programmatically, you can use the remaining function in the **Help** subpalette.

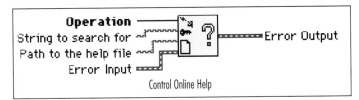

Control Online Help

Manipulates an external help file. You can display the contents, the index, or jump to a specific part of the help file.

Pointers and Recommendations for a "Wow!" Graphical Interface

It's the attention to detail that often makes a graphical interface so startling that people say, "Wow!" To achieve this, we've collected a "bag o'tricks" over the years from experience and from observing other people's fancy VIs. Many VIs start out as a scratchpad with no concern for order or aesthetics, but never get cleaned up later. Believe me, it is worth the effort; even if you're not out to impress anybody, you and others will have an easier time working with your own VI. What if you have to modify that graphical spaghetti you threw together last year?

The following list is just a set of recommendations for a cool GUI; they don't necessarily always apply to your particular situation—perhaps you'll find better ones—but it should help you get started.

■ Panel Layout

◆ If possible, make your top-level front panel and window size such that it fills up the whole screen, as long as you know that's the monitor size and resolution it will be used on.

◆ Try to use neatly aligned square decorations as "modules" on the front panel to group objects.

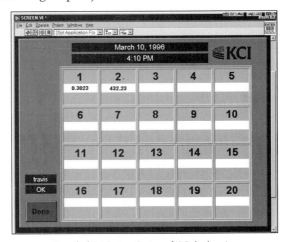

Biomedical Test System. Courtesy of VI Technology, Inc.

◆ If you have empty space, consider filling it up with a decoration "module" as something you could use later; or fill it up with your company's logo or some cool picture.

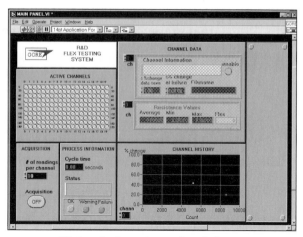

Cable Stress Test System. Courtesy of VI Technology, Inc.

◆ Label the decoration "modules."

■ Text, Fonts, and Color

◆ Use color, but don't overdo it and be *consistent* with a coloring scheme. If you use red for a "hi" alarm in one panel, use the same red for the "hi" alarms in other panels (and please, no puke-green or other weird colors...).

◆ Choose a different background color than your decoration box colors. Often a dark gray or black background seems to fit nicely to add a border effect. Alternatively, if you don't use the decoration modules, you may find that white or a soft, light color works best for the background.

◆ Select the "transparent" option for object label boxes from the **Preferences...** command under the **Edit** menu. Labels generally look much better without their gray box.

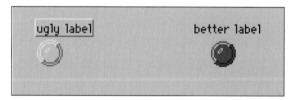

◆ For many labels in general, and especially for numeric indicators and controls, coloring the background black and making the text a vivid green or yellow makes them stand out. Choosing bold for the text usually helps as well.

◆ Use bright, highlighted colors only for items that are very important, such as alarms or error notifications.

◆ LabVIEW has three standard fonts (*Application, System,* and *Dialog* Font). In general, stick to these if you are writing a cross-platform application. A custom font style or size you selected on a Windows machine may look very different (usually too big or too small) on a Mac.

■ Graphs and Imported Pictures

◆ Make your panel look truly customized by importing your company's logo picture somewhere onto the front panel.

◆ Be aware that pictures, graphs, and charts can significantly slow down the update of your VI, especially if you have placed any controls or indicators on top of a picture or graph indicator.

◆ In general, you will want to select the **Use Smooth Updates** option from the **Preferences...** command to avoid the nasty "flickering" that would otherwise appear on graphs and other image-intensive operations.

■ Miscellaneous

◆ Include units on numerical controls if appropriate. Remember LabVIEW has built-in units, if you choose to use these. You can show the unit label for numeric control or indicator by popping up on it and selecting **Show Unit Label.** See Chapter 13 for a discussion of units.

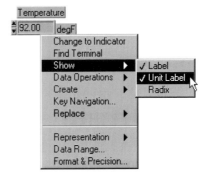

◆ Include custom controls when you can for neat visual effects.

◆ If you want to get really creative, for multimedia effects, use sounds and attribute nodes such as the position attribute.

◆ Use attribute nodes to hide or "gray out" controls when you want to indicate that these controls shouldn't be used.

◆ The **VI Setup** Window options discussed in Chapter 13 are very useful for making your front panel fill the screen, center itself, etc.

◆ Never forget to spend enough time on the icon editor—make cool icons!

◆ When you see a good front panel design, plagiarize! (Lawyers: Please ignore!)

■ Some Final Words on the GUI...

Most of you at some time or another have probably had to give a presentation to your manager or some other audience about how your software works. The success of this "dog and pony show," in our experience, usually depends far more on the razzle and dazzle of the GUI than on whether the software is efficient or even works properly. Perhaps it shouldn't be this way, but that's often the way projects get accepted or rejected. The story goes that at one large semiconductor manufacturer, some engineers had recently decided to revise a piece of software written in C that had taken two years to develop. Another engineer put together a LabVIEW front panel that demonstrated what the new software should look like. One of the top executives in the company was so impressed by the GUI that he immediately approved the engineer's proposal for all their worldwide plants—despite the fact that the front panel was just a demo and didn't do anything yet!

The moral of the story is, spend some time making the best, coolest graphical interface you can. You never know how far it might carry you.

How Do You Do That in LabVIEW?

This section is more about programming solutions than about LabVIEW features. Common problems arise in developing many applications and we show you some of them along with their solution in the following pages. Some of the solutions are very simple, some are quite ingenious. By no means do all or even most programming challenges appear here; rather, these examples have been collected from a variety of sources to give you a jump start on some applications, or at least food for thought.

You might take some of these problems as a challenge—try to come up with a programming solution on your own first before looking at the answer!

■ Where is the "Undo" in LabVIEW?

Umm . . . there isn't one, at least as of version 4.0. As simple as it seems, it's really not, and canceling the last thing you did in LabVIEW is not possible. The best you can do for now is revert to the last saved version of your VI (using the **Revert...** command under the **File** menu). Save your VI often!

■ How do you acquire data from different channels at different sampling rates?

Things work fine when you're sampling all your data channels at the same rate. It's more tricky when one channel has to be sampled at 10 kHz, and another channel only requires sampling at 50 Hz. There's no magical solution to the multiple-sampling rate requirements, but you have a couple of options:

1. If you have more than one DAQ board, try assigning the channels so each board can handle the channels with the same sampling rate.

2. Sample everything at the highest rate, and then throw away the extra points on the lower-rate channels. One efficient way to do this is to use the **Decimate Array** function (**Array** palette).

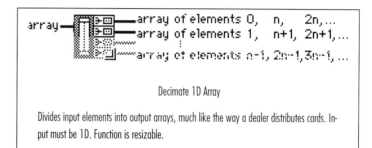

Decimate 1D Array

Divides input elements into output arrays, much like the way a dealer distributes cards. Input must be 1D. Function is resizable.

■ How do you allow a user to print data programmatically?

See the last section in Chapter 14 on creating a dummy VI for custom printing.

■ How can you create a set of Boolean controls so that only one can be true at any time?

This is the common "Radio Buttons" or "Menu Bar" problem. You'll find this solution extremely useful in many applications. The basic problem setup goes somewhat like this: you want to give the user a list of options (Boolean controls) that will each go to a different subVI or subroutine. For example, you might want a top-level screen for a test program that will allow the user to run a test, edit the test parameters, view a test result, or exit. The user should only be able to select one of these options at a time. What is an efficient way to do this? You certainly don't want four parallel While Loops, or some complicated Boolean logic diagram. Fortunately, one solution to this problem comes with the LabVIEW examples. It's found in `examples\general\ controls\booleans.llb` library, called **Simulating Radio Buttons.vi**.

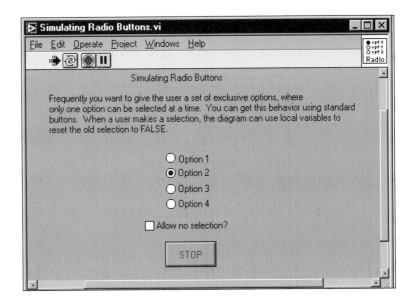

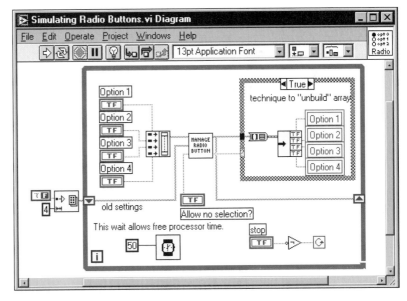

The heart of the previously shown VI is the subVI **Manage Radio Buttons**, which is located in the same library in the full version of LabVIEW. The block diagram of this VI is shown to give you an idea of how it works; however, we recommend just using this VI directly instead of reinventing the wheel.

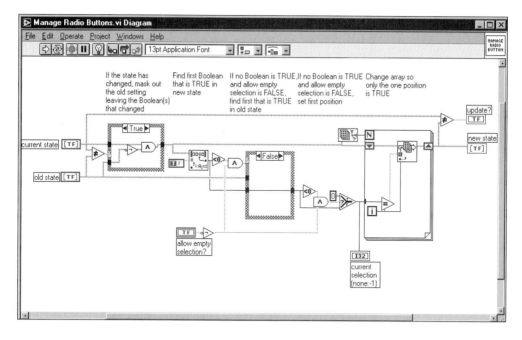

How do I clear a chart programmatically?

Wire an empty array to the History Data attribute of a chart. The data type of this empty array should be the same as the data type wired to the chart. There is an excellent example that illustrates this technique in `Examples\General\Graphs\Charts.llb\`**How to Clear Charts and Graphs.vi.**

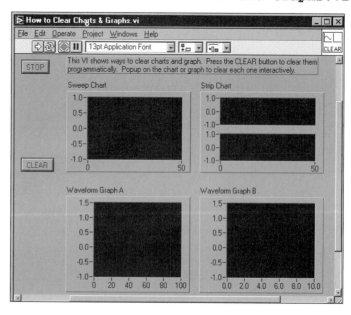

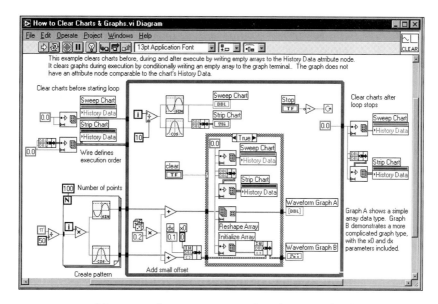

You can also pop up on the chart in the front panel and select **Clear Chart.**

How do I create a 3D plot in LabVIEW?

LabVIEW does not have any built-in capabilities for 3D graphs. Some third-party vendors provide add-on packages for 3D graphics and more, such as *SurfaceView* (Metric Systems) or *FastDraw* (Dixon DSP Design). You can, however, create one type of graph that represents data in three dimensions: the intensity charts and graphs (covered in Chapter 8).

How can I display a waterfall plot?

You can purchase the Picture Control Toolkit from National Instruments, or a third-party graphics add-on as mentioned above, or you can put a little more work into your program and make your own waterfall graph.

The basic idea to creating a waterfall plot is to build a multiplot graph, where you create the z-axis effect by successively adding a constant offset to each plot.

Is there a way to put more than one y-axis on a graph for two or more plots?

Yes, but it's tricky. LabVIEW has no built-in way to do this. A common work-around is to put one graph on top of another, with the first one being colored transparent. The two plots will appear to be on the same graph, but the scaling for each plot can be independently controlled.

My graphs flicker every time they are updated. It's getting really annoying. Is there any way to fix this?

Fortunately, yes. From the **Edit** menu, choose **Preferences...** Select the **Front Panel** item. Check the box that says "Use smooth updates during drawing." This will turn off the flicker; the trade-off being that more memory will be used.

Is there any way to jump to any given frame in a sequence structure (like a state machine)?

This is a very interesting problem whose solution can be a challenge to the novice user, but nevertheless presents us with a powerful programming structure—the state machine.

The problem statement goes somewhat like this: suppose you have a sequence structure, which you arrange to normally perform some operations in the expected sequence. However, suppose you occasionally need to skip a frame if a certain Boolean was true. Not a problem, you say; just add a case statement. But suppose it began to get more complicated; perhaps you actually need to go from frame 4 to frame 2, again under certain conditions. This is why it's called a state machine: you need a structure with a finite number of *states* it can be in, determined by your algorithm. Any time you have a chain of events where one operation is dependent upon the operation that occurred previously, a state machine is a good way to do the job.

Although LabVIEW doesn't have a built-in state machine structure, you can create one easily enough by using a Case Structure inside a While Loop and adding a shift register.

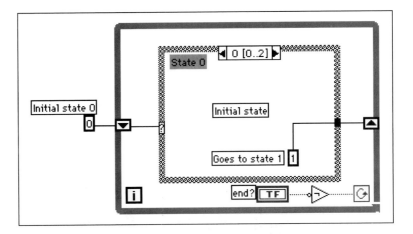

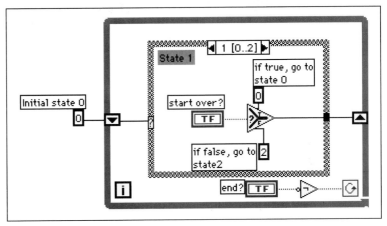

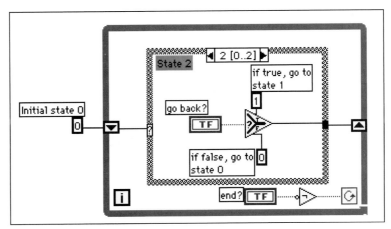

Each frame of the state machine's Case Structure can transfer control to any other frame that is allowed according to the algorithm on the next iteration, or cause the While Loop to terminate. By placing each "state" in a separate case, you can jump around to different states in any order you wish.

▩ How can you create a toolbar-type feature that will pop up different sets of controls and indicators on the same window?

This is a pretty neat trick. The idea is to create clusters of controls or indicators, one for each toolbar button. The clusters should be the same size, and then you literally stack them on top of each other. By using attribute nodes, you make all clusters invisible except the one that is selected. Toolbar buttons should be set up to work as radio buttons (only one can be selected at a time).

This is an excellent alternative when your front panel requires many controls or indicators, but it would be too confusing to display them all at once. The toolbar buttons serve as "menus" that appear to bring up a logically grouped set of objects.

▩ Can I access the parallel port in LabVIEW?

Yes. Use the serial port VIs (never mind that parallel is the opposite of serial). Just as you use port 0 for COM1, port 1 for COM2, etc., in LabVIEW for Windows, port 10 is LPT1, port 11 is LPT2, and so on. You can even send data to a printer connected to a parallel port, using **Serial Port Write**, although you may need to know the printer's special control characters. Another good use for accessing the parallel port is to do some digital I/O without a plug-in board: you get eight digital lines for free! (But a hardware buffer is recommended to protect your computer.)

Memory, Performance, and All That

Becoming a better LabVIEW programmer means you know how to make applications that are mean and lean. Sure, if you have oodles of RAM and a dual terahertz processor, you may not need to worry about these sort of things very much. But on average computers, critical or real-time applications are going to work better if you follow some simple guidelines for increasing performance and reducing memory consumption. Even if you don't foresee your application needing to conserve memory and processor speed, programmers who never take these sort of issues into account are, well, just plain sloppy. Nobody likes debugging sloppy programs.

◼ Curing Amnesia and Slothfulness

Let's face it—LabVIEW does tend to make applications gobble memory, but you can make the best of it by knowing some tips. Memory management is generally an advanced topic, but quickly becomes a concern when you have large arrays and/or critical timing problems. Read anything you can find about improving LabVIEW performance and saving memory. Here is a summary of the tips:

◆ Are you using the proper data types? Extended precision (EXT) floats are fine where the highest accuracy is needed, but they waste memory if a smaller type will do the job. This is especially important where large arrays are involved.

◆ Globals use a significant amount of memory. Minimize not only the creation of globals, but the amount of times you read or write to them.

◆ Don't use complicated, hierarchical data types (such as an array of clusters of arrays) if you need more memory efficiency and speed.

◆ Avoid unnecessary data type coercion (the gray dots on terminals). Coercion indicates that the expected data type is different than the data type wired to the terminal—LabVIEW does an astounding job of accepting the data anyway in most cases (polymorphism), but the result is a loss of speed and increased memory usage because copies of the data must be made. This is especially true of large arrays.

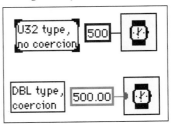

◆ How are you handling arrays and strings? Are you using loops where you don't have to? Sometimes there is a built-in function that can do the job, or several loops can be combined into one.

Watch out for putting unnecessary elements into loops, as shown in the following figure:

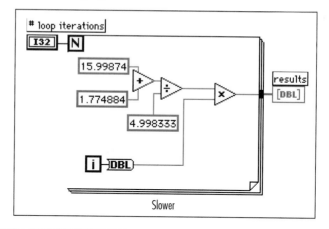

Slower

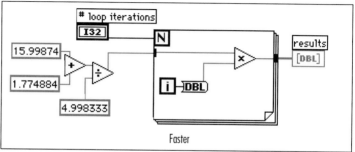

Faster

◆ Where possible, avoid using **Build Array** inside loops, thus avoiding repetitive calls to the Memory Manager. Every time you call the **Build Array** function, a new space in memory for the whole "new" array is allocated. Use auto-indexing or **Replace Array Element** with a pre-sized array instead. Similar problems occur with **Concatenate Strings**.

Obviously, this is not an option if you need to display the array data in real-time as it is being built. In this case, initialize the array to the maximum size first, and use the **Replace Array Element** function instead of the **Build Array**.

◆ Try to use For Loops instead of While Loops when arrays are involved. The advantage of a For Loop is that you can tell Lab-VIEW ahead of time how many elements to allocate in an array. With a While Loop, LabVIEW has to change the array size each time it goes around the loop.

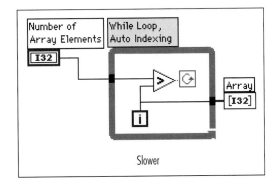

Slower

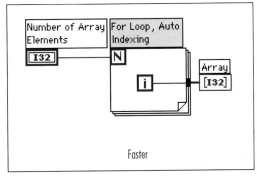

Faster

◆   Consider indicator overhead. Minimize the use of graphics (especially graphs and charts) when speed is extremely important.

◆   Update controls and indicators outside of loops whenever possible; that is, whenever it is not critical to view the object's value until the loop is finished executing.

■ The Declaration of Independence

LabVIEW does just an amazing job of porting VIs between different platforms. For the most part, you can take a VI created on a Sun and use it on a MacOS or Win95 system (as long they use the same LabVIEW version or a later one). However, not all parts of a block diagram are portable. If you want to design your VIs to be platform-independent, there are a few things you should know about portability.

◆   Be aware that the LabVIEW *application* and everything included with it (such as all the VIs in the vi.lib directory) are not portable. What the LabVIEW for each OS does is to *convert* VIs from other platforms, using its internal code, to VIs for its own platform.

◆ Some of the VIs in the **Advanced** palette (such as the **System Exec** VI in LabVIEW for Windows), as well as VIs in the **Communication** palette (like AppleEvents or DDE), are system-specific and thus are not portable.

◆ VIs that contain CINs are not immediately portable. But if you write your source code to be platform-independent and recompile it on the new operating system, the CIN should work fine.

◆ Keep in mind such things as filenames that have their own special rules for each OS (such as the infamous eight-character limit for Windows 3.1), and don't use characters such as [: / \], which are path delimiters in different operating systems.

End of Line Constant

◆ The end-of-line (EOL) character is different on each platform (\r for MacOS, \r\n for Windows, and \n for Sun). The easiest solution is to use LabVIEW's **End of Line** constant, from the **String** palette.

◆ Fonts can be a real mess when porting between systems. If you can, stick to the three standard LabVIEW font schemes (Application, Dialog, System), because custom fonts that look fine on one platform may look huge, tiny, or distorted on another.

◆ Screen resolution can also be a nuisance. Some people recommend sticking to using a 640 by 480 screen resolution, which will make VI windows fit fine on most monitors.

◆ Remember the colors available on one machine may not be available on another. If you used a 256-color palette, some colors may be dithered on a 16-color machine, which looks pretty bad sometimes.

Don't forget about the Profile window, discussed in Chapter 13, if you want to track your VI's memory usage and speed performance.

Programming with Style

Programming really is an art, and it can be especially fun in LabVIEW! The following section is a collection of final reminders and guidelines for writing a *good* LabVIEW application. You can find some of these concepts and many more in "The LabVIEW Style Guide: A Guide to Better LabVIEW Applications," written by Gary W. Johnson and Meg F. Kay. This document is available on the National Instruments Web page (`http://www.natinst.com`) or by request from National Instruments.

■ Modularize and Test Your VIs

Although it's theoretically possible to be *too* modular in designing your program, this rarely happens. Make subVIs out of all but the most simple and trivial functions and procedures. This gives you a chance to test each of your individual pieces of code before working with the big piece. It also lets you easily reuse code, keep yourself organized, and make the block diagram size manageable. Don't forget to test each subVI as a top-level VI—and be thorough; test all sorts of strange input combinations. If you know all your subVIs are working, it should be very easy to debug any problems with your top-level VI.

One useful tip: often LabVIEW programmers won't test certain VIs because they require DAQ hardware or other externally-generated inputs. Don't wait until you have the hardware. Write a simple "dummy data" VI to pass data to these VIs in the meantime so you can at least test part of their functionality.

■ Document as You Go Along

Please document your work! Many programmers shun documentation, thinking it will be done later or not needed, until, two years later, a user (possibly yourself) is trying to figure out how a VI works. Take advantage of LabVIEW's built-in documentation capabilities:

1. **VI Info**. At the very least, write a short description for each VI you create. This is extremely valuable when someone is looking at the subVI icons in the block diagram and needs an idea of what they do.

2. **Descriptions**. Ideally, write a help statement for each control and indicator using the pop-up **Description...** command. These invaluable comments will appear on the Help window if a user points to the control or indicator in question.

3. **VI History**. This option, available from the **Windows** menu, is a more sophisticated tool for larger projects. It allows you to enter

comments about the changes you've made to a VI along the way. The History window can be quite helpful when more than one person works on a project, since it keeps track of the user, time, and date.

4.   **Front Panel Text**. For important indications, just write some text (perhaps with a bold or large font) on the front panel itself. Users can't miss that one!

■ One More Time: Dataflow!

As you get more comfortable with LabVIEW, you will begin to take more advantage of the way dataflow programming works. Some tips to remember are:

◆   As you know, LabVIEW data is carried across wires. *Dataflow* means that a subVI or function will only start executing after data is available to ALL of its input terminals.

◆   Two or more objects or groups of objects on the block diagram that are not connected have no specific execution sequence. Many people new to LabVIEW have a feeling that execution should take place left-to-right or top-to-bottom. That is not true! There is no way to predict in what order two or more pieces of block diagram will occur unless it's specified by dataflow.

◆   When you need to "force" an execution sequence, you have the option of using an artificial data dependency structure, as shown in the figure.

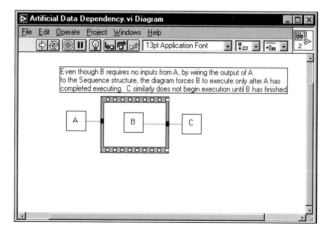

◆ You've noticed that many LabVIEW functions have common "threads": refnums, taskID, error clusters. These threads are designed to string several common VIs together in a natural dataflow sequence. You can also use threads for your own VIs, such as error clusters. Using threads will reduce the need for Sequence Structures with awkward sequence locals.

Wrap It Up!

This chapter gave you some instructions on the *art* of graphical programming. We looked first at the front panel end—suggestions, guidelines, and reasons for making an exciting graphical user interface (GUI). Secondly, we focused on the block diagram—programming solutions, performance, memory, and style.

Creating a good GUI is important for "selling" your program to your customer or your boss, as well as making it much easier to use. LabVIEW's "art" enhancements include decoration modules, align and distribution commands, layered objects, and the ability to import pictures.

Custom controls and indicators can add value to your GUI by providing graphical simulation and animation tools. The Control Editor lets you modify standard LabVIEW controls and indicators and import picture files to represent the new objects.

The Help window is not just for you, the programmer. The end user can use the Help window to examine front panel objects descriptions. You can open or close the Help window programmatically and set up custom help as well.

Many common questions and problems arise when you are programming in LabVIEW. Some of the solutions were discussed in this chapter. You can find more ideas to solve your specific problem by consulting some of the resources listed in Appendix A or the common questions listed in Appendix B.

When you need to improve the speed and/or memory performance of LabVIEW, follow the guidelines discussed in this chapter. Watch out always for unnecessary operations inside loops, especially with arrays.

Although for the most part LabVIEW VIs are platform independent, you do need to be aware of a few obstacles that can creep up—such as CINs or system-specific functions.

Finally, to be a good LabVIEW programmer, you need to be systematic in making modular VIs that are tested thoroughly. Good documentation is essential to making a quality, maintainable piece of software. Last but not least, learn to use LabVIEW's distinctive hallmark unknown to other languages—dataflow— to your advantage.

Concluding Remarks

This is the end of the book! (Well, there are some appendices, and if you're really bored, you can read the index.) By now, you've gained a solid understanding of how LabVIEW works. You've also begun to see how it can work for you, whether your application is teaching an electrical engineering class or building a process control system for a large plant.

Where do you go from here? More than anything else, hands-on experience is the best teacher. Experiment with a VI. Build a prototype. Look at examples. Be creative. Above all, have fun. If you decided to buy the full version of LabVIEW, don't be afraid to go through some of the manuals. They can be an invaluable reference for details on your application.

Good luck!

High-Speed Remote Process Control

By Steve Conquergood, President, Advanced Measurements, Inc.

The Challenge

Developing a reliable, high-speed, remote PC-based process monitoring and control system.

The Solution

Using LabVIEW—with a watchdog timer for reliability—to control the system using DAQ, GPIB, and RS-232 for I/O.

Field Test Site

Introduction

Using LabVIEW running on three networked PCs, we developed a high-speed system for remote monitoring and control of a proprietary oilfield process. We monitor process signals using National Instruments SCXI, DAQ, and GPIB hardware. We use GPIB and RS-232 outputs for process control. The system uses virtually every kind of function in LabVIEW, from DAQ, GPIB, and serial port I/O to advanced analysis techniques such as multivariable regression and matrix manipulation.

Requirements

The process developers required a control system that would be powerful and flexible, yet affordable and expandable. We chose LabVIEW software because of its rapid development environment and its easy integration capabilities.

The objectives for the system are as follows:

◆ Gather input signals every 33 ms (milliseconds) from 13 process measurement points located 5,000 feet from the computer.

◆ Use this updated information in real time to calculate the process control adjustments required for safe operation, then output new setpoints.

◆ Display process variables, setpoints, and status information for the operator, then log data at one-second intervals.

◆ Perform advanced analysis to determine progress toward process completion.

◆ Perform emergency shutdown if necessary.

Additional requirements that make this system unique:

◆ The process hardware will be destroyed in less than 200 ms if control is lost.

◆ Component failures, such as insulation breakdown, can result in 5,000 V being applied at the DAQ inputs.

◆ The entire system must be mobile, able to withstand extreme environmental conditions, and capable of continuous operation for several days.

System Hardware

Three 486/66 personal computers running Windows are networked together using Ethernet for high-speed exchange of all the required process information. Although the distance is short, we use fiber-optic cable between the DAQ PC and the other two PCs for safety in case the 5,000 V source is ever accidentally shorted to the inputs.

Signal inputs are connected to an SCXI-1100 multiplexer because of its fast settling times and ability to measure and compensate for amplifier offset. We digitize signals with an AT-MIO-16X board. The process control hardware is controlled with an AT-GPIB/TNT board cabled to the device with a 10 meter cable.

> The system uses virtually every kind of function in LabVIEW, from DAQ, GPIB, and serial port I/O, to advanced analysis techniques such as multivariable regression and matrix manipulation.

One of the main concerns with using PCs for process control is the potential for failure. To reduce the consequences of software or hardware failure, we installed a watchdog timer in the controller PC. This timer is reset by the LabVIEW program every iteration (33 ms), which keeps the con-

DAQ Computer and SCXI chassis

troller in the operating mode. If the program fails to reset the watchdog timer for any reason, it opens its relay and the process goes safely off-line.

System Software

We developed the software using LabVIEW 3.1, but we run compiled executables during operation. Virtual memory and power-saver features are disabled and the Monitor PC uses a graphics accelerator card for improved speed.

The DAQ software is essentially a data server that can have up to two clients logged in to receive the current data. The software must acquire voltage data, scale it to engineering units, test the data for emergency conditions, and transmit the data to the clients at a desired 33 ms iteration time. This high-speed requirement is dictated by the 200 ms time before destruction of the equipment if the process goes out of control.

We chose the overall sampling period as 33 ms to minimize 60 Hz noise because it represents two periods of 60 Hz. During the 33 ms window, we acquire and average 20 samples to generate each data point (sampling rate is 600 samples per second for each channel).

We also use double-buffered acquisition to assure continuous sampling at a fixed rate. But we could not use the normal AI Read function because it is a code interface node (CIN) that effectively stops other operations during data sampling. For such high-speed process control, the loop iteration time requires that data conversion, testing, and transmission occur concurrently with sampling. So we used another DAQ approach, with occurrences providing the timing cues for the AI Read function. An occurrence is a LabVIEW mechanism for coordinating parallel processes. Other functions in the loop proceed simultaneously as samples are placed in the buffer; when the required number of samples are present in the buffer, AI Read is complete.

> **O**ne of the greatest assets of LabVIEW is that you can easily substitute new portions of code in the future, without major redevelopment of the remaining code.

The VIs for the three PCs in the system use a state-machine architecture. For this application, the state machine provides a very efficient way to control the execution order in "normal" mode, while maintaining fast response in case an emergency

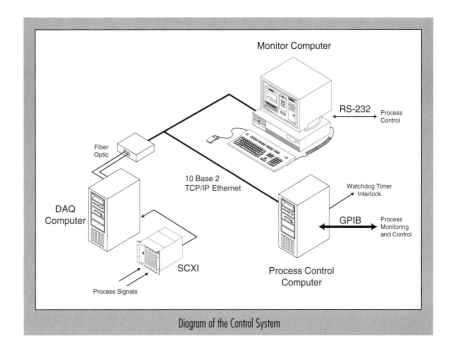

Diagram of the Control System

515

condition requires alternative actions. Another technique widely used in this application is circular data buffering. Data is stored in circular buffers, giving the operator the ability to review long-term trends in the process.

Process Control

We achieved the high-speed setpoint updates required for safety with a relatively simple control algorithm but without more expensive hardware because of the inherent stability of the parameters. However, the software is constructed so that we can easily substitute more sophisticated algorithms such as PID, as needed. One of the greatest assets of LabVIEW is that you can easily substitute new portions of code in the future, without major redevelopment of the remaining code.

Summary

We developed an inexpensive, sophisticated system for remote process control using readily available hardware and software—a fully open system with easy expansion for the future. Although nearly every functional area of LabVIEW is employed in the system, the graphical front panels mean that the software is not overwhelming to use.

For More Information Contact

Advanced Measurements, Inc.

5510 3rd St. SE, Bay 2

Calgary, AB T2H 1J9, Canada

tel (403) 571-7273

fax (403) 571-7279

Appendix A: National Instruments Contact Information, Resources, and Toolkits

How to Contact National Instruments

Corporate Office
National Instruments
6504 Bridge Point Parkway
Austin, TX 78730-5039 USA
Tel. (512) 794-0100
Fax. (512) 794-8411
E-mail: info@natinst.com
Web Address: http://www.natinst.com

Australia
National Instruments Australia
P.O. Box 466
Ringwood, Victoria 3134
Australia
Tel. (03) 9879 5166
Fax. (03) 9879 6277
E-mail: info.australia@natinst.com
Web Address: http://www.natinst.com/austral

Austria
National Instruments Ges.m.b.H.
Plainbachstr. 12
5101 Slazburg-Bergheim
Austria
Tel. 43 662 457990-0
Fax. 43 662 069959
Web Address:
http://www.natinst.com/austria

Belgium
National Instruments Belgium
Leuvensesteenweg 613
B-1930 Zaventem
Belgium
Tel. 32 2 757 00 20
Fax. 32 2 757 03 11
E-mail: info.belgium@natinst.com
Web Address:
http://www.natinst.com/belgium

Canada
National Instruments Canada
P.O. Box 42252
128 Queen Street South
Mississauga, ON
Canada, L5M 4Z0
Tel. (905) 785-0085
Fax. (905) 785-0086
E-mail: info@natinst.com
Web Address:
http://www.natinst.com/canada

Denmark
National Instruments Danmark
Christianshusvej 189
2970 Hørsholm
Danmark
Tel. 45 45 76 26 00
Fax. 45 45 76 26 02
E-mail: ni.danmark@natinst.com
Web Address:
http://www.natinst.com/denmark

Finland
National Instruments Finland Oy
PL 2, Sinimäentie 14 C
02631 Espoo, Finland

Tel. 358-0-527-2321
Fax. 358-0-502-2930
E-mail: ni.finland@natinst.com
Web Address:
http://www.natinst.com/finland

France
National Instruments France
Centre d'Affaires Paris-Nord
Immeuble Le Continental - BP 217
93 153 Le Blanc-Mesnil CEDEX
France
Tel. 33 01 48 14 24 24
Fax. 33 01 48 14 24 14
Web Address:
http://www.natinst.com/france

Germany
National Instruments Germany GmbH
Konrad-Celtis Str. 79
81369 München
Germany
Tel. 49 89 74 13 13 0
Fax. 49 89 7 14 60 35
E-mail: nig.cs@natinst.com
Web Address:
http://www.natinst.com/german

Hong Kong
National Instruments Hong Kong Limited
Unit 10, 6/F Block B
New Trade Plaza
6 On Ping Street
Shantin, N.T.
Hong Kong
Tel. 852 2645 3186
Fax. 852 2686 8505
E-mail: general@nihk.com.hk
Web Address:
http://www.natinst.com/hongkong

Israel
National Instruments Israel, LTD.
P.O. Box 184
Givatayim 53101
Israel
Tel. 972 357 34815
Fax. 972 3f57 34816

Web Address:
http://www.natinst.com/israel

Italy

National Instruments Italy S.r.l.
Via Anna Kuliscioff, 22
20152 Milan, Italy
Tel. 39 2 413 091
Fax. 39 2 4130 9215
E-mail: ni.italy@natinst.com
Web Address:
http://www.natinst.com/italy

Japan

Nihon National Instruments K.K.
Shuwa Shiba Park Bldg. B-5F
Shibakoen 2-4-1, Minato-ku
Tokyo, Japan 105
Tel. 81 3 5472 2970
Fax. 813 5472 2977
Web Address: http://www.natinst.com/nni

Korea

National Instruments Korea Limited
Room #201, Seocho B/D
1680-3, Seocho-dong, Seocho-ku
Seoul, Korea 137-070
Tel. 822 596-7456
Fax. 822 596 7455
Web Address:
http://www.natinst.com/korea

Mexico

National Instruments Mexico
Galileo, 31B
Suite 570
Col. Polanco
11560 Mexico, D.F.
Tel. 525 202 2544
Fax. 525 520 3282
E-mail: info@natinst.com
Web Address:
http://www.natinst.com/mexico

Netherlands

National Instruments Netherlands BV
Pompmolenlaan 25
3447 GK Woerden

The Netherlands
Tel. 31 348 433466
Fax. 31 348 430673
Email: info.netherlands@natinst.com
Web Address:
http://www.natinst.com/netherl

Norway

National Instruments Norge
Industrigt. 15
Postboks 592
3412 Lierstranda
Norway
Tel. 47 32 84 84 00
Fax. 47 32 84 86 00
Email: ni.norway@natinst.com
Web Address:
http://www.natinst.com/norway

Singapore

National Instruments Singapore Pte Ltd
138 Cecil Street, #05-03
#05-03 Cecil Court
Singapore 069538
Tel. 65 226 5886
Fax. 65 226 5887
Email: natinst@singnet.com.sg
Web Address:
http://www.natinst.com/singapor

Spain

National Instruments Spain, S.L.
Europa Empresarial
C/ Rozabella N°. 6
Edf. París, 2° Planta, Oficina N° 8
28230 - Las Rozas, Madrid, Spain
Tel. (34) 1 640 0085
Fax. (34) 1 640 0533
E-mail: ni.spain@natinst.com
Web Address:
http://www.natinst.com/spain

Sweden

National Instruments Sweden AB
Box 2004
Råsundavägen 166
171 02 Solna
Sweden

Tel. 46 8 7304970
Fax. 46 8 7304370
E-mail: ni.sweden@natinst.com
Web Address:
http://www.natinst.com/sweden

Switzerland
National Instruments Switzerland
Sonnenbergstr. 53
CH-5408 Ennetbaden
Switzerland
Tel. 41 56 200 5151
Fax. 41 56 200 5155
E-mail: ni.switzerland@natinst.com
Web Address:
http://www.natinst.com/switzerl

Taiwan
National Instruments Taiwan
3F, NO. 97-1

Ho Ping East Road Section 3
Taipei, Taiwan
R.O.C.
Tel. 8862 377 1200
Fax. 8862 737 4644
Web Address:
http://www.natinst.com/taiwan

United Kingdom
National Instruments Corporation (UK) Ltd.
21 Kingfisher Court
Hambridge Road
Newbury, Berkshire RG14 5SJ
United Kingdom
Tel. 44 1635 523 545
Fax. 44 1635 523 154
E-mail: info.uk@natinst.com
Web Address: http://www.natinst.com/uk

Resources to Help You

You have many options for getting help and finding additional information about LabVIEW and virtual instrumentation. Read on to discover the vast array of help and information available.

■ LabVIEW Documentation and Online Help

The LabVIEW manuals provide excellent tutorials to get you up to speed with LabVIEW. They also contain plenty of reference information. You can find the answers to almost all of your questions in these manuals. In addition, LabVIEW has extensive online help to assist you as you build your application.

■ EXAMPLES Directory

The LabVIEW EXAMPLES directory, which ships with the full version of LabVIEW, contains hundreds of LabVIEW programs that you can run as is or modify to suit your application. You can learn a lot about LabVIEW programming just by browsing through these examples and looking at the techniques used. You can get a description of each example by opening it and then selecting **Get Info...** from the **File** menu.

■ Info-LabVIEW

Info-LabVIEW is a user-sponsored Internet mailing list that you can use to communicate with other LabVIEW users. You can post messages containing questions, answers, and discussions about LabVIEW on this mailing list. These messages will be sent to LabVIEW users worldwide.

If you want to subscribe to Info-LabVIEW, send an e-mail message to `info-labview-request@pica.army.mil` requesting that your e-mail address be added to the subscription list. All messages posted to Info-LabVIEW will then be forwarded to your e-mail address. You can cancel your subscription by sending a message to the above address requesting that your e-mail address be removed from the list.

If you want to post a message, send e-mail to `info-labview@pica.army.mil`. Although Info-LabVIEW is user-sponsored, many National Instruments engineering, marketing, and sales personnel read the postings. Your discussions, questions, and product requests will be heard by National Instruments.

■ FaxBack Information Retrieval System

You can have product data sheets, answers to common questions, technical and application notes, user solutions, and other information sent to you from National Instruments' automatic FaxBack information retrieval system. FaxBack is available 24 hours a day, seven days a week, from a touch-tone telephone. Simply call (800) 329-7177 or (512) 418-1111 and follow the instructions provided. You can request indexes of product information, technical support information, and User Solutions; these indexes will list the documents available to you on FaxBack.

■ World Wide Web

INSTRUMENTATION WEB™
The Internet Home for Instrumentation Information

You can access the National Instruments World Wide Web site at `http://www.natinst.com`. The Instrumentation Web site contains up-to-date online information about National Instruments products, services, developer programs, and many other topics.

■ LabVIEW Bulletin Board Service and Internet FTP Site

You can connect to the LabVIEW Bulleting Board Service (BBS) using a modem or to the Internet FTP site to receive the following services:

- ◆ *Software and technical support*
- ◆ *Electronic correspondence*
- ◆ *Common questions*
- ◆ *Technical publications, including application notes*
- ◆ *Software updates and the latest hardware drivers*
- ◆ *Utility VIs for specialized applications*
- ◆ *Free instrument drivers*
- ◆ *Educational VI Exchange*

National Instruments offers free application software for the Macintosh (NovaTerm) or PC (RipTerm) that you can use to connect to its BBS. This software features easy-to-use graphical user interfaces (GUIs) that simplify sending mail and transferring files. You can download these applications from the BBS.

BBS (Modem)	
Telephone	(512) 794-5422
Baud Rate	14,400
Data Bits	8
Stop Bits	1
Parity	None

You will create an account the first time you log on.

Internet FTP Site	
Address	ftp.natinst.com
Login	anonymous
Password	your e-mail address

■ Educational VI Exchange

The Educational VI Exchange (EVE) gives you access to LabVIEW VIs and VI libraries that are used in colleges and universities around the world. The Exchange is an excellent way for users to see new ways to apply LabVIEW in their fields. You can download these VIs to see educational uses of LabVIEW in many disciplines and then run them in your own applications. You can also upload your programs to share with others who can benefit from your ideas.

The Exchange is located on the National Instruments FTP site in the `contrib\labview\education` directory. You can also access it from the worldwide web. Inside the education directory, you will find three folders: `windows`, `mac`, and `all_platforms`. Most VIs should be placed in the `all_platforms` directory. If your VIs contain code interface nodes (CINs) or platform specific code such as Apple Event or dynamic data exchange (DDE) calls, then place the VIs in the appropriate Mac or Windows directory. Files compressed with ZIP and self-extracting archives should also be placed in the appropriate platform-specific directory. Virtual instruments and VI libraries should be transferred in binary mode.

National Instruments encourages you to put VIs written in both the professional version and student edition on the Exchange. They request that you provide a detailed description in the **Get Info...** section of your top-level VIs so that others can easily understand what you are doing. You should include the following information:

◆ *Description of the VI*

◆ *Course(s) or setting in which you use the VI*

◆ *Hardware and other equipment you use*

◆ *Whether the VI was written in the professional version or student edition*

◆ *Any special instructions or recommendations*

◆ *Your name and institution*

◆ *Your e-mail address, regular mail address, and phone number (optional)*

Also, please upload an accompanying "readme" file containing the same information so that users can browse through your descriptions. The "readme" file should have the same name as your VI or VI library file, with a `.txt` extension.

■ Application and Technical Notes

National Instruments has many application and technical notes available to you at no cost. These notes are available from our BBS, Internet site, and FaxBack system.

■ Technical Support

National Instruments provides top-quality technical support to LabVIEW users. You can get help using e-mail, fax, and of course, the telephone. To make your life easier, be sure to compose a detailed description of the problem before you contact them.

In the U.S. and Canada	
Fax	(512) 794-5678
E-mail	lv.support@natinst.com
Telephone	(512) 795-8248

International	
Contact your local branch office	

■ National Instruments Customer Education

National Instruments offers hands-on customer education classes worldwide on LabVIEW, DAQ, GPIB, VXI, and LabWindows/CVI. These courses help you get up to speed quickly. For more information, call National Instruments at (512) 794-0100 in the United States and Canada or contact the local branch office in your country.

■ LabVIEW Basics—Interactive: A Multimedia Training CD-ROM

LabVIEW Basics—Interactive is a self-paced multimedia course on CD-ROM that teaches LabVIEW concepts through an interactive user interface. Experience high quality video and audio effects as a National Instruments applications engineer builds sample VIs on screen. Based on material from the highly acclaimed LabVIEW Basics course, LabVIEW Basics—Interactive progresses from LabVIEW fundamentals and the construction of simple VIs to developing data acquisition and instrument control applications.

If you are new to LabVIEW, you can take advantage of the step-by-step instructions included with each hands-on exercise. If you are already familiar with certain LabVIEW programming concepts, you may choose to advance at a quicker pace. You can pause, review, or skip lessons to suit your level of expertise and learning style. You can also add bookmarks for noteworthy material, so that you (and other users) can exit the course and pick up where you left off. An index feature provides direct access to content using key-word searches.

LabVIEW Basics—Interactive ships on a single CD-ROM that will run on PCs under Windows 95, Windows 3.1 or greater, and Macintosh/Power Macintosh running System 7.1 or greater. LabVIEW version 4.0 or greater must be installed separately.

■ Training Videos

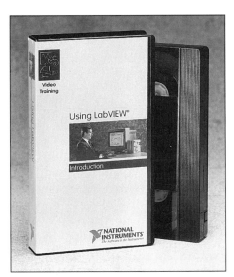

You can learn LabVIEW at your own pace using an instructional video available from National Instruments. This two-hour video teaches engineers and scientists how to use LabVIEW to develop data acquisition and control applications. For more information, call National Instruments at (512) 794-0100 in the United States and Canada or contact your local branch office.

■ National Instruments Catalogue

The *National Instruments Instrumentation Reference and Catalogue* contains detailed information about all National Instruments hardware and software products.

■ LabVIEW Graphical Programming

LabVIEW Graphical Programming is a book about LabVIEW written by an avid and versatile LabVIEW user of many years, Gary Johnson. It contains valuable information on LabVIEW, data acquisition, instrument control, and specific application configurations, presented in a user-friendly manner . To order *LabVIEW Graphical Programming*, contact McGraw-Hill at (800) 822-8158 and reference ISBN# 0-07-032692. Outside the United States, call (717) 794-2191.

■ Seminars

National Instruments regularly holds seminars around the world. The seminars are designed to keep you abreast of the latest technologies and trends in measurement and instrumentation. The material is usually presented in a hands-on format, with equipment donated by leading instrumentation and computer manufacturers.

■ Instrupedia™ 96

Instrupedia™ 96, the encyclopedia of instrumentation, is an interactive and searchable reference CD-ROM that you can request from National Instruments for no charge. It includes a glossary of common instrumentation terms, a summary of books on instrumentation and measurement, collections of actual LabVIEW user applications, and application notes covering specific topics in GPIB, VXI, DAQ, data analysis, and software. Instrupedia 96 also features National Instruments' online catalog, a technical support reference library, demonstration programs, and configuration-advising programs that help you design your system. It's available for Windows platforms only.

■ Software Showcase

The National Instruments Software Showcase is a CD-ROM that provides one-stop shopping for anyone interested in demos of their software. The Software Showcase will teach users about the concept of virtual instrumentation and will allow them to investigate the virtual instrumentation software tools of their choice. Available for Windows 3.1, Windows 95, WindowsNT, Macintosh, and Power Macintosh, the Software Showcase will also discuss existing user applications, common technical uses, and available complementary products. To request one, contact National Instruments.

■ The Instrumentation Newsletter

The *Instrumentation Newsletter*, produced quarterly by National Instruments, features User Solutions along with updates on the latest products for instrumentation. Academic, Radio Frequency/Microwave, Automotive, and international versions are available. You can request this free newsletter by contacting National Instrument.

■ LabVIEW Technical Resource

LabVIEW Technical Resource is a quarterly newsletter that provides technical information for LabVIEW systems developers. It offers solutions to common problems, programming tips, tools, and techniques, and each issue comes with a disk containing LabVIEW programs. For more information about *LabVIEW Technical Resource*, call (214) 827-9931 or write to LTR Publishing, 5614 Anita, Dallas, TX 75206

■ National Instruments Alliance Members

NATIONAL INSTRUMENTS
ALLIANCE
PROGRAM

National Instruments' Alliance Program is comprised of several hundred third-party companies that provide consulting services, turnkey solutions, and valuable add-on products. Located all over the world, Alliance members can help you design a system or write a specific application to meet your needs. For a complete listing of Alliance member companies and their services, request a *Solutions* guide from National Instruments.

LabVIEW Add-On Toolkits

You can purchase the following special add-on toolkits from National Instruments to increase LabVIEW's functionality. In addition, new toolkits are created frequently, so if you have a particular goal, it's worthwhile to check and see if a toolkit already exists to accomplish it. If one doesn't exist, National Instruments will appreciate your suggestions on what to develop in the future!

Many third-party developers, often Alliance Members, also make add-ons to LabVIEW to do all sorts of things. We suggest posting to the `info-labview` user forum (described in the previous section of this appendix) if you have a specific task and you want to know if someone's already done it.

■ Application Builder

The LabVIEW Application Builder is an add-on package for creating stand-alone, executable applications. When accompanied by the Application Builder, a LabVIEW system can create VIs that operate as stand-alone applications. You can run the executable file but cannot edit it. Stand-alone applications minimize RAM and disk requirements by saving only those resources needed for execution. The application contains the same execution resources as those in the LabVIEW Full Development System, so VIs will execute at the same high-performance rates.

■ Test Executive Toolkit

The LabVIEW Test Executive is a library of virtual instruments used to create test sequences and control test execution for production and manufacturing automated test applications. The toolkit delivers the Test Executive in high-level modules, making it easy to modify according to your application needs. The Test Executive includes an intuitive, multilevel user interface for displaying test results, operator and UUT (unit under test) serial number prompts, pass/fail banners, and run-time error reporting.

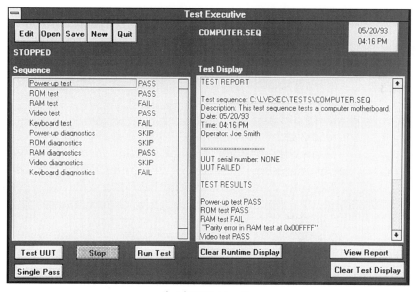

The LabVIEW Test Executive

■ JTFA Toolkit

The Joint Time-Frequency Analysis (JTFA) Toolkit is a LabVIEW add-on library and stand-alone executable that enhances the signal processing capabilities of LabVIEW software on nonstationary signals. It is useful for precise signal analysis of data whose frequency content changes with time. The toolkit features the award-winning and patented Gabor Spectrogram, which is ideal in applications of speech processing, sound analysis, sonar, radar, vibration analysis, and dynamic signal monitoring.

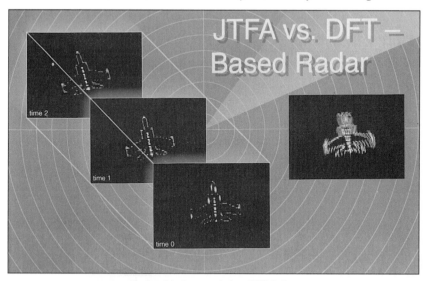

The Joint Time-Frequency Analysis (JTFA) Toolkit

■ Digital Filter Design Toolkit

The Digital Filter Design Toolkit is a ready-to-run virtual instrument for interactively designing Finite Impulse Response (FIR) filters and Infinite Impulse Response (IIR) filters. Outputs include pole zero plots, magnitude and phase plots, impulse response, and step response.

■ Third Octave Analyzer Toolkit

The Third Octave Analyzer Toolkit, combined with National Instruments DAQ hardware, is a ready-to-run virtual instrument. The Third Octave Analyzer provides a standard instrument interface used widely in acoustics and vibration analysis. The toolkit includes LabVIEW source code so that you can make modifications for custom applications.

■ Signal Processing Suite

The Signal Processing Suite gives users ready-to-run signal processing capabilities. Scientists and engineers can use the JTFA Toolkit, the Digital Filter Design Toolkit, the Virtual Bench/DSA dynamic signal analyzer, and the Third Octave Analyzer Toolkit in applications such as acoustics, analog telephony, radar, seismology, remote sensing, and vibration analysis.

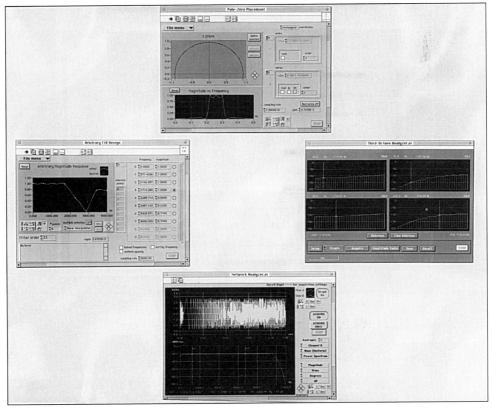

Signal Processing Suite includes the JTFA Toolkit, the Digital Filter Design Toolkit, the Virtual Bench/DSA dynamic signal analyzer, and the Third Octave Analyzer Toolkit.

■ Picture Control Toolkit

The LabVIEW Picture Control Toolkit is a versatile graphics add-on package for creating arbitrary front panel displays. The toolkit adds the Picture control and a library of VIs to your

LabVIEW system. You can create diagrams using a set of VIs to describe the drawing operations to build these images dynamically. With the toolkit, you can create new front panel displays, such as specialized bar graphs, pie charts, and Smith charts. You can also display and animate arbitrary objects such as robot arms, test equipment, a unit under test, or a real-world process.

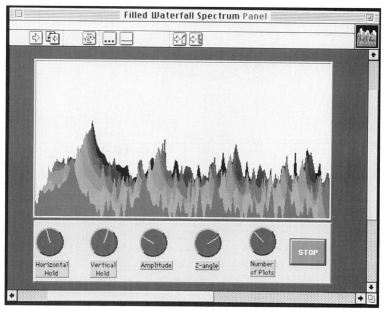

The Picture Control Toolkit

■ SPC Toolkit

The LabVIEW SPC Toolkit is a VI library for statistical process control (SPC) applications. Using the SPC Toolkit, you can use SPC methods to analyze and track process performance. In addition to subVIs that perform the SPC computations, the toolkit contains an extensive collection of example and graph VIs that demonstrate how to incorporate typical SPC methods and displays into LabVIEW applications. The toolkit addresses three areas of SPC: Control Charts, Process Statistics, and Pareto Analysis.

■ PID Control Toolkit

The PID Control Toolkit adds sophisticated control algorithms to LabVIEW. With this package, you can quickly build data acquisition and control systems for your own control application. By combining the PID Control Toolkit with the math and logic functions in LabVIEW, you can quickly develop programs for control.

■ SQL Toolkit—DatabaseVIEW

National Instruments is distributing DatabaseVIEW from Ellipsis Inc. as the SQL Toolkit—DatabaseVIEW. DatabaseVIEW is a collection of LabVIEW VIs for direct interaction with any SQL local or remote database. High-level Access VIs simplify database access by intelligently encapsulating common database operations into easy-to-use VIs. Low-level Interface VIs directly access the database, operating on columns and records in database tables.

Appendix B: Troubleshooting and Common Questions

This appendix will help you fix those hard-to-spot problems in your VIs and will also answer some common LabVIEW questions. Sometimes you'll find the answers before you even come up with the question, so you might want to skim through it even if you don't have problems or questions right now. You can also find some very helpful information describing possible errors and common reasons for bad wires in the LabVIEW User Manual.

Debugging Tips

■ Most Common Reasons For Broken VIs

◆ A function terminal requiring an input is unwired. You cannot leave unwired functions on the diagram while you run a VI to try out different designs.

◆ The block diagram contains a bad wire due to a data type mismatch or a loose, unconnected end, which may be hidden under something or be so tiny you can't see it. Remove these using the **Remove Bad Wires** command from the **Edit** menu.

◆   A subVI is broken, or you edited its connector after placing
    its icon on the diagram.

◆   You have a problem with an object that is disabled, invisi-
    ble, or altered using an attribute node.

■ Debugging Techniques for Executable VIs—When Your Results Aren't Right

If your program executes but does not produce the expected
results, use these suggestions as a reference to locate the problem.

◆   *Eliminate any warnings* your VI may have. You can view
    them by selecting the **Show Warnings** option in the Error
    List window.

◆   *Check wire paths* to ensure that the wires connect to the prop-
    er terminals. Triple-clicking on the wire with the
    Positioning tool highlights the entire path. A wire that
    appears to emanate from one terminal may in fact emanate
    from another, so look closely to see where the end of the
    wire connects to the node.

◆   *Use the Help window* (from the Help menu) to make sure that
    functions are wired correctly.

◆   *Verify that the default value is what you expect* if functions or
    subVIs have unwired inputs.

◆   *Use breakpoints, execution highlighting, and single-stepping* to
    determine if the VI is executing as you planned. Make sure
    you disable these modes when you do not want them to
    interfere with performance.

◆   *Use the probe* to observe intermediate data values. Also
    check the error output of functions and subVIs, especially
    those performing I/O.

◆   *Observe the behavior of the VI or subVI with various input val-
    ues.* For floating-point numeric controls, you can enter the
    values NaN (not a number) and ±Inf (infinity) in addition
    to normal values.

◆   *Make sure execution highlighting is turned off* in subVIs if the
    VI runs more slowly than expected. Also, close subVI win-
    dows when you are not using them.

◆ *Check the representation of your controls and indicators* to see whether you are getting an overflow; you may have converted a floating-point number to an integer, or an integer to a smaller integer.

◆ *Check the data range and range error actions* of controls and indicators. They may not be taking the error action you want.

◆ *Check for For Loops that may inadvertently execute zero iterations* and produce empty arrays.

◆ *Verify that you initialized shift registers properly*, unless you specifically intend them to save data from one execution of the loop to another.

◆ *Check the order of cluster elements* at the source and destination points. Although LabVIEW detects data type and cluster size mismatches at edit time, LabVIEW does not detect mismatches of elements of the same type. Use the **Cluster Order...** option on the cluster shell pop-up menu to check cluster order.

◆ *Check the node execution order*. Nodes that are not connected by a wire can execute in any order. The spatial arrangement of these nodes does not control the order. That is, unconnected nodes do not execute from left to right or top to bottom on the diagram like statements do in textual languages.

◆ *Check that your diagram does not contain any extraneous subVIs* that do unexpected things. Unlike functions, unwired subVIs do not always generate errors (unless you configure an input to be required or recommended).

◆ Check that you don't have hidden VIs or other objects. You can inadvertently hide something by dropping it directly on top of another node, by decreasing the size of a structure without keeping it in view, or by placing it off the main diagram area (off your screen).

Common LabVIEW Questions

We've answered some common questions about LabVIEW here, organized by topic: Charts and Graphs, DAQ, Error Messages and Crashes, File I/O, GPIB, Installation, Miscellaneous, Platform Issues and Compatibilities, Printing, and Serial I/O. Our answers come from National Instruments LabVIEW Common Questions

List and assume you are working with the full version of LabVIEW, not the sample software. A few of the questions only apply to a specific platform, so watch for the appropriate icon.

Charts and Graphs

How do I wire data to a graph or chart?

Show the Help window and move the wiring tool across a graph terminal in the block diagram to see a brief description of how to wire basic graph types. You can find excellent examples on graphs and charts in the directory EXAMPLES\GENERAL\GRAPH.LLB.

What is the basic difference between graphs and charts?

Graphs and charts differ in the way they display and update data. VIs with graphs usually collect the data in an array and then plot it on the graph, similar to a spreadsheet that first stores the data, then generates a plot of it. In contrast, a chart appends new data points to those already in the display. Using the chart, you can see the current reading or measurement in context with data previously acquired. You can set the length of the chart history buffer by using a pop-up option on the chart. It is possible, of course, to implement a history buffer with the graph indicators as well. However, you need to do this in the block diagram, while this feature is already built into the chart.

How do I update a graph without clearing it?

Graphs always clear before writing new data; however, it is a simple procedure to keep track of the data previously written to the graph and append new data with each write. The example **Separate Array Values.vi,** located in EXAMPLES\GENERAL\ARRAYS.LLB, demonstrates the technique of using the **Build Array** function to append new values to an existing array.

Why do the scales flip on a graph when I am changing the limits?

National Instruments intentionally gave scales the ability to flip, so you can flip them if you need to. Whenever the minimum of any scale is set to a value greater than or equal to the maximum, the scales flip from their current orientation.

■ How can I create polar plots and Smith charts?

The LabVIEW Picture Control Toolkit contains examples with routines to create polar plots and Smith charts. The toolkit is a versatile graphics package for creating arbitrary front panel displays. The Picture VI Library, which is shipped with the toolkit, implements a common set of drawing commands for building the graphic images.

■ DAQ

■ Where (besides this book) is the best place to get up to speed quickly with data acquisition and LabVIEW?

Read the *LabVIEW Data Acquisition Basics Manual.* Then look at the RUN_ME.LLB examples (in EXAMPLES\DAQ) included with LabVIEW.

■ What are the advantages/disadvantages of reading AI Read's backlog rather than a fixed amount of data?

Reading the backlog is guaranteed not to cause a synchronous (uninterruptable) wait for the data to arrive. However, it adds more delay until the data is processed (because the data really was available on the last call) and it can require constant reallocation or size adjustments of the data acquisition buffer in LabVIEW.

■ How can I tell when a continuous data acquisition operation does not have enough spare capacity?

The backlog rises with time, either steadily or in jumps, or takes a long time to drop to normal after an interrupting activity like mouse movement. If you can open a VI during the operation without receiving an overrun error, you should have adequate capacity.

■ I want to group two or more ports using my DIO32F, DIO24, or DIO-96 board, but I don't want to use handshaking. I just want to read one group of ports once. How can I set it up in software?

Use **Write to Digital Port** or **Read from Digital Port** to set several ports in the port list. Whatever data you try to output to each port of your *group* will correspond to each element of the data array. The same principle applies for input. You can also use the Advanced Digital VIs (**DIO Port Config**, **DIO Port Write**, or **DIO Port Read**) to set several ports in the port list.

■ I want to use the OUT1, OUT 2, OUT3, and IN1, IN2, IN3 pins on my DIO32F board. How do I address those pins using the Digital VIs in LabVIEW?

You can use **Write to Digital Port** or **Read from Digital Port**. The OUT1, OUT 2, OUT3, and IN1, IN2, IN3 pins are addressed together as port 4. OUT1 and IN1 are referred to as bit 0, OUT2 and IN2 are referred to as bit 1, and OUT3 and IN3 are referred to as bit 2. Only the NB-DIO-32F has three pins for each direction. If you use the **Write To Digital Port** VI, you will output on the OUT pins, and if you use the **Read From Digital Port** VI, you will input from the IN pins.

■ When are the DAQ boards initialized?

All DAQ boards are initialized automatically when the first DAQ VI is loaded in on a diagram when you start LabVIEW. In the professional version, you can also initialize a particular board by calling the **Device Reset** VI.

■ I bought LabVIEW and also have a slightly older DAQ board from National Instruments. I installed the whole LabVIEW package. Should I go ahead and install my NI-DAQ® drivers that I got with the board?

No. LabVIEW installs a set of DAQ driver files that are guaranteed to work with LabVIEW, whereas if you happen to install an older version of the drivers, you may run into problems.

■ Error Messages and Crashes

■ Why does LabVIEW tend to crash randomly or when printing? These crashes may be general protection faults or LabVIEW failure messages that include a source code file and line number.

Often, random crashes involve problems with the video driver on the machine. To establish that this is the case, use standard VGA as the video driver. If the crashes do not occur with the standard VGA driver, then you should get an update to your video driver.

▦ While running LabVIEW, why do I receive a memory parity error followed by a general protection fault?

Memory parity errors are the result of bad memory in your machine. This memory might be virtual memory, indicating a problem on your hard disk, or physical memory, indicating a bad SIMM. To check for problems with your hard disk, use a standard disk utility package such as Norton Utilities. To check for problems with physical RAM, you will need to physically rotate the SIMMs in your machine, rebooting the PC after each rotation. When the bad SIMM is placed in the lowest bank of memory, the machine will not boot at all. You should replace the SIMM.

■ File I/O

▦ When should I wire the count input of Read File and what should I wire to it?

Wire the count input of **Read File** when you want to read a string from a byte-stream file, several records of a single data type from a byte-stream file, or several records from a datalog file.

To read a string from a byte-stream file, do not wire the type input, but wire the count input to a numeric whose value is the number of bytes that you want to read.

To read several records of a single data type from a byte-stream file, wire the type input to the type of the records that you want to read, and wire the count input to either a numeric or a cluster of numerics. If you wire the count input to a numeric whose value is n, you will read n records of the type provided and set the data output to a 1D n-element array of this type. If you wire the count input to a cluster of numerics whose m elements are (in cluster order) $n1, n2, \ldots, nm$, you will read $n1 \times n2 \times \ldots \times nm$ records of the type provided and set the data output to an m-D array of these records whose dimensions are $n1$ by $n2$ by $\ldots$ by nm.

To read several records from a datalog file, do not wire the type input (because the record type was already provided when the file was opened, and the refnum retains this record type information), but wire the count input to either a numeric or a cluster of numerics. Otherwise, the data read in this case is determined as for the previous case, reading several records of a single data type from a byte-stream file.

How can I find out how long a file is?

Use the **EOF** function. This function is used to both set and get a file EOF (End Of File). If you wire either or both of the **pos mode** and **pos offset** inputs, **EOF** will set the file size and then return the new file size. If you do not wire either the **pos mode** or **pos offset** inputs, **EOF** simply returns the current file size. In addition, **Open/Create/Replace File.vi** returns file size in bytes when it opens a file.

What is a file refnum and how do I pass it to a subVI?

A file refnum (reference number) is an identification number that links all of the VIs that operate on that file (for example, **Open File**, **Write File**, **Close File**, and so on). It is a special LabVIEW data type. If you probe a file refnum, you will see that it looks like a hexadecimal number. Use this special data type for platform-independent operations. When you open or create a file, a unique refnum is created, and should be passed to all subsequent VIs that operate on that file. Because the refnum is a special data type, you cannot display it in a digital indicator with a hexadecimal radix. To pass the refnum to another VI, use the file refnum control/indicator located in the **Path & Refnum** subpalette of the **Controls** palette.

Whenever I try to create a new file, I get a manager argument error, error number 1. Why?

This error occurs for a variety of reasons, the most common being that the path is in a format that the operating system does not recognize. Make sure not to use these special characters in the filename: * : \ / ~ ?. These are interpreted differently on each platform, and LabVIEW will not accept these characters as part of a filename.

■ **I am using** Write to Spreadsheet File.vi **to write an array to disk. This VI is inside a loop and seems to take an extremely long time to write to disk. How can I improve performance?**

Each time **Write to Spreadsheet File.vi** is called, it opens the file, appends information to the file, and then closes the file. If you intend to write to a spreadsheet file within a loop and have concerns about performance, store the data generated by the loop, then write it to a file after the loop completes. Or, you can use more low-level VIs to open the file before entering the loop, write to the file in the loop, and then close the file after the loop has completed execution. **Write to Spreadsheet File.vi** is best suited for a one-shot operation when you only write to the file once.

■ **A VI that performs file I/O works well on one platform but indicates unprintable characters on other platforms when I look at the text file the VI produces. Why?**

Check if you are using any <CR> constants in your VI. The line termination character on the Macintosh is a <CR> whereas on the PC it is <CR><LF>. So, you will need to replace all instances of <CR> with <CR><LF>. An easier way is to use the **End of Line** constant, which is platform independent. The **End of Line** constant represents <CR> on the Macintosh and <CR><LF> on the PC.

■ GPIB

■ **Why can I communicate with my GPIB instrument with a LabVIEW VI running in execution highlighting mode but not when it is running full speed?**

You most likely have a timing problem. VIs run much slower with execution highlighting enabled than they would otherwise. Your instrument may need more time to prepare the data to send. Add a delay function such as **Wait (ms)** or use service requests before **GPIB Read.vi** to give the instrument enough time to generate the data it needs to send back to the computer.

■ Why can I write successfully to my instrument but can't read back from it?

When the **GPIB Write.vi** executes, a message is sent from the computer to a device, over the GPIB bus. When **GPIB Read.vi** executes, the device may not respond for the one of the following two reasons: The device may not have understood the command sent across the GPIB bus or the command was not properly terminated. You can refer to the manual that came with the device to confirm the syntax of the command as well as the necessary termination. The 'mode' terminal on the **GPIB Write.vi** is used to control how the GPIB Write is terminated. Generally, devices terminate their reads on <CR><LF> or assertion of the EOI (End Or Identify) line on the GPIB bus.

Use the GPIB configuration utility for your platform to change the termination character.

■ Installation

■ I just received a new DAQ or GPIB board and a copy of LabVIEW. What do I have to install and in what order?

You need to only install the LabVIEW package. The LabVIEW installation includes the necessary files to control DAQ and GPIB boards with LabVIEW. See the LabVIEW release notes for further details.

■ Why does the installation process abort prematurely?

On Macintosh and Windows, virus protection software can prevent installation of the LabVIEW package; the error message you get depends on the virus protection software. First check the disks for viruses and then turn off the virus-protection software while installing LabVIEW.

On any platform, this could indicate a problem with the disk. Make sure you have enough space available on your disk for the installation.

■ Miscellaneous

■ How can I dynamically load and run VIs?

When a subVI exists in the block diagram of another VI, the subVI is loaded into memory as soon as the calling VI is loaded into memory. For memory considerations, you may want LabVIEW to dynamically load and unload VIs from memory during the execution of your program. The **Advanced►VI Control** subpalette of the **Functions** palette contains VIs that let you programmatically open, run, and close other VIs.

In addition to the platform-independent VIs, you can use the Apple Events VIs (located under **Functions►Communication► AppleEvent**) if you have a Macintosh to dynamically load and run VIs.

■ Why do I have problems loading two VIs that call two distinct subVIs with the same name?

Assume you have **Main1.vi** and **Main2.vi**, calling different subVIs which are both named **SubVI.vi**. **Main1.vi** is loaded into memory and **SubVI.vi** is loaded because it is called by **Main1.vi**.

When you try to load **Main2.vi** into memory, linker information tells LabVIEW to load **SubVI.vi**. LabVIEW realizes that it already has **SubVI.vi** loaded into memory and tries to link **Main2.vi** with this **SubVI.vi**. If the connector pane is exactly the same, it will probably link fine (but the behavior may be very wrong). If the connector pane is different, you will get a bad linkage error.

Solution: Always give subVIs distinct names.

■ When loading and executing several VIs, how is the priority of execution assigned to the VIs?

With the LabVIEW execution system, you can execute several VIs or subdiagrams in parallel and assign priorities to VIs. Priorities range from 0 (low) to 3 (high), with an additional (highest) level of subroutine. Only one section of code executes at a given time. Other sections wait on an execution queue. If the code does not complete in a given time period, LabVIEW returns it to the queue and the front-most queue element starts running. In this fashion, execution passes round-robin between VIs or subdiagrams.

You can alter the priority of a VI through **VI Setup:** Note that a VI with high priority automatically bumps its subVIs to the same level of priority. For example, if **main.vi** has priority 3 (high) and **subVI.vi** has priority 1 (not as high), then **main.vi** calls **subVI.vi** and **subVI.vi** has priority 3.

A potential danger in using VI priority is that low-priority items will never execute if higher level tasks are always available to run. Also, priority does not specify which of the two items in parallel gets executed first. If something with high priority is put on the queue, it will execute as soon as the currently executing lower priority node checks the queue.

■ With nested While Loops, how do I stop both loops without anything in the outer loop executing after the inner loop stops?

Put the code you don't want to execute on the last iteration into a Case Structure. The condition terminal of the inner loop should be connected to the selection terminal of the Case Structure.

■ How do I hide the menu bars of a LabVIEW VI?

All VI attributes, including whether the menu bar and execution palettes are displayed, are set through the **VI Setup...** menu item.

■ How do I disable mouse interrupts to improve performance?

Interrupts caused by moving or clicking the mouse take CPU time. In some extremely time-critical applications, this interrupt activity can result in a loss of data or some other problem with the application. There is no known way to disable mouse interrupts while an application is running, unless you want to unplug your mouse. Just make sure that the mouse does not move during the time-critical portions of your application.

■ Platform and Compatibility Issues

■ What is required to transfer VIs between platforms?

Once the file is brought to another platform, no conversion is necessary for LabVIEW to read the VI (assuming the LabVIEW version number is the same). When LabVIEW opens the VI, it will recognize that the VI has been compiled for a different platform and then recompile the VI for the new platform. However, you must include the block diagram of a VI if you want to take the VI to another platform.

You can move VIs between platforms through networks, modems, and disks. LabVIEW VIs are saved in the same file format on all platforms. If you transfer VIs across a network via FTP or modem, make sure that you specify a binary transfer.

If disks are the method of transfer, you need disk conversion utilities to read the disks from other platforms. Conversion utilities change the format of files stored on disk because each platform saves files to disk in a different format. Most file conversion utilities not only read files from another platform, but also write files in the disk format of that platform. For example, there are utilities such as MacDisk and TransferPro available for the PC that transfer Macintosh disks to PC format and vice versa. On the Macintosh, DOS Mounter and PC Exchange are two utilities that convert files on DOS-formatted disks to Macintosh format and vice versa.

You will need to reload the code resources for VIs containing CINs. You should eliminate platform-specific functions (for example, Apple Events for Macintosh, DDE for Windows, and so on) before you port a VI to another platform.

■ Printing

■ When using automatic (programmatic) printing while executing a VI, why is the information not printed until the VI stops executing?

By design, programmatic printing does not print a VI or subVI until it stops execution. For printing during execution, either:

◆ Print from the **File** menu to print the panel manually.

◆ Use a call to a subVI that accepts the input data to its controls (the front panel might be identical to that of the executing VI) and prints its results with programmatic printing.

■ How do I print a single control from the front panel (for example, a graph)?

To print only a graph from the front panel, create a subVI with a graph on its front panel. Change the graph from an indicator to a control. Put the subVI in print mode by selecting its print mode button. Assign the subVI a connector and pass the data from the graph on the main VI to the graph on the subVI. Every time your main VI calls the subVI, it will automatically print the graph.

■ How can you print a string from LabVIEW?

Use the Serial I/O VIs: **Serial Port Init.vi** to initialize the port where the printer is connected (`LPT1`, `LPT2`, and so on, on the PC, or the printer port on the Macintosh) and then **Serial Port Write.vi** to write the string to the initialized port. The printer will see the data at the port and print it.

On the Macintosh, the VI **AESend Print Document** will tell the Finder to print any document. The VI is located in **Communication▶AppleEvents** in the **Functions** palette. On other platforms, you can use the **System Exec.vi** to print a file through a command-line function; the VI is located in the **Communication** palette. You can also use LabVIEW's programmatic printing option.

■ How can I scale a VI before I print it?

Select **Print Documentation** from the **File** menu and check the **Scale Front Panel to Fit** or **Scale Block Diagram to Fit** box.

■ Why is the text on labels and front panel controls clipped when the VI prints?

This clipping occurs when the size of the font for the printer does not match the size of the font on the monitor. There are three things you can try to remedy the situation:

1. Enlarge the labels and front panel controls so that when the text expands because of the font mismatch, the text still fits in the control or label boundary.

2. Find a font that is the same size on both the monitor and the printer.

3. On Windows, try Postscript or bitmap printing.

■ I'm having trouble getting LabVIEW screens to print in other applications. How do you take a screenshot of LabVIEW?

Sometimes you can simply copy and paste a selected area into another application. On Windows, you might also try hitting <alt>-<print screen> to copy the selected screen to the clipboard. However, these pictures do not always print properly, so you may have to use a screen capture utility, such as Capture for Macintosh or HiJaak for Windows.

■ How do I select bitmap printing?

Go to **Preferences** under the **Edit** menu and select **Printing**. The very last option on the screen is **Bitmap Printing**.

■ Why does LabVIEW crash when printing?

This crash may be related to the video driver. Please see the *Error Messages and Crashe*s section of this appendix for more information.

■ Serial I/O

■ Why doesn't my instrument respond to commands sent with Serial Port Write.vi?

Many instruments expect a carriage return or line feed to terminate the command string. The **Serial Port Write.vi** in LabVIEW sends only those characters included in the string input; no termination character is appended to the string. Many terminal emulation packages (for example, Windows Terminal) automatically append a carriage return to the end of all transmissions. With LabVIEW, you will need to include the proper termination character in your string input to **Serial Port Write.vi** if your instrument requires it.

Some instruments require a carriage return (\r); others require a line feed (\n). When you enter a return on the keyboard (on PC keyboards, this is the <enter> key on the main alphanumeric keypad), LabVIEW inserts a \n (line feed). To insert a carriage return, use the **Concatenate Strings** function to append a **Carriage Return** constant to the string, or manually enter \r after selecting '\' **Codes Display** from the string pop-up menu.

Make sure that your cable works. Many technical support questions are related to bad cables. In computer to computer communication with serial I/O, use a null-modem to reverse the receive and transmit signals.

See the example **LabVIEW <-> Serial.vi** to establish communication with your instrument. It is located in EXAMPLES\ SERIAL\SMPLSERL.LLB. The VI also demonstrates the use of **Bytes at Serial Port.vi** before reading data back from the serial port.

How do I reset or clear a serial port?

Use **Serial Port Init.vi** to reinitialize the port. This will automatically clear the serial port buffers associated with the port number.

How do I access the parallel port?

In LabVIEW for Windows, port 10 is LPT1, port 11 is LPT2, and so on. To send data to a printer connected to a parallel port, use the **Serial Port Write.vi**. See the *Printing* section of this appendix for more details. For Windows 95 and NT, see the *LabVIEW Instrument I/O VI Reference Manual*.

Glossary

Prefix	Meaning	Value
m-	milli-	10^{-3}
μ-	micro-	10^{-6}
n-	nano-	10^{-9}

Symbols

∞   Infinity.

π   Pi.

Δ   Delta. Difference. Δx denotes the value by which x changes from one index to the next.

A

absolute path	File or directory path that describes the location relative to the top of level of the file system.
active window	Window that is currently set to accept user input, usually the frontmost window. The title bar of an active window is highlighted. You make a window active by clicking on it, or by selecting it from the Windows menu.
A/D	Analog-to-digital conversion. Refers to the operation electronic circuitry does to take a real-world analog signal and convert it to a digital form (as a series of bits) that the computer can understand.
ADC	See A/D.
ANSI	American National Standards Institute.
array	Ordered, indexed set of data elements of the same type.
array shell	Front panel object that houses an array. It consists of an index display, a data object window, and an optional label. It can accept various data types.
artificial data dependency	Condition in a dataflow programming language in which the arrival of data, rather than its value, triggers execution of a node.
ASCII	American Standard Code for Information Interchange.
asynchronous execution	Mode in which multiple processes share processor time. For example, one process executes while others wait for interrupts during device I/O or while waiting for a clock tick.
auto-indexing	Capability of loop structures to disassemble and assemble arrays at their borders. As an array enters a loop with auto-indexing enabled, the loop automatically disassembles it with scalars extracted from one-dimensional arrays, one-dimensional arrays extracted from two-dimensional arrays, and so on. Loops assemble data into arrays as they exit the loop according to the reverse of the same procedure.
autoscaling	Ability of scales to adjust to the range of plotted values. On graph scales, this feature determines maximum and minimum scale values, as well.

autosizing	Automatic resizing of labels to accommodate text that you enter.

B

block diagram	Pictorial description or representation of a program or algorithm. In LabVIEW, the block diagram, which consists of executable icons called nodes and wires that carry data between the nodes, is the source code for the VI. The block diagram resides in the block diagram window of the VI.
Boolean controls	Front panel objects used to manipulate and display or input and
and indicators	output Boolean (TRUE or FALSE) data. Several styles are available, such as switches, buttons and LEDs.
breakpoint	A pause in execution. You set a breakpoint by clicking on a VI, node, or wire with the Breakpoint tool from the Tools palette.
Breakpoint tool	Tool used to set a breakpoint on a VI, node, or wire.
broken VI	VI that cannot be compiled or run; signified by a broken arrow in the run button.
Bundle node	Function that creates clusters from various types of elements.
byte stream file	File that stores data as a sequence of ASCII characters or bytes.

C

case	One subdiagram of a Case Structure.
Case Structure	Conditional branching control structure, which executes one and only one of its subdiagrams based on its input. It is the combination of the IF, THEN, ELSE, and CASE statements in control flow languages.
channel	Pin or wire lead to which an analog signal is read from or applied.
chart	See scope chart, strip chart, and sweep chart.
CIN	See Code Interface Node.

cloning	To make a copy of a control or some other LabVIEW object by clicking the mouse button while pressing the \<ctrl\> (Windows); \<option\> (Macintosh); \<meta\> (Sun); or \<alt\> (HP-UX) key and dragging the copy to its new location.
	(Sun and HP-UX) You can also clone an object by clicking on the object with the middle mouse button and then dragging the copy to its new location.
cluster	A set of ordered, unindexed data elements of any data type including numeric, Boolean, string, array, or cluster. The elements must be all controls or all indicators.
cluster shell	Front panel object that contains the elements of a cluster.
Code Interface Node (CIN)	Special block diagram node through which you can link conventional, text-based code to a VI.
coercion	The automatic conversion LabVIEW performs to change the numeric representation of a data element.
coercion dot	Glyph on a node or terminal indicating that the numeric representation of the data element changes at that point.
Color tool	Tool you use to set foreground and background colors.
Color Copy tool	Copies colors for pasting with the Color tool.
compile	Process that converts high-level code to machine-executable code. LabVIEW automatically compiles VIs before they run for the first time after creation or alteration.
conditional terminal	The terminal of a While Loop that contains a Boolean value that determines whether the VI performs another iteration.
connector	Part of the VI or function node that contains its input and output terminals, through which data passes to and from the node.
connector pane	Region in the upper right corner of a front panel window that displays the VI terminal pattern. It underlies the icon pane.
constant	See universal constant and user-defined constant.

continuous run	Execution mode in which a VI is run repeatedly until the operator stops it. You enable it by clicking on the continuous run button.
control	Front panel object for entering data to a VI interactively or to a subVI programmatically.
control flow	Programming system in which the sequential order of instructions determines execution order. Most conventional text-based programming languages, such as C, Pascal, and BASIC, are control flow languages.
Controls palette	Palette containing front panel controls and indicators.
conversion	Changing the type of a data element.
count terminal	The terminal of a For Loop whose value determines the number of times the For Loop executes its subdiagram.
CPU	Central Processing Unit.
current VI	VI whose front panel, block diagram, or Icon Editor is the active window.
custom PICT controls	Controls and indicators whose parts can be replaced by graphics and indicators you supply.

D

D/A	Digital-to-analog. The opposite operation of an A/D.
data acquisition (DAQ)	Process of acquiring data, usually by performing an analog-to-digital (A/D) conversion. Its meaning is sometimes expanded to include data generation (D/A).
data dependency	Condition in a dataflow programming language in which a node cannot execute until it receives data from another node. See also artificial data dependency.
data flow	Programming system consisting of executable nodes in which nodes execute only when they have received all required input data and produce output automatically when they have executed. LabVIEW is a dataflow system.
data logging	Generally, to acquire data and simultaneously store it in a disk file. LabVIEW file I/O functions can log data.

data storage formats	The arrangement and representation of data stored in memory.
data type descriptor	Code that identifies data types, used in data storage and representation.
datalog file	File that stores data as a sequence of records of a single, arbitrary data type that you specify when you create the file. While all the records in a datalog file must be of a single type, that type can be complex; for instance, you can specify that each record is a cluster containing a string, a number, and an array.
DC	Direct Current. The opposite of AC (Alternate Current). Refers to a very low-frequency signal, such as one that varies less than once a second.
device	A plug-in DAQ board
device number	Number assigned to a device (DAQ board) in the NI-DAQ configuration utility.
Description box	Online documentation for a LabVIEW object.
destination terminal	See sink terminal.
dialog box	An interactive screen with prompts in which you specify additional information needed to complete a command.
differential measurement	Way to configure a device to read signals in which the inputs need not be connected to a reference ground. The measurement is made between two input channels.
dimension	Size and structure attribute of an array.
DMA	Direct Memory Access. A method by which you can transfer data to computer memory from a device or memory on the bus (or from computer memory to a device) while the processor does something else. DMA is the fastest method of transferring data to or from computer memory.
drag	To drag the mouse cursor on the screen to select, move, copy, or delete objects.

E

empty array	Array that has zero elements, but has a defined data type. For example, an array that has a numeric control in its data display window but has no defined values for any element is an empty numeric array.
EOF	End-of-File. Character offset of the end of file relative to the beginning of the file (that is, the EOF is the size of the file).
execution highlighting	Feature that animates VI execution to illustrate the data flow in the VI.

F

FFT	Fast Fourier transform.
file refnum	An identifier that LabVIEW associates with a file when you open it. You use the file refnum to specify that you want a function or VI to perform an operation on the open file.
flattened data	Data of any type that has been converted to a string, usually, for writing it to a file.
For Loop	Iterative loop structure that executes its subdiagram a set number of times. Equivalent to conventional code: For i=0 to n-1, do ....
Formula Node	Node that executes formulas that you enter as text. Especially useful for lengthy formulas that would be cumbersome to build in block diagram form.
frame	Subdiagram of a Sequence Structure.
free label	Label on the front panel or block diagram that does not belong to any other object.
front panel	The interactive user interface of a VI. Modeled from the front panel of physical instruments, it is composed of switches, slides, meters, graphs, charts, gauges, LEDs, and other controls and indicators.

function	Built-in execution element, comparable to an operator, function, or statement in a conventional language.
Functions palette	Palette containing block diagram structures, constants, communication features, and VIs.

G

G	The LabVIEW graphical programming language.
global variable	Non-reentrant subVI with local memory that uses an uninitialized shift register to store data from one execution to the next. The memory of copies of these subVIs is shared and thus can be used to pass global data between them.
glyph	A small picture or icon.
GPIB	General Purpose Interface Bus. Also known as HP-IB (Hewlett-Packard Interface Bus) and IEEE 488.2 bus (Institute of Electrical and Electronic Engineers standard 488.2), it has become the world standard for almost any instrument to communicate with a computer. Originally developed by Hewlett-Packard in the 1960s to allow their instruments to be programmed in BASIC with a PC, now IEEE has helped define this bus with strict hardware protocols that ensure uniformity across instrument.
graph control	Front panel object that displays data in a Cartesian plane.
ground	the common reference point in a system; i.e., ground is at 0 volts.

H

Help window	Special window that displays the names and locations of the terminals for a function or subVI, the description of controls and indicators, the values of universal constants, and descriptions and data types of control attributes. The window also accesses LabVIEW's Online Reference.

hex	Hexadecimal. A base-16 number system.
hierarchical palette	Menu that contains palettes and subpalettes.
Hierarchy window	Window that graphically displays the hierarchy of VIs and subVIs.
housing	Nonmoving part of front panel controls and indicators that contains sliders and scales.
Hz	Hertz. Cycles per second.

I

icon	Graphical representation of a node on a block diagram.
Icon Editor	Interface similar to that of a paint program for creating VI icons.
icon pane	Region in the upper right corner of the front panel and block diagram that displays the VI icon.
IEEE	Institute for Electrical and Electronic Engineers.
indicator	Front panel object that displays output.
Inf	Digital display value for a floating-point representation of infinity.
instrument driver	VI that controls a programmable instrument.
I/O	Input/Output. The transfer of data to or from a computer system involving communications channels, operator input devices, and/or data acquisition and control interfaces.
iteration terminal	The terminal of a For Loop or While Loop that contains the current number of completed iterations.

L

label	Text object used to name or describe other objects or regions on the front panel or block diagram.
Labeling tool	Tool used to create labels and enter text into text windows.

LabVIEW	Laboratory Virtual Instrument Engineering Workbench.
LED	Light-emitting diode.
legend	Object owned by a chart or graph that display the names and plot styles of plots on that chart or graph.
line	The equivalent of an analog channel—a path where a single digital signal is set or retrieved.

M

marquee	A moving, dashed border that surrounds selected objects.
matrix	Two-dimensional array.
MB	Megabytes of memory.
menu bar	Horizontal bar that contains names of main menus.
modular programming	Programming that uses interchangeable computer routines.

N

NaN	Digital display value for a floating-point representation of not a number, typically the result of an undefined operation, such as log(–1).
NI-DAQ	Driver software for National Instruments DAQ boards and SCXI modules. This software includes a configuration utility to configure the hardware, and also acts as an interface between LabVIEW and the devices.
nodes	Execution elements of a block diagram consisting of functions, structures, and subVIs.
nondisplayable characters	ASCII characters that cannot be displayed, such as newline, tab, and so on.
not-a-path	A predefined value for the path control that means the path is invalid.
not-a-refnum	A predefined value that means the refnum is invalid.

numeric controls and indicators	Front panel objects used to manipulate and display or input and output numeric data.
NRSE	Nonreferenced single-ended.
NRSE measurement	All measurements are made with respect to a common reference. This reference voltage can vary with respect to ground.
Nyquist frequency	One-half the sampling frequency. If the signal contains any frequencies above the Nyquist frequency, the resulting sampled signal will be aliased, or distorted.

O

object	Generic term for any item on the front panel or block diagram, including controls, nodes, wires, and imported pictures.
Object pop-up menu tool	Tool used to access an object's pop-up menu.
Octal	Numbering system, base-eight.
Operating tool	Tool used to enter data into controls as well as operate them. Resembles a pointing finger.

P

palette	Menu of pictures that represent possible options.
platform	Computer and operating system.
plot	A graphical representation of an array of data shown either on a graph or a chart.
polymorphism	Ability of a node to automatically adjust to data of different representation, type, or structure.
pop up	To call up a special menu by clicking (usually on an object) with the right mouse button (on Window, Sun, and HP-UX) or while holding down the command key (on the Macintosh).
pop-up menus	Menus accessed by popping up, usually on an object. Menu options pertain to that object specifically.

port	A collection of digital lines that are configured in the same direction and can be used at the same time.
Positioning tool	Tool used to move, select, and resize objects.
probe	Debugging feature for checking intermediate values in a VI.
Probe tool	Tool used to create probes on wires.
programmatic printing	Automatic printing of a VI front panel after execution.
pseudocode	Simplified language-independent representation of programming code.
pull-down menus	Menus accessed from a menu bar. Pull-down menu options are usually general in nature.

R

reentrant execution	Mode in which calls to multiple instances of a subVI can execute in parallel with distinct and separate data storage.
representation	Subtype of the numeric data type, of which there are signed and unsigned byte, word, and long integers, as well as single-, double-, and extended-precision floating-point numbers, both real and complex.
resizing handles	Angled handles on the corner of objects that indicate resizing points.
ring control	Special numeric control that associates 32-bit integers, starting at 0 and increasing sequentially, with a series of text labels or graphics.
RS-232	Recommended Standard #232. A standard proposed by the Instrument Society of America for serial communications. It's used interchangeably with the term "serial communication," although serial communications more generally refers to communicating one bit at a time. A few other standards you might see are RS-485, RS-422, and RS-423.
RSE	Referenced single-ended.
RSE measurement	All measurements are made with respect to a common ground; also known as a grounded measurement.

S

sample	A single analog input or output data point.
scalar	Number capable of being represented by a point on a scale. A single value as opposed to an array. Scalar Booleans and clusters are explicitly singular instances of their respective data types.
scale	Part of mechanical-action, chart, and graph controls and indicators that contains a series of marks or points at known intervals to denote units of measure.
scope mode	Mode of a waveform chart modeled on the operation of an oscilloscope.
Scroll tool	Tool used to scroll windows.
SCXI	Signal Conditioning eXtensions for Instrumentation. A high-performance signal conditioning system devised by National Instruments, using an external chassis that contains I/O modules for signal conditioning, multiplexing, etc. The chassis is wired into a DAQ board in the PC.
sequence local	Terminal that passes data between the frames of a Sequence Structure.
Sequence Structure	Program control structure that executes its subdiagrams in numeric order. Commonly used to force nodes that are not data-dependent to execute in a desired order.
shift register	Optional mechanism in loop structures used to pass the value of a variable from one iteration of a loop to a subsequent iteration.
sink terminal	Terminal that absorbs data. Also called a destination terminal.
slider	Moveable part of slide controls and indicators.
source terminal	Terminal that emits data.
state machine	A method of execution in which individual tasks are separate cases in a Case Structure that is embedded in a While Loop. Sequences are specified as arrays of case numbers.

string controls and indicators	Front panel objects used to manipulate and display or input and output text.
strip mode	Mode of a waveform chart modeled after a paper strip chart recorder, which scrolls as it plots data.
structure	Program control element, such as a Sequence, Case, For Loop, or While Loop.
subdiagram	Block diagram within the border of a structure.
subVI	VI used in the block diagram of another VI; comparable to a subroutine.
sweep mode	Similar to scope mode—except a line sweeps across the display to separate old data from new data.

T

terminal	Object or region on a node through which data passes.
tool	Special LabVIEW cursor you can use to perform specific operations.
Toolbar	Bar containing command buttons that you can use to run and debug VIs.
Tools palette	Palette containing tools you can use to edit and debug front panel and block diagram objects.
top-level VI	VI at the top of the VI hierarchy. This term distinguishes the VI from its subVIs.
trigger	a condition for starting or stopping a DAQ operation.
tunnel	Data entry or exit terminal on a structure.
typecast	To change the type descriptor of a data element without altering the memory image of the data.
type descriptor	See data type descriptor.

U

| universal constant | Uneditable block diagram object that emits a particular ASCII character or standard numeric constant, for example, pi. |

user-defined constant	Block diagram object that emits a value you set.

V

VI	See virtual instrument.
VI library	Special file that contains a collection of related VIs for a specific use.
virtual instrument	LabVIEW program; so called because it models the appearance and function of a physical instrument.

W

While Loop	Loop structure that repeats a section of code until a condition is met. Comparable to a Do loop or a Repeat-Until loop in conventional programming languages.
wire	Data path between nodes.
Wiring tool	Tool used to define data paths between source and sink terminals.

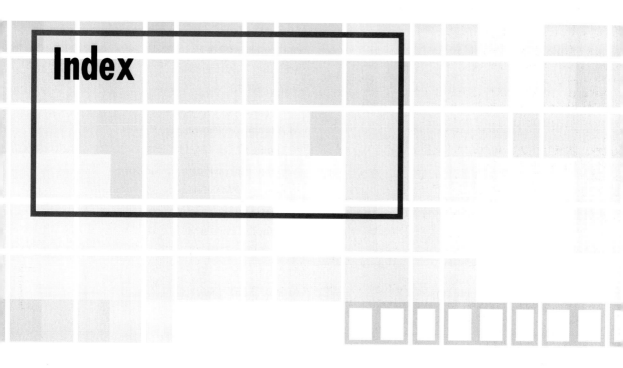

Index

H

I